THE COGNITIVE NEUROSCIENCE OF SECOND LANGUAGE ACQUISITION

Marianne Gullberg and Peter Indefrey, Editors

© 2006 Language Learning Research Club, University of Michigan

Blackwell Publishing
350 Main Street
Malden, MA 02148 USA

Blackwell Publishing, Ltd.
9600 Garsington Road
Oxford OX4 2DQ
United Kingdom

All rights reserved. Except for the quotations of short passages for the purpose of criticism and review, no part of this publication may be reproduced, stored in a retrieval system, or transmitted, in any form or by any means, electronic, mechanical, photocopying, recording or otherwise, without the prior written permission of the publisher.

Library of Congress Cataloging-in-Publication Data

The cognitive neuroscience of second language acquisition / Marianne Gullberg and Peter Indefrey, editors.
    p. ; cm. – (The Language Learning–Max Planck Institute for Psycholinguistics cognitive neuroscience series)
  Includes bibliographical references and index.
  ISBN-13: 978-1-4051-5542-7
  ISBN-10: 1-4051-5542-6
    1. Cognitive neuroscience—Congresses. 2. Second language acquisition—Congresses. 3. Education, Bilingual—Congresses. I. Gullberg, Marianne. II. Indefrey, Peter. III. Language Learning–Max Planck Institute for Psycholinguistics. IV. Series.
    [DNLM: 1. Language Development—Congresses. 2. Neurophysiology—Congresses. 3. Brain—physiology—Congresses. 4. Learning—physiology—Congresses. 5. Multilingualism—Congresses. 6. Psycholinguistics–Congresses. WL 102 C67575 2006]

QP360.5.C64. 2006
612.8'233–dc22
                                                         2006015711

Contents

Foreword — v–vi

Peter Indefrey and Marianne Gullberg
   Introduction — 1–8

David Birdsong
   Age and Second Language Acquisition and
   Processing: A Selective Overview — 9–49

Peter Coopmans
   L2 Acquisition, Age, and Generativist
   Reasoning. Commentary on Birdsong — 51–58

H. B. M. Uylings
   Development of the Human Cortex and the
   Concept of "Critical" or "Sensitive" Periods — 59–90

Peter Hagoort
   What We Cannot Learn From Neuroanatomy
   About Language Learning and Language
   Processing. Commentary on Uylings — 91–97

David W. Green, Jenny Crinion, and Cathy J. Price
   Convergence, Degeneracy, and Control — 99–125

Kees de Bot
   The Plastic Bilingual Brain: Synaptic Pruning
   or Growth? Commentary on Green et al. — 127–132

A. Rodriguez-Fornells, R. De Diego Balaguer,
and T. F. Münte
   Executive Control in Bilingual Language
   Processing — 133–190

Ton Dijkstra and Walter van Heuven
   On Language and the Brain—Or on
   (Psycho)linguists and Neuroscientists?
   Commentary on Rodriguez-Fornells et al. — 191–197

Lee Osterhout, Judith McLaughlin, Ilona Pitkänen, Cheryl Frenck-Mestre, and Nicola Molinaro
Novice Learners, Longitudinal Designs, and Event-Related Potentials: A Means for Exploring the Neurocognition of Second Language Processing 199–230

Doug Davidson
Strategies for Longitudinal Neurophysiology. Commentary on Osterhout et al. 231–234

Jutta L. Mueller
L2 in a Nutshell: The Investigation of Second Language Processing in the Miniature Language Model 235–270

Monique J. A. Lamers
Cracking the Nutshell Differently. Commentary on Mueller 271–277

Peter Indefrey
A Meta-analysis of Hemodynamic Studies on First and Second Language Processing: Which Suggested Differences Can We Trust and What Do They Mean? 279–304

Laurie A. Stowe
When Does the Neurological Basis of First and Second Language Processing Differ? Commentary on Indefrey 305–311

John H. Schumann
Summing up: Some Themes in the Cognitive Neuroscience of Second Language Acquisition 313–319

Glossary of Neuroanatomical Terms and Phrases 321–329

Author Index 331–342

Subject Index 343–348

Foreword

The articles in this volume are the result of a conference held at the Max Planck Institute for Psycholinguistics in Nijmegen, the Netherlands, in September 2005. The Conference, entitled "The Cognitive Neuroscience of Second Language Acquisition," was the first in a newly established conference series, The A. Guiora Roundtable Conference Series in the Cognitive Neuroscience of Language, jointly initiated and cosponsored by the journal *Language Learning* and by the Max Planck Institute for Psycholinguistics.

In October 2003, Alexander "Shonny" Guiora contacted some people in Nijmegen about an idea that he had. In characteristically enthusiastic and engaging terms, he expressed his conviction that cognitive neuroscience is the "next frontier for the language sciences," offering not only new methodological options but new explanatory possibilities. Shonny's idea was to create a permanent forum for discussing and taking stock of the present stage of the neurosciences of language behavior. This idea led to the conference series, which now bears Shonny's name in honor of his foresight and initiative. True to his initial idea, the aim of the conference series is to establish a forum that goes beyond a mere exchange of data and that stimulates participants to work toward a broader view of the neural implementation of language.

We wish to express our thanks to the former and the current board of *Language Learning* and to the directorate of the Max Planck Institute for Psycholinguistics for their generous support for this enterprise. We also wish to thank the other members of the steering committee of the Conference series, Alexander Guiora, *Language Learning*; Peter Hagoort, director of the FC Donders Centre for Cognitive Neuroimaging, Nijmegen; and Wolfgang Klein, director of the Max Planck Institute for Psycholinguistics. We gratefully acknowledge the help provided by Nanjo Bogdanowicz and Evelyn Giering in organizing the conference.

Finally, we owe a debt of gratitude to Frauke Hellwig and Arna van Doorn for editorial help with indices, the creation of artwork, and meticulous proofreading.

Nijmegen and Los Angeles, February, 2006
Marianne Gullberg and Peter Indefrey
John H. Schumann, Series Editor

Introduction

Peter Indefrey
Max Planck Institute for Psycholinguistics
and
F. C. Donders Centre for Cognitive Neuroimaging

Marianne Gullberg
Max Planck Institute for Psycholinguistics

This volume is a harvest of articles from the first conference in a series on the cognitive neuroscience of language. The first conference focused on the cognitive neuroscience of second language acquisition (henceforth SLA). It brought together experts from as diverse fields as second language acquisition, bilingualism, cognitive neuroscience, and neuroanatomy. The articles and discussion articles presented here illustrate state-of-the-art findings and represent a wide range of theoretical approaches to classic as well as newer SLA issues. The theoretical themes cover age effects in SLA related to the so-called Critical Period Hypothesis and issues of ultimate attainment and focus both on age effects pertaining to childhood and to aging. Other familiar SLA topics are the effects of proficiency and learning as well as issues concerning the difference between the end product and the process that yields that product, here discussed in terms of convergence and degeneracy. A topic more related to actual usage of a second language once acquired concerns how multilingual speakers control and regulate their two languages.

Correspondence concerning this article should be addressed to Peter Indefrey, Max Planck Institute for Psycholinguistics, Box 310, 6500 AH Nijmegen, The Netherlands. Internet: peter.indefrey@mpi.nl

Age Effects

The first important issue concerns the nature and origin of age effects in SLA. Although it is uncontroversial that SLA becomes, on average, less successful with increasing age of acquisition, why this is the case is much debated. A frequently held assumption is that of a *critical period* for language acquisition, after which SLA becomes fundamentally different in one way or another.

The notion of a critical period is borrowed from biology and refers to the phenomenon that certain abilities, such as stereoscopic vision or a particular type of birdsong, might only be acquired if appropriate external stimulation is provided during a certain time window in the ontogenetic development of the individual. If this time window is missed, acquisition of the ability in question becomes impossible or at least imperfect (in the latter case, one speaks of a *sensitive* period; see Knudsen, 2004). Although it is not clear exactly how the animal brain responds to the external stimulation by developing the neural structures subserving a particular skill, it seems clear that at the end of a critical period, there must be some change in the brain that stops its responsiveness.

Researchers who assume that an analogous brain-based mechanism underlies language acquisition in humans tend to view the imperfect attainment of a second language following late exposure as an indication that the mechanism has ceased to function. Of course, not all brain changes have the temporal (and functional) properties of critical periods. Age effects might also be due to changes of brain structure and function over the lifetime that are not specific to language acquisition but that, nevertheless, affect it. David Birdsong critically discusses different versions of the critical period hypothesis and provides an overview of general and regional age-related brain changes as well as cognitive changes that might account for age effects in SLA without assuming a critical period. Harry Uylings discusses the notion of a critical period from the perspective of the neuroanatomist and distinguishes different types of critical periods. He provides in-depth information about neuroanatomical prenatal and postnatal changes in the development of the cortex. Peter Hagoort argues that structural neuroanatomy as such provides fewer insights into language acquisition and processing than the investigation of functional brain activation patterns.

Age effects have not only been observed for performance in a second language but also in the brain activation patterns during second language processing. Peter Indefrey presents a meta-analysis of hemodynamic studies that examined within-subject differences between the first language (L1) and a second language (L2). He finds that L2 onset seems to play a role in activation differences related to syntactic/sentence processing, but less so for other processing levels. Jutta Mueller presents event-related potential (ERP) studies investigating age effects in a task that was designed to achieve comparable proficiency in L1 and L2 participants using a miniature version of Japanese. She concludes that the observed differences in electrophysiological responses to grammatical violations might be interpreted as supporting the notion of a critical period. Monique Lamers cautions against equating performance on a small subset of Japanese and proficiency. If proficiency is understood in the broader sense of everything that the participants know about Japanese, the two subject groups are no longer comparable in this respect and the age-of-acquisition variable might be a disguised competence variable.

Proficiency

The second important variable influencing L2 performance is proficiency. More recently, proficiency has also been linked to functional (Chee, Soon, Lee, & Pallier, 2004) and structural (Mechelli et al., 2004) brain changes. It is important to note, however, that not all designs allow for a distinction between changes that occur as a consequence of becoming proficient in a second language and preexisting anatomical or functional neural differences that might cause some individuals to acquire languages better than others. It is also problematic that, at least in the language domain, we lack a real understanding of the functional significance of differences at the neural level. Observed differences tend to be interpreted in a circular manner starting from the (plausible) assumption that whatever is found in the more proficient speakers must be more effective, be it an increase or decrease of hemodynamic activation, an increase or decrease of gray matter. In a second step, more or less convincing speculations as to why it might be more effective are added. The chapters by Indefrey and Laurie Stowe both contribute to this type of speculation, but they attempt to narrow the interpretation space by searching for commonalities across studies.

Proficiency is mostly correlated with age of acquisition (AoA). Therefore, its effect can only be assessed if the effect of AoA is in some way controlled. One way to achieve such control is to study L2 learners with a common starting point (typically zero L2 knowledge) longitudinally while they are learning a new language. Indefrey and Lee Osterhout and colleagues report data from hemodynamic and electrophysiological studies using this approach. Osterhout et al. are particularly interested in what it means to become proficient and suggest a possible operationalization of proficiency in terms of online processing ability. Doug Davidson points out that even in groups of learners with a common starting point, individual differences might have substantial effects and suggests methodological approaches to this problem.

Convergence

The term *convergence* refers to a process whereby the representations of two languages become more similar either as a function of increasing proficiency or time—for instance, duration of L2 learning (cf. Bullock & Toribio, 2004; Clyne, 2003). However, the term is used in a number of different ways as applied to different situations and phenomena. Some authors might use this term in reference to a processing level, whereas others use it when referring to the neural level. To make things worse, the term might also be used to describe the relationship between the first and second language within an individual speaker or to describe a situation between speakers in contact where a given language is one speaker's L1 and another speaker's L2. Obviously, this could easily lead to confusion, especially because the term might be used to describe opposite findings at the neural level. Mueller, for example, uses the term to describe her finding that a specific neural response (the so-called P600) is similar for L1 and L2 processing in proficient L2 speakers. By contrast, David Green and colleagues discuss a concept of convergence that assumes that over time, the processing and the neural representation of a L2 in L2 speakers will become similar to the representation of this language in L1 speakers due to the processing requirements of the language itself. As a consequence, the neural correlates of the L1 and L2 within a bilingual speaker should become different. Indefrey's meta-analysis of hemodynamic studies comparing L1 and L2 within subjects speaks directly to the latter prediction, but finds no evidence for it. Stowe warns that a failure to observe reliable L1/L2 differences might at least in part be due to a lack of statistical power and individual anatomical differences.

According to Green et al., convergence to a new language also underlies structural brain differences found in L2 speakers (Mechelli et al., 2004). Kees de Bot points out that it is not clear whether structural or functional differences at the neural level are a consequence of L2 acquisition or are in fact preexisting differences that might have influenced L2 acquisition.

Degeneracy

A crucial matter for research on L2 acquisition is the relationship between the observed behavior and the underlying processes. David Birdsong and Peter Coopmans both warn against the common assumption that the product necessarily reflects the underlying process in a straightforward manner. In other words, it might not be warranted to infer that differences in performance in L1/L2 processing imply different underlying processes or, conversely, that similarities in performance imply similar processes. The latter consideration is lately more and more referred to with the term *degeneracy*, thoroughly discussed by Green et al. Note that when taking the neural level into account, degeneracy can refer to two different situations that should be distinguished: (a) The same behavior/performance/goal can be achieved by different underlying processes (which, in turn, have different neural correlates); (b) the same behavior/performance/goal can be achieved by the *same* underlying processes, which, however, might have different neural correlates because more than one brain system is capable of subserving this process. Mueller provides a nice example of a degeneracy (type a) explanation when suggesting that her L2 speakers perform at the same level as L1 speakers in grammaticality judgments but use a different, prosody-based mechanism to do so.

Control

Whereas the previous issues consider the two languages of a bilingual speaker separately, the issue of control concerns the interaction of the two languages. What needs to be explained is how the bilingual speaker manages to speak or comprehend one language at a time without more than the occasional interference of the other language. Our understanding of this ability is still very limited and the full range of possible mechanisms suggested by common sense has been proposed. These range from the assumption that the two languages have separate lexicons that do not interact to full lexical integration. The more the representations of the two languages are integrated, the more additional mechanisms are needed to regulate the appropriate language output in a given communicative situation. Suggested control mechanisms might increase the activation level of the target language, inhibit the other language, or affect the selection of L1 or L2 lexical items based on a high-level goal representation.

Antoni Rodriguez-Fornells and colleagues discuss executive/cognitive control and regulation of language choice in production (both in the sense of interference suppression and response inhibition), favoring a language-neutral view of the mechanism. They review models of control for speech production and propose a model for control inhibition that combines a top-down regulatory function associated with the prefrontal cortex and a more local bottom-up mechanism regulating the activation level of the nontarget language. Ton Dijkstra and Walter van Heuven point out that there is massive evidence for language nonselective lexical access and that different kinds of cognitive control are at work and should be distinguished (e.g., monitoring and conflict resolution). They argue strongly in favor of an integrative approach combining neurocognitive studies with behavioral studies and computational modeling to provide theoretical coherence. Green et al. also suggest a network of control, corresponding to different levels of control (goal maintenance, management of competing tasks), language selection, and response selection. Birdsong sees working memory as the crucial control instance and emphasizes its executive role in the suppression of irrelevant information. He suggests that age effects on these components in L2 use are likely to be more pronounced than in the L1 case, due to a relatively low degree of *automaticity* in L2 processing.

Concluding Remarks

The articles in this volume chart the current state of the art in a rapidly growing field of study, touching on a range of theoretical questions. They also outline new venues of research in a new approach to understanding language, the acquisition of language in adults, and the usage of several languages in the minds and brains of multilingual speakers. We hope that the articles in this volume will convey some of the excitement and sense of momentum from the conference to a broader audience.

References

Bullock, B. E., & Toribio, A. J. (2004). Introduction: Convergence as an emergent property in bilingual speech. *Bilingualism: Language and Cognition*, 7, 91–93.

Chee, M. W. L., Soon, C. S., Lee, H. L., & Pallier, C. (2004). Left insula activation: A marker for language attainment in bilinguals. *Proceedings of the National Academy of Sciences of the United States of America, 101*, 15,265–15,270.

Clyne, M. G. (2003). *Dynamics of language contact: English and immigrant languages*. Cambridge: Cambridge University Press.

Knudsen, E. I. (2004). Sensitive periods in the development of the brain and behavior. *Journal of Cognitive Neuroscience, 16*, 1412–1425.

Mechelli, A., Crinion, J. T., Noppeney, U., O'Doherty, J., Ashburner, J., Frackowiak, R. S., et al. (2004). Structural plasticity in the bilingual brain: Proficiency in a second language and age at acquisition affect grey-matter density. *Nature, 431*, 757.

Age and Second Language Acquisition and Processing: A Selective Overview

David Birdsong
University of Texas at Austin

This article provides a selective overview of theoretical issues and empirical findings relating to the question of age and second language acquisition (L2A). Both behavioral and brain-based data are discussed in the contexts of neurocognitive aging and cognitive neurofunction in the mature individual. Moving beyond the classical notion of "deficient" L2 processing and acquisition, we consider the complementary question of learner potential in postadolescent L2A.

The outcome of second language acquisition (L2A) among adults is demonstrably different in many respects from the outcome of first language acquisition (L1A) among children. Departing from this basic observation, researchers attempt to understand the various sources of age-related effects in L2A.

The present article is an overview of facts and theoretical issues concerning age and L2A. This contribution considers both behavioral data and brain-based processing data. The review includes findings and controversies in the areas of neurocognitive development and aging, and cognitive neurofunction in the mature brain.

A comprehensive treatment of the facts and issues is not possible in the space available. It is hoped, nevertheless, that this selective offering provides useful scaffolding for other articles in

---

Correspondence concerning this article should be addressed to David Birdsong, Department of French and Italian, University of Texas, 1 University Station B7600, Austin, TX 78712-0224. Internet: birdsong@ccwf.cc.utexas.edu

this volume that examine cognitive and neural aspects of L2 use and acquisition.

Background and Terminology

Over the past 20 or so years, a great deal of empirical research on the age question in L2A has focused on the end state of L2A, not on rates of attainment or on stages of L2 development. The developmental literature and comparative rate (adult vs. child) literature are certainly not without interest, and overviews of this research can be found in Klein (1995), Marinova-Todd, Marshall, and Snow (2000), and Pienemann, Di Biase, Kawaguchi, and Hakansson (2005).

However, it is essential that the end state receive its share of attention, because it is evidence from the end state that determines the upper limits of L2 attainment. Knowing the potential of the learner permits inferences about the nature of putative constraints on acquisition, including their relative strength and ultimate impact on learning (see Long, 1990, pp. 253–259). Accordingly, the end state is the focus of the present article. Both as a matter of logic and as a matter of theoretic adequacy, it is important to recognize that when comparing L1A and L2A, a superficial difference in ends does not necessarily imply an underlying difference in means. Nor does similarity of ends/products necessarily imply similar means/processes. Thus, for example, with respect to the question of Universal Grammar's (UG) mediating role in L2A, we understand that nativelikeness at the L2A end state does not always imply access to UG.[1] By the same token, it is clear that nonnativelike linguistic behaviors are not necessarily evidence of lack of access to UG. Researchers must be wary of linking end-state differences in L1A and L2A exclusively to a loss of general learning ability or exclusively to some erosion of any putative mechanism(s) responsible for successful L1A. Thus, linkages between product and process are to be established only with due caution.

In the literature, the terms *end state*, *final state*, *steady state*, *ultimate attainment*, and *asymptote* are used more or less interchangeably to refer to the outcome of L2A. Note that "ultimate attainment" has occasionally and erroneously been used as a synonym for nativelike proficiency. However, the term properly refers to the final product of L2A, whether this be nativelike attainment or any other outcome. For divergent views of the construct of "end state" in L2A, see Larsen-Freeman (2005) and White (2003). For discussion of operationalizing the L2A end state, see Birdsong (2004).

Researchers have explored several biographical variables that might be predictive of L2A outcomes. Age of acquisition (AoA) is understood as the age at which learners are immersed in the L2 context, typically as immigrants. This landmark is distinct from age of first exposure (AoE), which can occur in a formal schooling environment, visits to the L2 country, extended contact with relatives who are L2 speakers, and so forth. Researchers tend to equate the terms *late L2A*, *postadolescent L2A*, and *postpubertal L2A*; these are typically operationalized as AoA of >12 years. Length of residence (LoR) refers to the amount of time spent immersed in the L2 context. Because residence does not guarantee exposure to and use of the L2, researchers quantify the actual amount of contact L2 learners have with the L2 (in spoken and written modalities) and the relative use of the L1 versus the L2 in day-to-day activities. Other experiential variables include amount of formal training in the L2 as a foreign language (e.g., grammar courses, corrective phonetics) as well as amount of exposure to the L2 in so-called content courses, where nonnatives are enrolled in high school, vocational, or university classes in the L2 country.

Endogenous variables of interest to L2A researchers include the following: motivation (with several subtypes relating to outcome, e.g., motivation to pass for a native, motivation to acquire lexico-grammatical accuracy), psycho-social integration with the L2 culture, aptitude (with several presumed components, including imitative ability, working memory capacity, metalinguistic awareness, etc.), and learning styles and strategies. These are understood to be continuous, not all-or-nothing, variables. For overviews of these variables, see Dörnyei and Skehan (2003) and Doughty (2003).

AoA and L2 Ultimate Attainment

It is widely recognized that AoA is predictive of L2A outcomes, in the simple sense that AoA is observed to significantly correlate negatively with attained L2 proficiency at the end state. This conclusion is based on the results of more than two dozen experimental studies; see Birdsong (2005) and DeKeyser and Larson-Hall (2005) for overviews. The areas of language most commonly investigated are morphosyntax and pronunciation. Typically, morphosyntax errors in production or grammaticality judgments increase with advancing AoA, as does degree of judged nonnative accent.

Across many studies that examine AoA and other factors that might be related to L2 success, it has emerged that, of all the above-mentioned experiential variables, AoA is reliably the strongest predictor of ultimate attainment. This is not to say that other variables, indeed some that are confounded with AoA, are not predictive. In many cases, variables such as LoR and AoE are controlled statistically or included as factors in the experimental design.

The Age Function

From the actual behavioral data, a recurrent finding is that a linear function captures the relationship between AoA and outcome over the span of AoA (i.e., when considering aggregate data from both early- and late-AoA subjects). In 10 surveyed studies, the range of correlations is .45 to .77, with a median of about .64 (all absolute values).[2] The slope of the age function varies (i.e., it is steeper or shallower) as a function of such factors as L1-L2 pairing, amount of L2 use, task, education in the L2, and so on. It is also not surprising to find, given what is known about learning and cognitive performance over the life span (Schaie, 1994; Weinert & Perner, 1996), that there is less intersubject variation in outcome among early arrivals than among late arrivals.

When data from early- and late-AoA subjects are disaggregated, inconsistent results are obtained, producing a clouded picture of the timing and geometry of the age function. For example, DeKeyser (2000) studied 57 Hungarian L1 English L2 subjects with AoA ranging from 1 to 40 years, all with at least 10 years of U.S. residence. On a grammaticality judgment test using some items from Johnson and Newport (1989) along with some novel items, a significant correlation of AoA with scores was obtained ($r = -.63, p < .001$). However, when DeKeyser broke out the data by early- and late-arriving subjects, neither set of data yielded a significant correlation with AoA (early arrivals $n = 15, r = -.24$, ns; late arrivals $n = 42$, r $= -.04$, ns).

Another illustration of the disparate results of analyses of aggregate versus disaggregated data is seen in the comparison of the results of Johnson and Newport (1989) and Birdsong and Molis (2001). Johnson and Newport looked at accuracy on a 276-item grammaticality judgment by a group of Chinese and Korean natives ($n = 46$) with English as their L2. The Birdsong and Molis study was a strict replication of Johnson and Newport, but in this case, the subjects were Spanish natives ($n = 61$). Over all subjects and AoAs, Johnson and Newport found a strong linear relationship between AoA and accuracy ($r = -.77, p < .01$). This finding was reproduced by Birdsong and Molis ($r = -.77, p < .0001$). However, when the subjects were divided into AoA groups of ≤16 years and >16 years, the analyses produced divergent results. Figure 1 represents these differences.

The pattern of results seen in Johnson and Newport (1989) is a decline in scores with increasing AoA for early arrivals ($r = -.87, p < .01$) and an essentially random distribution of scores for the older-arriving group ($r = -.16$, ns). A quite different pattern was obtained by Birdsong and Molis (2001). For early arrivals,

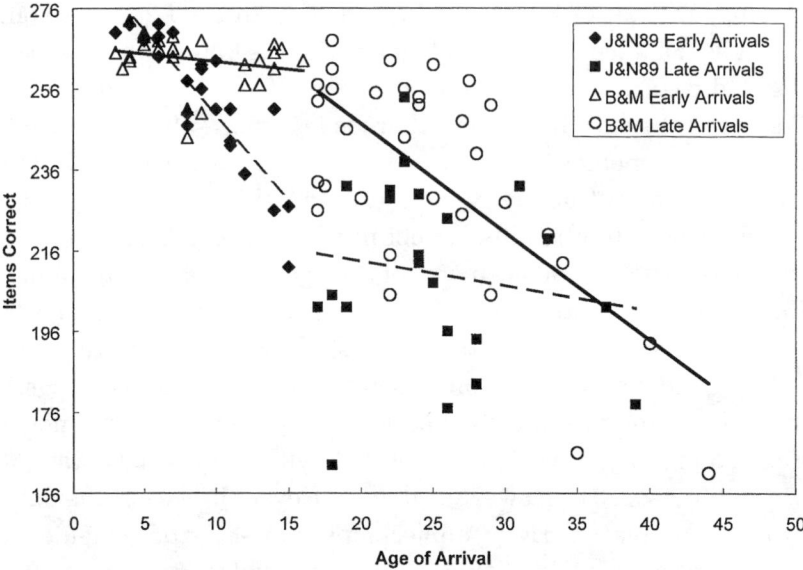

Figure 1. Plot of accuracy over AoA from Birdsong and Molis (2001, p. 240). Solid regression lines are fit to the Birdsong and Molis data; dashed lines are fit to the Johnson and Newport (1989) data. Early and late AoA groups are divided at 16 years.

the correlation of scores with age is not significant ($r = -.24, p = .22$), as this subgroup performed at ceiling. For late arrivals, the correlation is strongly negative ($r = .69, p < .0001$).

In a reexamination of the Johnson and Newport (1989) data, Bialystok and Hakuta (1994) moved the cutoff point separating early- and late-arriving groups to 20 years. For late learners, the subsequent correlation reached significance ($r = -.50, p < .05$). Birdsong and Molis (2001) conducted a similar reanalysis of their data, placing the cutoff at various ages between 15 years and 27.5 years; all correlations reached significance.

The meta-analysis by Birdsong (2005) of L2 end-state morphosyntactic and pronunciation behavioral research arrives at three main conclusions: (a) In all analyses of pooled data from early and late arrivals, age effects persist indefinitely across the span of surveyed AoA (i.e., they are not confined to a circumscribed period); (b) In analyses of disaggregated samples (and in studies that look only at late AoA), most studies find significant AoA effects for the late learners, indicating postmaturational declines in attainment; (c) in analyses of early-arrival data alone, AoA effects are inconsistent: Some are flat, some are random, and some are monotonically declining.

Do Observed AoA Effects Suggest a Maturationally Based Critical Period?

We can now take a step back and consider whether observed AoA effects can be interpreted as critical period effects.[3] If what we are dealing with is in fact a period, the age effects observed in the data must be confined to a finite time span; see Bornstein (1989) for a further discussion of characteristics of a critical period. Moreover, if the effects are maturational in nature, then the age function prior to the end of maturation should look different from the age function after the end of maturation.

Taken together, the requirement of finite age effects and a discontinuity in the age function synchronized with the end of maturation permute into three basic patterns (see Figure 2). One is a stretched "L" or hockey stick shape, with age-related declines ceasing at a point of articulation that coincides with the end of maturation. The second is an upside-down mirror image of the stretched "L," resembling a stretched "7." The flat portion at the top left of the image is the period where success is guaranteed. A third possibility, laid out by Johnson and Newport (1989) and expanded by Pinker (1994), specifies a causal role of brain

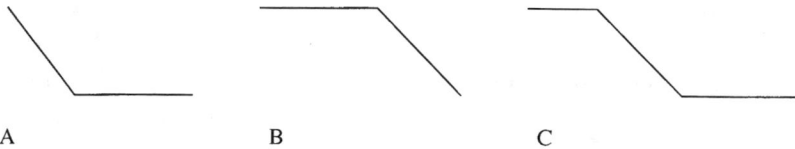

Figure 2. Three patterns of bounded age effects: (A) stretched "L" shape; (B) stretched "7" shape; (C) stretched "Z" shape.

maturation in L2A age effects, with the end of age effects synchronized with the completion of brain maturation. This version combines features of the first two possibilities to produce the image of a stretched "Z." The function begins with a period of ceiling effects, followed by a decline that ceases at the end of maturation, after which the age function flattens and no further age effects are seen.

Let us consider the third possibility first. The stretched "Z" shape (Figure 2C) includes two finite periods. At the upper left portion of the image, where performance is at ceiling, we indeed observe a bounded period, which is actually a period during which age effects are absent, as there is no downward slope in the age function. The next segment is a bounded downward slope; the age effect begins prepubertally and ends at the completion of maturation. The third segment, which is unbounded, captures the hypothesized bottoming out or flattening of the age function. Johnson and Newport (1989) purport to have produced findings consistent with the timing and geometric features just described. However, instead of an orderly array of scores parallel to the x-axis—that is, the hypothesized floor effect—one finds a random dispersion of points. In other words, the crucial flattening feature of the function, whose beginning should coincide with the end of maturation, is in fact not present in the data.

Moreover, as mentioned earlier, if following Bialystok and Hakuta (1994), one moves the cutoff point to 20 years, the late-arrivals data in Johnson and Newport (1989) start to look a bit more orderly. The result of the ensuing analysis is neither a random distribution nor a floor effect, but a significant negative correlation of AoA and performance for the late-arriving group.

The stretched "L" or hockey-stick representation (Figure 2A) incorporates a sloping segment on the left that would satisfy the requirement of a bounded period during which AoA is negatively correlated with outcomes. It also contains a flattened segment, the beginning of which coincides with the end of maturation. A review of the literature (see Birdsong, 2005; DeKeyser & Larson-Hall, 2005) reveals that several analyses of disaggregated data show prematurational declines—the left portion of the stretched L. However, for later learners (i.e., those whose performance would be represented by the right segment of the stretched "L"), there is no evidence of a flat function or floor effect. Instead, for late-learner groups, there is either a random array of scores (e.g., DeKeyser, 2000; Patkowski, 1990) or a persistent decline in performance with increasing AoA (e.g., Bialystok & Miller, 1999; Birdsong, 1992). Returning to Figure 2A, we note that the *appearance* of a stretched "L" shape (i.e., the two rightward segments of the "Z") is obtained for the Johnson and Newport (1989) data when linear functions are applied separately to early- and late-arrival data. A systematic performance decline over AoA is indeed observed for early arrivals ($r = -.87$). However, as we saw earlier, for late arrivals, a flat segment is a misleading representation of the correlation coefficient in this instance ($r = -.16$), as the best-fitting near-horizontal regression line actually goes through a random array of scores, not through an orderly set of points that are parallel to the x-axis.

The final scenario by which age effects would be considered critical period effects is the mirror image of the one just discussed, a stretched "7" or upside-down hockey stick shape with the "blade" at the top left (Figure 2B). This is an unconventional, although often implicitly invoked, notion of a critical period function (see Birdsong, 2005, for discussion of conventional and unconventional conceptions; based on Bornstein, 1989). The leftmost part of the function is flat, with performance at ceiling. On the right portion of the image, the age gradient (i.e., the decline in ultimate attainment with advancing AoA) is not bounded. What is bounded is the left segment of the image, the period of peak attainment, which is often referred to as a "window of opportunity"—the temporal span during which sensitivity or learning potential is at its highest and full attainment is guaranteed. Such a period has been observed in at least one study: Birdsong and Molis (2001).[4] As seen in Figure 1, a roughly flat function at ceiling is generated by the performance of the early-arriving AoA group of Spanish L1 speakers. This "age noneffect" is confined to a limited span, thus satisfying the geometric criterion and corresponding to the unconventional "window of opportunity" version of the critical period. However, because of the apparent duration of the window of opportunity, the temporal features do not conform to a maturational account of AoA effects. For their L2 learners' results, Birdsong and Molis (pp. 241–242) conducted a series of post hoc piecewise regression analyses that included the inflection point (i.e., the terminus of the period) as a free parameter. Under these conditions, the best-fitting function placed the end of the ceiling period, and thus the beginning of the decline, at 27.5 years. In other words, the period of peak performance extends 10 or more years beyond the end of maturation. Thus, although the Birdsong and Molis results reveal a stretched "7" shape and its circumscribed period of full attainment, the temporal parameters do not mesh with a maturational-effects account of L2 ultimate attainment.

Divergent Conceptualizations of "Critical Period"

Singleton (2005) examined several proposals for the timing of the "end of the critical period." In most cases, these proposals made reference to the end of the period of peak sensitivity; that is, they invoked the "window of opportunity" notion of critical period. In the studies that Singleton surveyed, hypothesized beginnings of declines ranged from near birth to late adolescence. Some proposals made distinct timing claims for phonetics/phonology versus other areas of linguistic knowledge and performance. Such so-called "multiple critical period" accounts of attainment in various language domains were advanced by Long (1990) and Seliger (1978) for the L2 context and are consistent with current neurobiological thinking about critical periods in other contexts (Knudsen, 2004).

The proposals of Johnson and Newport (1989), Lenneberg (1967), Long (1990), Pinker (1994), Scovel (1988), and Seliger (1978) signaled changes that occur around puberty. Significantly, in some cases, this maturational milestone is thought to be the point at which declines in performance *begin* (i.e., the unconventional notion of critical period), and in other cases, this maturational milestone is thought to be the point at which performance declines *cease* (the conventional notion). Thus, a serious conceptual issue confronts proponents of a maturational account of constraints on L2A attainment: Does maturation determine the beginning of age effects or the end of age effects?

Empirically, neither account of the timing of maturational effects fares very well. As discussed earlier, it is now understood (e.g., Birdsong, 2005; Hyltenstam & Abrahamsson, 2003) that the behavioral data are generally inconsistent with either a period of peak sensitivity whose end coincides with the end of maturation or with a leveling off of sensitivity whose beginning coincides with the end of maturation. For additional commentary on the timing of age effects, see Moyer (1999).

Incidence of Nativelike Attainment in Late L2A

Like the facts about the age function, the facts relating to nativelike attainment in L2A do not lend themselves to simple generalization. Moreover, as was the case with the age function, the interpretation of these facts is not without controversy.

Historically, research in L2A has been guided by what has been termed the *deficit model*. Characterizing the end state of L2A as a "lack of success," research in this tradition looks to explain the "near-universal failure" of adults to reach attainment comparable to that observed in L1A (Bley-Vroman, 1989). The prevailing view was that nativelikeness, if ever observed, was so rare as to be of no relevance to L2A theory (e.g., Bley-Vroman; Selinker, 1972). Estimates of a 0–5% incidence of nativelikeness were more a matter of guesswork than experimentation and might have referred to a population that included foreign language learners and others who were not at the L2A end state. More recently, however, a number of studies have targeted immigrants with sufficient LoR and contact with natives to qualify for end-state status and have scrupulously attempted to ascertain the rate of nativelikeness in the sample. The findings of these studies suggest that nativelikeness in late L2A is not typical, but neither is it exceedingly rare.

More than 20 studies have reported the rate of nativelikeness among late (AoA ≥ 12 years) L2A learners. In these studies, the incidence of nativelikeness ranges from 0% to 45.5%. Higher rates of nativelikeness in the area of morphosyntax are associated with certain L1-L2 pairings (e.g., Cranshaw, 1997), with increased L2 use (e.g., Flege, Yeni-Komshian, & Liu, 1999), and with L2 dominance (e.g., Flege, MacKay, & Piske, 2002). In the area of pronunciation, those learners who are taken for natives by native judges tend to be those with high levels of L2 practice, motivation to sound like a native, and L2 phonetic training (e.g., Bongaerts, 1999).

Anecdotal evidence, along with some research, suggests that nativelikeness is attested less often in the domain of pronunciation than in other performance domains. However, nativelike pronunciation is not impossible, as studies by Birdsong (2003) and Bongaerts and colleagues (see Bongaerts, 1999, for summaries of their studies) have shown. The perceptual abilities underlying unaccented L2 pronunciation have proved to be amenable to training in some studies (e.g., Bradlow, Pisoni, Akahane-Yamada, & Tohkura, 1997; McCandliss, Fiez, Protopapas, Conway, & McClelland, 2002; McClelland, Fiez, & McCandliss, 2002) but resistant to training in others (e.g., Takagi, 2002; see Darcy, Peperkamp, & Dupoux, in press, for an overview).

Domains of Nativelikeness

There exists a widespread belief (Hyltenstam & Abrahamsson, 2000, 2003; Long, 1990; Scovel, 1988) that nativelike attainment by late L2 learners, if observed at all, will be confined to one or a few tasks and that an individual will not display nativelikeness across a variety of linguistic behaviors (or experimental performances). The coinage "Joseph Conrad effect" captures this notion. However, recent work suggests that the attainment of broad nativelikeness among late L2 learners is in fact possible. In a study of end-state L2 English acquisition, Marinova-Todd (2003) recruited 30 late learners (AoA > 16 years; mean = 11 years) with at least 5 years' residence (mean = 11 years) in an English-speaking country. These subjects had been informally screened for high English proficiency and, like the 30 native controls, were college educated. Nine tasks targeted an array of linguistic performance. Two tasks related to pronunciation, one for spontaneous speech and one for read-alouds; three tasks tested morphosyntactic accuracy in both online and offline performance; two tasks probed lexical knowledge in oral descriptions; and two tasks involved language use in narrative and discourse. Of the 30 late learners, 3 performed to nativelike criteria across all nine tasks. Six others were indistinguishable from natives on seven tasks. The results of this study are of particular interest because the performances tested included not only the core areas of grammar and pronunciation but also lexical diversity and narrative and discourse competence. Moreover, some of the tasks used by Marinova-Todd did not involve reflection and metalinguistic analysis, thus muting the argument that nonnativelikeness will inevitably be ferreted out in spontaneous language use (Hyltenstam & Abrahamsson, 2000, 2003). See Birdsong (to appear) and Ioup, Boustagui, El Tigi, and Moselle (1994) for additional evidence of broad nativelikeness in late L2A. Where nativelikeness is perhaps least likely to be observed is in certain domains of language processing. Differences between highly proficient late L2 learners and monolingual natives have been noted in the areas of lexical retrieval, structural ambiguity resolution, and detection of acoustic distinctions in the areas of syllable stress, consonant voicing, and vowel length (e.g., Clahsen & Felser, 2006; Dussias, 2004; Dupoux & Peperkamp, 2002; Papadopoulou & Clahsen, 2004). The observed behavioral differences appear to be both quantitative (speed, accuracy) and qualitative (parsing in a structurally shallow manner; mishearing segments) in nature. Other types of sentence processing difference between natives and learners are revealed in event-related potential (ERP) and eye-tracking studies; see Frenck-Mestre (2005) for a review.

Use of Evidence of (Non-)Nativelikeness

There is ongoing discussion about the relevance to L2A theory of behavioral evidence showing end-state nativelikeness and nonnativelikeness. Should researchers dig around for any soupçon of nonnativelikeness and declare this to be proof that learning mechanisms are rendered defective by aging? Consider the use of a nonnative lexical item—say, an exclamation in a moment of passion or pain. Does this departure from nativelikeness constitute evidence of defective L2 learning?

Now consider small quantitative differences between the L2 and native L1 (e.g., shorter-than-native-norm voice onset time [VOT] values averaged over subjects). In bilingualism, L2 VOT values tend to move toward L1 VOT values; at the same time, L1 VOT values of bilinguals move toward L2 values (Flege & Hillenbrand, 1984; Mack, Bott, & Boronat, 1995). L2 effects in the L1 have been observed in such diverse domains as collocations (Laufer, 2003), middle-voice constructions (Balcom, 2003), syntactic processing (Cook, Iarossi, Stellakis, & Tokumaru, 2003), and lexical decision (Van Hell & Dijkstra, 2002). Rather than invoke deficiencies in learning (which could not apply to changes in the L1), it is more reasonable to argue that minor quantitative departures from monolingual values are artifacts of the nature of bilingualism, wherein each language affects the other and neither is identical to that of a monolingual. For further discussion, see Cook (2002), Flege (2002), and Grosjean (1989).

L2 Dominance

To conclude this consideration of nativelikeness in late L2A, I would like to suggest that investigations of the upper limits of attainment in late L2A could profit by targeting an underrepresented group, namely late L2 learners who are L2-dominant. Flege et al. (2002) presented an illustration of the possible benefits of such investigations. The researchers looked at the English pronunciation of three groups of Italian

L1/English L2 bilinguals: L1-dominants, balanced bilinguals, and L2-dominants. They found that both L1-dominants and balanced bilinguals spoke with detectable accents, whereas the pronunciation of L2-dominant bilinguals was indistinguishable from that of native controls. Flege et al. speculated that, in the area of pronunciation at least, L2-dominants are less likely to be subject to interference effects from the L1. Whether or not this speculation proves tenable, researchers should recognize the possibility that data from L1-dominants, high-L2-proficients, and even balanced bilinguals might not give the full picture of the capacities of late L2 learners.[5]

To take the discussion of L2 dominance to a logical extreme, consider the case of the adoptees studied by Pallier et al. (2003). These eight individuals were removed from their native Korea at ages ranging from 3 to 8 years and were placed in homes in the Paris area. With no subsequent contact with Korean into their adult years, the adoptees' L2 (French) became their dominant language. Behavioral measures revealed no trace of residual Korean knowledge, and functional magnetic resonance imaging (fMRI) scans showed no specific activation when listening to Korean. Informal measures of their French speaking showed that the Korean adoptees behaved like French natives, and they performed like native French speakers on formal tests of French grammatical knowledge as well (Ventureyra, 2005). For determining the upper limits of L2A as a function of AoA, it would potentially be revealing to study larger numbers of such individuals, covering a larger range of age of adoption. Under such a design, the confounding effects of L1 representational entrenchment (on L1 entrenchment, see Kuhl, 2000; MacWhinney, 2005a) and L1 use would be minimized.

Age and Nativelikeness in Brain-Based Measures

As a complement to linguistic and metalinguistic data, brain-based evidence illuminates important dimensions of the question of age and L2A. A number of recent reviews gave more breadth and depth to discussion of relevant research than space permits here (e.g., Abutalebi, Cappa, & Perani, 2005; Indefrey, this volume; Stowe & Sabourin, 2005). See also Green (2005) and Paradis (2004, 2005) for discussions of the limitations of localization research using imaging techniques.

Comparing L1 Processing and L2 Processing

The basic research issue addressed in this area of cognitive neuroscience is whether processing in the L2 is accomplished in the same way as processing in the L1. The degree of observed similarity hinges on three principal factors: the age at which L2 acquisition is begun, the level of L2 proficiency, and the type of task demanded of the subjects. As one would expect with any complex cognitive activity, some of the most revealing results relate to interactions among these factors.

When comparing L1 and L2 processing, we might be referring to the psychology of cognition (e.g., automatic vs. controlled processes; implicit vs. explicit knowledge), the nature of mental representations (e.g., symbolic vs. subsymbolic representations; encapsulated vs. distributed representations), the general area of the brain that is activated (e.g., involvement of cortical vs. subcortical regions; left hemisphere vs. right hemisphere), or, within a given region of the brain, the particular neuronal circuits engaged in language processing.

AoA and Proficiency: Imaging Studies. Early research (e.g., Kim, Relkin, Lee, & Hirsch, 1997) showed cortical activation differences between late and early bilinguals in L2 production tasks. It was tempting to conclude from these findings that later AoA results in nonnativelike brain activity patterns. However, this conclusion is not supported in subsequent investigations that have controlled for or manipulated the factor of L2 proficiency.

In studies of production, it is L2 proficiency level, not AoA, that emerges as the strongest predictor of degree of similarity between late learners and monolingual natives. This generalization must be qualified, however, as the degree of similarity varies from study to study. Moreover, what is meant by "production" is also quite variable, with tasks ranging from word repetition (Klein, Zatorre, Milner, Meyer, & Evans, 1994), to (typically cued) word generation (Chee, Tan, & Thiel, 1999; Klein et al., 1995), to sentence generation (Kim et al., 1997), to cognate and noncognate naming (De Bleser et al., 2003). Finally, from study to study, there are exposure differences and degree of proficiency differences that make comparisons and generalizations difficult.

In comprehension studies (of which there are relatively few), similar issues of incommensurability must be taken into account. Still, a coherent pattern of sorts emerges. For story listening tasks, two studies (Perani et al., 1996 [PET; positron-emission tomography]; Dehaene et al., 1997 [fMRI]) found differential activation between natives and low-proficient late learners. However, when Perani et al. (1998) compared high-proficiency late and early bilinguals on story listening (PET), overlapping patterns of brain activity were found.

Two fMRI investigations involving comprehension, then judgment, are worthy of note. Chee, Hon, Lee, and Soon's (2001) fMRI study of high- and low-proficiency bilinguals found that highly proficient subjects (AoA $\geq$ 12 years) had relatively reduced brain activity in left prefrontal and parietal areas. The fMRI study of Wartenburger et al. (2003) involved semantic and grammar judgments by three groups of Italian-German bilinguals divided by AoA and proficiency (early acquisition/high proficiency; late acquisition/high proficiency; late acquisition/low proficiency). Activation for grammar judgments in the L2 was found to be related to AoA: The two high-proficiency groups with different AoAs showed different activations, the activations being more extensive across Broca's and other areas for the later learners. However, the authors point out that some differences might have been related to proficiency, as the nominally equal proficiency groups actually differed in grammaticality judgment accuracy. On the L2 semantic judgment task, similar activations were found for early and late high-proficients (i.e., irrespective of the AoA difference among the groups). Comparisons of L1 processing versus L2 processing were also carried out in within-group analyses. On the grammatical task, for both late-acquiring groups (i.e., both low and high proficiency), there was more extensive activation in Broca's and subcortical regions in L2 processing than in L1 processing. On the semantic task, the early acquisition/high-proficiency group did not exhibit differences in processing the L1 versus the L2. However, on this task, both of the late-acquiring groups showed greater bilateral activation in inferior frontal areas for L2 versus L1 processing.

Studies of word-level meaning and reference (Chee et al., 2000; Ding et al., 2003; Xue, Dong, Jin, Zhang, & Wang, 2004) have shown similarities in areas of activation in L1 and L2. In the case of Xue et al. (2004), where subjects were asked to judge whether pairs of words were related, relatively late Chinese learners of English (age of exposure between 8 and 10) with rather low proficiency (subjects had had 2 years of English study and no other exposure or practice) showed activations in both L1 and L2 in the fusiform gyrus, Broca's area, and left parietal lobe.

Although there is a general congruence of brain areas activated in the L1 and L2 by proficient late bilinguals, the degree of activation might be different. Specifically, more neuronal activity in a given area is sometimes seen in L2 versus L1 processing, as indicated either by more voxels in a given area being activated or by more signal change for the same voxels. This pattern has been observed for early bilinguals as well as late bilinguals. These indexes correspond to increased neural activity in a specified area, and the extra activity could be viewed as evidence that the L2 is being processed with more effort than the L1 (Stowe & Sabourin, 2005).

For discussion of the relationship of extent of activation to proficiency in the two languages, see Wartenburger et al. (2003, pp. 167–168). For a discussion of the extent of activation pattern changes with aging, see Park and Gutchess (2005, pp. 237–238).

AoA and Proficiency: ERP Studies. Broadly speaking, the ERP literature relating to the When of language processing is consistent with the fMRI and PET literature that speaks to the Where question; that is, the timing components of high-proficient L2 use are by and large similar to those of L1 use, even when acquisition of L2 was begun at age 12 or later (e.g., Hahne, 2001; Hahne & Friederici, 2001; Ojima, Nakata, & Kakigi, 2005; Proverbio, Cok, & Zani, 2002; Stowe & Sabourin, 2005). From the ERP studies, as with imaging studies, it appears that there is general support for the "convergence hypothesis" articulated by Green (2005), which states that as L2 proficiency increases, the processing profile in the L2 becomes more similar to that of native L1 use.

Recent research suggests that the similarities appear earlier in the course of adult L2 learning than had been previously thought; for example, McLaughlin, Osterhout, and Kim (2004) found P600 effects for syntactic violations after just 4 months of L2 learning, and Osterhout, McLaughlin, Kim, and Inoue (2004) demonstrated nativelike word versus pseudo-word N400 effects after only 14 hr of instruction. The corresponding behavioral results for the subjects were not nativelike, suggesting that the amount of L2 learning taking place might be understated in behavioral data (see also Indefrey, Hellwig, Davidson, & Gullberg, 2005; Mueller, Hahne, Fujii, & Friederici, 2005).

On the other hand, the study by Sabourin (2003) suggests that behavioral measures might overstate similarities, whereas ERP might pick up on differences, at least among mid-proficiency L2 learners of different L1 backgrounds. Subjects were late learners (>12 years AoE) of Dutch from German, Romance, and English native-language backgrounds. On grammaticality judgments of verb feature agreement, all three learner groups performed at about 90% accuracy (native controls' accuracy = 97.4%). However, the groups differed in terms of ERP signals recorded as the judgments were being made. The German group showed roughly nativelike N400 and P600 responses, whereas the Romance and English groups displayed no early negativity and their P600 was delayed and smaller relative to native control data.

The Aging Brain

The next descriptive component in our consideration of age and L2A consists of facts about the aging brain, with which explanatory accounts of age-related differences in ultimate attainment must be compatible. Neurocognitive features of aging are amenable to investigation at various organizational and analytic levels. Those relevant to language learning and use include the functional/processing level (lexical encoding and retrieval, processing speed and depth, concatenation and coordination of grammatical units in real time, etc.), the functional/learning level (Hebbian learning, declarative memory and procedural memory, etc.), the brain structure level (hippocampus, striatum, etc.), and the cellular level (neurotransmission, regional volumetric decline, neurogenesis, etc.). The basic consideration is the degree and locus of age effects at these various levels of analysis.

L2 and Cognitive Aging

From the work of Bäckman and colleagues (e.g., Bäckman, Small, & Wahlin, 2001), Park (e.g., Park, 2000), Salthouse (e.g., Salthouse, 1996), and others, we have come to recognize several general patterns in cognitive aging. In tasks that tap working memory and episodic memory, there is an observed performance decline over age, starting in young adulthood. Declines in associative memory and incremental learning also appear to begin in young adulthood. On tasks involving priming, recent memory, procedural memory, and semantic memory, age-related effects, when observed, are comparatively mild (Craik, 2000, pp. 78ff). Age effects are also comparatively mild for implicit memory tasks versus explicit memory tasks relating to lexical recall (Park, pp. 7–8).

Researchers have identified three principal components of cognitive aging (Park, 2000): decreases in processing speed, deficits in working memory, and decreases in suppression (i.e., the ability to focus attention on relevant material that some link to working memory; see also Rogers, 2000). Each of these abilities is involved in some stages of L2 acquisition and routinely in language use (L1 and L2).[6] With increasing age, both L1 and L2 use are affected via declines in these areas of language processing. In L2 use, age effects in these domains are likely to be more pronounced than in the L1 case, due to a relatively low degree of automaticity in L2 processing (Segalowitz & Hulstijn, 2005).

On tasks where speed and efficiency demands are made and when relatively new information is involved, two features of the age gradient stand out. First, the onset of performance decline begins in early adulthood (around age 20). Second, the decline across the adult life span is generally linear and, in all cases, continuous (Bäckman & Farde, 2005, p. 68). Note that within the general trends in cognitive performance, there is a range of variation among individuals. These should play out in L2A as interindividual differences in ultimate attainment.

Age, Brain Volume, and L2

In this subsection we speculatively explore the possibility of a connection between brain volume decreases in aging and declines in L2 acquisition and processing. The volumetric decreases are known to begin in the twenties or later, indicating that if there were a link between brain volume and L2A, it would clearly be biological in nature, but not maturational.

In vivo studies using magnetic resonance imaging (MRI) reveal that, as a general rule, brain volume decreases with advancing age (see Raz, 2005, for a review). The degree of shrinkage varies from brain structure to brain structure, as do the details of timing of the onset of decline. In all cases surveyed, the declines, once begun, are typically linear and are consistently continuous, with no leveling off at the end.

Starting at the coarsest level of investigation, *in vivo* studies reveal that gray matter volume declines in a linear fashion beginning in childhood (e.g., Pfefferbaum et al., 1994; Courchesne et al., 2000). (Postmortem studies reveal a slightly different trajectory; see Miller, Alston, & Corsellis, 1980; cited in Raz, 2005.) In contrast, white matter volume enjoys a linear increase until the early twenties. An ensuing plateau continues into the sixties, after which there is a linear decline into old age. The inverted U shape for white matter volume over age has been replicated in many but not all studies. Declines are minimized in healthy subjects and are heightened in subjects with cardiovascular disease (Raz, p. 22).

Looking now at specific regions of interest, the question driving a great deal of research is whether the volumes of some areas of the brain are more affected by age than others. The answer to this question is not straightforward, as differing results are obtained by different measurement techniques and in longitudinal and cross-sectional studies, with the latter typically underestimating the amount of shrinkage. However, a reasonably clear picture of age-related declines in regional brain volumes was offered by Raz (2005) in his survey of relevant studies. Results of cross-sectional studies reveal that the sites most affected by age are the prefrontal cortex, the putamen, the caudate nucleus, the hippocampus, and the temporal cortex. In longitudinal studies, we find that the four areas most susceptible to volumetric declines are the entorhinal cortex, the hippocampus, the caudate nucleus, and the frontal lobe, all with $\geq 1\%$ annual declines.

In addition to these data, consider the results of the Raz et al. (2003) study of 53 healthy adults between the ages of 20 and 77 years. Focusing on the striatum, the researchers found that the caudate nucleus volume declined at .83% per year, the putamen at .73% per year, and the globus pallidus at .51% per year. The shrinkage began in young adulthood. The observed declines were also linear; that is, the same rate of decline was observed for younger and older subjects. These volume declines in the striatal region go hand in hand with dopamine declines in this area (see next subsection).

Most studies, however, do not reveal the epochs at which declines begin and at which the slopes are most dramatic. However,

Raz (2005) sifted through the relevant studies to come to a few generalizations about timing and geometry of declines. First, volumes of the caudate nucleus, cerebellum, and cortical structures decline in a linear fashion that starts in adolescence and continues throughout the life span. Second, the entorhinal cortex and hippocampus appear to incur a greater annual shrinkage than other areas of the brain. These declines tend to begin in middle age to old age in the case of the hippocampus, and only in older age for the entorhinal cortex.

Whereas the relationship between brain volume and aging is typically linear and unbounded (bearing in mind that the age of onset of declines might vary from structure to structure), the relationship between brain volume and cognitive declines is apparently not linear in many cases. It has been suggested that cognitive deficits start to be expressed after structural deterioration reaches a certain threshold, but not before (Raz, 2000, p. 65). Consequently, it is challenging to connect regional morphological changes to specific cognitive deficits that might be related to L2 acquisition and processing. (Additional difficulties in making such connections arise from concerns relating to sampling, measurement, and methodological differences between studies, and interpretation of behavioral and imaging data.)

Two studies illustrate the challenges posed by this type of research. Golomb et al. (1994) found that declines in hippocampal volume predicted performance decrements on delayed declarative memory tasks (e.g., list recall, paragraph recall, and paired associates). On the other hand, Reuter-Lorenz (2000, pp. 101ff) observed that volumetric declines in the medial/temporal areas were not clearly paralleled by performance declines in episodic/associative (declarative) memory.[7] Given the present state of research, the fairest observation to be made is that neural resources, for which regional brain volume is a proxy, are reasonably good predictors of performance subserved by certain brain areas, but not others. (See related discussion in the following section.)

Age, Dopamine Systems, and L2A

The role of the nigrostriatal dopamine (DA) system in efficient motoric function is well known. In addition, DA appears to be involved in certain higher order cognitive functions, many of which are implicated in language learning and language processing, such as attention, motoric sequencing, and working memory (for a review, see Bäckman & Farde, 2005).

Schumann (1997, 2001) and colleagues (Schumann et al., 2004) have argued that DA is involved in basal ganglia functions in L2A, some of which are implicated in motivation to learn and learning reinforcement. These mechanisms are thought to contribute to proceduralization (i.e., the creation and strengthening of linguistic rules; Lee, 2004, pp. 66–67). The results of the study by Teichmann et al. (2005) of Huntington disease patients reinforce the notion that the striatum is involved in the processing of rules as opposed to words. Crosson et al. (2003) argued for a role of the basal ganglia (BG) in a variety of language production processes at the levels of syntax, lexicon, and phonology. For additional studies of BG involvement in language processing, see Friederici and Kotz (2003), Moro et al. (2001), Newman, Pancheva, Ozawa, Neville, and Ullman (2001), and Ullman (in press).

Dopamine is likewise considered essential to defossilization, an undoing of automatized nontargetlike linguistic performance (Lee, 2004, pp. 68–71). Arguably, similar DA-mediated processes are also involved in minimizing L1 influence; for example, one could envision the role of DA in suppressing and supplanting L1 routines in syntax (e.g., association of noun-first clausal sequences with subject-initial canonical word order, when in fact the L2's canonical word order is object-initial) and routines in phonology (e.g., the representation of aspirate and nonaspirate stops as allophones, as in English, when the L2 represents them as separate phonemes, as in Korean).

In humans, D1 and D2 receptors are distributed throughout the neocortex, and there is dense innervation in the caudate nucleus and putamen. Damage to the DA system in humans results in deficits in executive function, verbal fluency, and perceptual speed. In rodent and monkey studies, destruction of dopaminergic pathways in the limbic system produces memory impairments and attentional deficits. Lesioning these pathways in the subthalamic nucleus results in deficits in attention, executive function, and motor sequencing. Pharmacological interventions in humans show increased performance on tasks that measure information processing speed, discrimination, and working memory. Both D1 and D2 receptors appear to be implicated in working memory modulation. Models of DA function converge on the notion that DA facilitates switching between attentional targets both within and between neural networks, with the effect of enhancing the ratio of incoming neural signal to background noise. For a review of effects on cognition of age-related changes in nigrostriatal DA, see Bäckman and Farde (2005).

Li, Lindenberger, and Sikström (2001) found that declines in D2 receptors begin in the early twenties and continue across the life span. These declines are observed not only in the BG but also in the hippocampal structures, frontal cortex, anterior cingulate cortex, and amygdala. Of particular interest is the suggestion by Li et al. (2001) that with increased age and DA loss, neural noise increases, resulting in less distinctive neural representations. This decrease is linked to age-related cognitive deficits across domains such as working memory and executive function (Bäckman & Farde, 2005, p. 61).

A few PET studies have looked at age-related declines in DA markers and associated cognitive declines. A familiar pattern of results emerges from these studies: Declines begin in early twenties and continue linearly throughout the life span. A representative study is that of Volkow et al. (1998), who determined by PET the striatal D2 binding potential in adults aged 24–86 years. Behavioral measures included executive, motoric, and perceptual speed. D2 receptor binding decreased with advancing age in the caudate nucleus ($r = -.62$) and putamen

($r = -.7$); similar correlations were obtained between age and task performance.

Thus, with respect to the geometry and timing of the DA age gradient and in terms of the cognitive functions mediated by DA, it would appear that DA declines are a plausible mechanism (among others) underlying age effects in L2 acquisition and processing. A similar conclusion could apply to stress- and age-related increases in cortisol, which have been linked to hippocampal atrophy (Lupien et al., 1994, 1998). Also, with adjustments in the temporal and geometric features of the age-related declines, the same might be said of fluctuations in estrogen levels over age, as forms of this hormone are known to mediate verbal memory, production, and processing (e.g., Kimura, 1995; Resnick & Maki, 2001). As was the case with respect to brain volume declines, the possible linkage to L2A of changes in dopamine, estrogen/testosterone, and acetylcholine metabolism (e.g., Freeman & Gibson, 1988) is understood to be biological in nature, but given that the changes do not begin until adulthood, a maturational explanation is ruled out.

Summary

In a nutshell, what do studies of the aging brain reveal about L2 acquisition and processing? From the cognitive literature, we learn that the associative memory and incremental learning elements of language learning are steadily compromised by age, as are the working memory and processing speed components of language processing and production. It appears that these declines are linear and that they begin in early adulthood and continue throughout the life span.

Second language use, at least among non-L2-dominants, is less automatic and less efficient than L1 use. As increasing demands are made on a finite-capacity functional system, performance declines are to be expected. For this reason, processing deficits are likely to show up earlier and to be more pronounced in typical L2 use than in L1 use.

For some areas of the brain, we see some evidence of linkages between age-related morphological changes and the cognitive processes mediating L2 learning, production, and processing; for example, age-related declines in working memory, attention, and speed of processing appear to be roughly correlated with volumetric declines in the frontal lobe and prefrontal cortex, the latter area being particularly susceptible to the effects of aging. A somewhat stronger case can be made for the relation of age-related dopamine declines to a variety of cognitive deficits that could undermine L2 processing and acquisition, as the research findings appear to be more straightforwardly interpretable than those associated with brain volume studies.

As for the timing of changes in the aging brain, none of the evidence from the cognitive, brain volume, or dopamine literature is consistent with a maturational account because the observed declines commence after the end of maturation. With respect to the geometry of declines, the literature in all three areas generally indicates linear declines. However, it has been suggested that for some brain regions, the actual expression of functional deficits does not begin at the onset of volumetric changes, but at a point later in life when a theorized threshold has been crossed.[8] Arguably, such a suggestion could extend to the connection between neurobiological/neurochemical/neuroanatomical states and cognitive processes in general.

Finally, it should be emphasized that in this section, we have mentioned only a few of the well-studied neural sources of age-related cognitive decline. For a comprehensive overview of the cognitive neuroscience of aging, see Cabeza, Nyberg, and Park (2005).

The Nature of Age Effects in L2A

To conclude this overview, let us step back and reflect briefly on the sources of age effects in L2A. In the literature, we find a multiplicity of candidate causal mechanisms—biological and experiential—and mediating factors—endogenous and exogenous—that underlie age effects in L2A. Singleton (2005) saw no less than 14 versions of the Critical Period Hypothesis as it applies to L2A (CPH/L2A). Birdsong (1999) cited six major variants of the CPH/L2A and pointed to numerous endogenous and exogenous factors that affect ultimate attainment in L2A. MacWhinney (2005b) identified 10 "concrete proposals" in the literature that relate AoA to ultimate L2A attainment, and to these were added two explanations for variability in L2A outcomes. The various hypothesized mechanisms relate to the biology of the species (in its neurobiological or neurocognitive dimensions), developmental aspects of cognition, L1 influence, use of the L1 and L2, and psycho-social/affective dimensions of individuals' personalities, including a person's motivation to learn, appear nativelike, or integrate into the L2 culture.

A summary and evaluation of these accounts would be impractical in the present context (for critical reviews, see Herschensohn, in press; Singleton & Ryan, 2004). A brief commentary must suffice.

There is an understandable tendency in discussions of the underlying sources of age effects in L2 learning and processing to isolate a single mechanism or to focus on one type of mechanism. Yet, this practice often simplifies the phenomena in question and polarizes stances on an extremely textured set of issues. It is arguably more reasonable to take the initial position that the identified factors and mechanisms that are not at odds with empirical findings are each potentially at work in some fashion in L2A. Some might account for more variance than others, and individual differences in L2 attainment and processing are to be expected (Bowden, Sanz, & Stafford, 2005; Dörnyei & Skehan, 2003; Skehan, 1989). Some factors trump others; for example, it is pointless to invoke neurobiological capacities (or deficiencies) in the context of an individual who has no interest in passing for a native (Klein, 1995; Moyer, 2004; Piller, 2002).

Ongoing research in L2 acquisition must account not only for the typical decline in L2 attainment with age but also for the nativelikeness that late learners are manifestly capable of. To do so adequately will require clear-eyed and open-minded attempts to integrate biological, cognitive, experiential, linguistic, and affective dimensions of L2 learning and processing.

Notes

[1] Innately specified linguistic knowledge given by UG is posited to account for the apparent gap between learners' knowledge of linguistic structure and what they have been exposed to in the linguistic input (e.g., Chomsky, 1975). In late L2A, learners have access to fully developed linguistic representations in their L1. With this knowledge, supplemented by domain-general learning procedures such as inference and analogical patterning—and of course, L2 input—nativelikeness in at least some areas of the grammar is undeniably possible. Thus, not all nativelikeness is evidence for access to UG. For particular abstract linguistic features or structures, access to UG is inferred from evidence that L2 learners' knowledge could not have been attained by L2 input and domain-general cognition. For elaboration on this point, see Coopmans (this volume).

[2] These figures are expressed as absolute values because some experiments correlate AoA with numbers of errors or degree of foreign accent—thus resulting in positive correlation coefficients—whereas others correlate AoA with numbers of correct items or degree of nativelike accent—thus yielding negative correlations.

[3] The hypothesis of a critical period for L2A has been formulated by different researchers in different ways and invoking a variety of explanatory mechanisms; see the final section of this article as well as Birdsong (1999), Herschensohn (in press), Singleton (2005), and Singleton and Ryan (2004) for overviews.

[4] For their early-arriving subjects, DeKeyser (2000) and Patkowski (1990) found near-zero correlations of attainment and age. However, in both studies, the data for early arrivals did not constitute flat functions at ceiling, which would be consistent with a window of opportunity for full attainment.

[5] It is important to recognize that "L2-dominant" is not a homogeneous category. Like proficiency, dominance is a continuous construct. The degree to which a person is L2-dominant, as operationalized by performance on quantitative psycholinguistic measures such as reading speed or numbers of words extracted from L1 versus L2 speech amid background noise, varies from one individual to another. Dominance is also a relative construct; it is expressed, for example, as the proportion of words read per minute in the L2 and L1. L2 dominance does not always equate to nativelikeness. For a given L2-dominant, the number of words read per minute in the L2 might

not necessarily fall in the range of reading speed of monolingual natives. Further, dominance is not to be confused with grammatical proficiency. The fact that a given bilingual is better at extracting signal from noise in a particular language does not necessarily mean that this person is highly proficient in that language. Finally, this section has concentrated on psycholinguistic definitions of dominance, as opposed to frequency of use or psycho-social identification with a given language. For discussion of various operationalizations of dominance, see Flege et al. (2002), Golato (1998), and Grosjean (1998).

[6]Among bilinguals, working memory might be involved in controlling the activation of the two languages (see Kroll & Tokowicz, 2005; Michael & Gollan, 2005), which might be tied to a general ability to suppress irrelevant information (see Anderson, 2003; Michael & Gollan). Scores on tests of working memory in the L2 correlate with L2 proficiency (Stafford, 2005; see also Bowden et al., 2005). Thus, in the typical case of nonnativelike L2 proficiency, working memory capacities in the L2 should be inferior to those in the L1.

[7]Associative memory is essential to connectionist accounts of language acquisition and use and to the representation and processing of irregular forms under the words-and-rules approach (e.g., Pinker & Ullman, 2002) in both L1 and L2 (Ullman, 2001).

[8]This review of the aging brain has not considered the relationship between neurocognitive processing resources that are affected by aging and the actual level of activation of neural tissue in the particular brain regions in question. For consideration of technical, empirical, and theoretical issues surrounding this question, see Park and Gutchess (2005).

References

Abutalebi, J., Cappa, S. F., & Perani, D. (2005). What can functional neuroimaging tell us about the bilingual brain? In J. F. Kroll & A. M. B. de Groot (Eds.), *Handbook of bilingualism: Psycholinguistic approaches* (pp. 497–515). New York: Oxford University Press.

Anderson, M. C. (2003). Rethinking interference theory: Executive control and the mechanisms of forgetting. *Journal of Memory and Language, 49,* 415–445.

Bäckman, L., & Farde, L. (2005). The role of dopamine systems in cognitive aging. In R. Cabeza, L. Nyberg, & D. Park (Eds.), *Cognitive neuroscience of aging: Linking cognitive and cerebral aging* (pp. 58–84). New York: Oxford University Press.

Bäckman, L., Small, B. J., & Wahlin, Å. (2001). Aging and memory: Cognitive and biological perspectives. In J. E. Birren & K. W. Schaie (Eds.), *Handbook of the psychology of aging* (5th ed., pp. 349–377). San Diego, CA: Academic Press.

Balcom, P. (2003). Cross-linguistic influence of L2 English on middle constructions in L1 French. In V. Cook (Ed.), *Effects of the second language on the first* (pp. 168–192). Clevedon, UK: Multilingual Matters.

Bialystok, E., & Hakuta, K. (1994). *In other words: The science and psychology of second-language acquisition*. New York: Basic Books.

Bialystok, E., & Miller, B. (1999). The problem of age in second language acquisition: Influences from language, task, and structure. *Bilingualism: Language and Cognition, 2*, 127–145.

Birdsong, D. (1992). Ultimate attainment in second language acquisition. *Language, 68*, 706–755.

Birdsong, D. (1999). Introduction: Whys and why nots of the Critical Period Hypothesis. In D. Birdsong (Ed.), *Second language acquisition and the Critical Period Hypothesis* (pp. 1–22). Mahwah, NJ: Erlbaum.

Birdsong, D. (2003). Authenticité de prononciation en français L2 chez des apprenants tardifs anglophones: Analyses segmentales et globales. *Acquisition et Interaction en Langue Étrangère, 18*, 17–36.

Birdsong, D. (2004). Second language acquisition and ultimate attainment. In A. Davies & C. Elder (Eds.), *The handbook of applied linguistics* (pp. 82–105). Malden, MA: Blackwell.

Birdsong, D. (2005). Interpreting age effects in second language acquisition. In J. F. Kroll & A. M. B. de Groot (Eds.), *Handbook of bilingualism: Psycholinguistic approaches* (pp. 109–127). New York: Oxford University Press.

Birdsong, D., & Molis, M. (2001). On the evidence for maturational effects in second language acquisition. *Journal of Memory and Language, 44*, 235–249.

Bley-Vroman, R. (1989). What is the logical problem of foreign language learning? In S. Gass & J. Schachter (Eds.), *Linguistic perspectives on second language acquisition* (pp. 41–68). Cambridge, UK: Cambridge University Press.

Bongaerts, T. (1999). Ultimate attainment in L2 pronunciation: The case of very advanced late learners. In D. Birdsong (Ed.), *Second language acquisition and the critical period hypothesis* (pp. 133–159). Mahwah, NJ: Erlbaum.

Bornstein, M. H. (1989). Sensitive periods in development: Structural characteristics and causal interpretations. *Psychological Bulletin, 105*, 179–197.

Bowden, H. W., Sanz, C., & Stafford, C. A. (2005). Individual differences: Age, sex, working memory, and prior knowledge. In C. Sanz (Ed.), *Mind and context in adult second language acquisition: Methods, theory, and practice* (pp. 105–140). Washington, DC: Georgetown University Press.

Bradlow, A. R., Pisoni, D. B., Akahane-Yamada, R. A., & Tohkura, Y. (1997). Training Japanese listeners to identify English /r/ and /l/: IV. Some effects of perceptual learning on speech production. *Journal of the Acoustical Society of America, 101*, 2299–2310.

Cabeza, R., Nyberg, L., & Park, D. (2005). *Cognitive neuroscience of aging: Linking cognitive and cerebral aging.* New York: Oxford University Press.

Chee, M. W. L., Hon, N., Lee, H. L., & Soon, C. S. (2001). Relative language proficiency modulates BOLD signal change when bilinguals perform semantic judgments. *NeuroImage, 13*, 1155–1163.

Chee, M. W. L., Tan, E. W. L., & Thiel, T. (1999). Mandarin and English single word processing studied with functional magnetic resonance imaging. *Journal of Neuroscience, 19*, 3050–3056.

Chee, M. W. L., Weekes, B., Lee, K. M., Soon, C. S., Schreiber, A., Hoon, J. J., et al. (2000). Overlap and dissociation of semantic processing of Chinese characters, English words, and pictures: Evidence from fMRI. *NeuroImage, 12*, 392–403.

Chomsky, N. (1975). *Reflections on language.* New York: Pantheon.

Clahsen, H., & Felser, C. (2006). Grammatical processing in language learners. *Applied Psycholinguistics, 27*, 3–42.

Cook, V. (2002). Background to the L2 user. In V. Cook (Ed.), *Portraits of the L2 user.* Clevedon, UK: Multilingual Matters.

Cook, V., Iarossi, E., Stellakis, N., & Tokumaru, Y. (2003). Effects of the L2 on the syntactic processing of the L1. In V. Cook (Ed.), *Effects of the second language on the first* (pp. 193–213). Clevedon, UK: Multilingual Matters.

Courchesne, E., Chisum, H. J., Townsend, J., Cowles, A., Covington, J., Egaas, B., et al. (2000). Normal brain development and aging: Quantitative analysis at in vivo MR imaging in healthy volunteers. *Radiology, 216*, 672–682.

Craik, F. I. M. (2000). Age-related changes in human memory. In D. C. Park & N. Schwarz (Eds.), *Cognitive aging: A primer* (pp. 75–92). Philadelphia: Psychology Press.

Cranshaw, A. (1997). *A study of Anglophone native and near-native linguistic and metalinguistic performance.* Unpublished doctoral dissertation, Université de Montréal, Montréal, Canada.

Crosson, B., Benefield, H., Cato, M. A., Sadek, J. R., Moore, A. B., Wierenga, C. E., et al. (2003). Left and right basal ganglia activity during language generation: Contributions to lexical, semantic and phonological processes. *Journal of the International Neuropsychological Society, 9*, 1061–1077.

Darcy, I., Peperkamp, S., & Dupoux, E. (in press). Plasticity in compensation for phonological variation: The case of late second language learners. *LabPhon 9*.

De Bleser, R., Dupont, P., Postler, J., Bormans, G., Speelman, D., Mortelmans, L., et al. (2003). The organization of the bilingual lexicon: A PET study. *Journal of Neurolinguistics, 16,* 439–456.

Dehaene, D., Dupoux, E., Mehler, J., Cohen, L., Paulesu, E., Perani, D., et al. (1997). Anatomical variability in the cortical representation of first and second languages. *NeuroReport, 8,* 3809–3815.

DeKeyser, R. M. (2000). The robustness of critical period effects in second language acquisition. *Studies in Second Language Acquisition, 22,* 499–533.

DeKeyser, R., & Larson-Hall, J. (2005). What does the critical period really mean? In J. F. Kroll & A. M. B. de Groot (Eds.), *Handbook of bilingualism: Psycholinguistic approaches* (pp. 88–108). New York: Oxford University Press.

Ding, G., Perry, C., Peng, D., Ma, L., Li, D., Xu, S., et al. (2003). Neural mechanisms underlying semantic and orthographic processing in Chinese-English bilinguals. *NeuroReport, 14,* 1557–1562.

Dörnyei, Z., & Skehan, P. (2003). Individual differences in second language learning. In C. J. Doughty & M. H. Long (Eds.), *The handbook of second language acquisition* (pp. 589–630). Malden, MA: Blackwell.

Doughty, C. J. (2003). Instructed SLA: Constraints, compensation, and enhancement. In C. J. Doughty & M. H. Long (Eds.), *The handbook of second language acquisition* (pp. 256–310). Malden, MA: Blackwell.

Dupoux, E., & Peperkamp, S. (2002). Fossil markers of language development: Phonological "deafnesses" in adult speech processing. In J. Durand & B. Laks (Eds.), *Phonetics, phonology, and cognition* (pp. 168–190). Oxford: Oxford University Press.

Dussias, P. E. (2004). Syntactic ambiguity resolution in L2 learners: Some effects of bilinguality on L1 and L2 processing strategies. *Studies in Second Language Acquisition, 25,* 529–557.

Flege, J. (2002). No perfect bilinguals. In A. James & J. Leather (Eds.), *New sounds 2000: Proceedings of the Fourth International Symposium on the Acquisition of Second-Language Speech* (pp. 132–141). University of Klagenfurt.

Flege, J. E., & Hillenbrand, J. (1984). Limits on phonetic accuracy in foreign language speech production. *Journal of the Acoustical Society of America, 76,* 708–721.

Flege, J. E., MacKay, I. R. A., & Piske, T. (2002). Assessing bilingual dominance. *Applied Psycholinguistics, 23,* 567–598.

Flege, J. E., Yeni-Komshian, G. H., & Liu, S. (1999). Age constraints on second-language acquisition. *Journal of Memory and Language, 41,* 78–104.

Freeman, G. B., & Gibson, G. E. (1988). Dopamine, acetylcholine, and glutamate interactions in aging. Behavioral and neurochemical correlates. *Annals of the New York Academy of Sciences, 515,* 191–202.

Frenck-Mestre, C. (2005). Ambiguities and anomalies: What can eye movements and event-related potentials reveal about second language sentence processing? In J. F. Kroll & A. M. B. de Groot (Eds.), *Handbook of bilingualism: Psycholinguistic approaches* (pp. 268–281). New York: Oxford University Press.

Friederici, A. D., & Kotz, S. A. (2003). The brain basis of syntactic processes: Functional imaging and lesion studies. *NeuroImage, 20*(Suppl. 1), S8–S17.

Golato, P. (1998). *Syllabification processes among French-English bilinguals: A further study of the limits of bilingualism.* Unpublished doctoral dissertation, University of Texas-Austin.

Golomb, J., Kluger, A., de Leon, M. J., Ferris, S. H., Convit, A., Mittelman, M., et al. (1994). Hippocampal formation size in normal human aging: A correlate of delayed secondary memory performance. *Learning and Memory, 1,* 45–54.

Green, D. W. (2005). The neurocognition of recovery patterns in bilingual aphasics. In J. F. Kroll & A. M. B. de Groot (Eds.), *Handbook of bilingualism: Psycholinguistic approaches* (pp. 516–530). New York: Oxford University Press.

Grosjean, F. (1989). Neurolinguists, beware! The bilingual is not two monolinguals in one person. *Brain and Language, 36,* 3–15.

Grosjean, F. (1998). Studying bilinguals: Methodological and conceptual issues. *Bilingualism: Language and Cognition, 1,* 131–149.

Hahne, A. (2001). What's different in second-language processing? Evidence from event-related brain potentials. *Journal of Psycholinguistic Research, 30,* 251–256.

Hahne, A., & Friederici, A. (2001). Processing a second language: Late learners' comprehension mechanisms as revealed by event-related brain potentials. *Bilingualism: Language and Cognition, 4,* 123–141.

Herschensohn, J. (in press). *Language development and age.* Cambridge: Cambridge University Press.

Hyltenstam, K., & Abrahamsson, N. (2000). Who can become native-like in a second language? All, some, or none? On the maturational constraints controversy in second language acquisition. *Studia Linguistica, 54,* 150–166.

Hyltenstam, K., & Abrahamsson, N. (2003). Maturational constraints in SLA. In C. J. Doughty & M. H. Long (Eds.), *The handbook of second language acquisition* (pp. 539–588). Malden, MA: Blackwell.

Indefrey, P., Hellwig, F., Davidson, D., & Gullberg, M. (2005). *Nativelike hemodynamic responses during sentence comprehension after six months of learning a new language.* Poster presented at the 11th Annual Meeting of the Organization for Human Brain Mapping, Toronto.

Ioup, G., Boustagui, E., El Tigi, M., & Moselle, M. (1994). Reexamining the critical period hypothesis: A case study of successful adult SLA in a naturalistic environment. *Studies in Second Language Acquisition, 16,* 73–98.

Johnson, J. S., & Newport, E. L. (1989). Critical period effects in second language learning: The influence of maturational state on the acquisition of English as a second language. *Cognitive Psychology, 21,* 60–99.

Kim, K. H. S., Relkin, N. R., Lee, K.-M., & Hirsch, J. (1997). Distinct cortical areas associated with native and second languages. *Nature, 388,* 171–174.

Kimura, D. (1995). Estrogen replacement therapy may protect against intellectual decline in postmenopausal women. *Hormones & Behavior, 29,* 312–321.

Klein, D., Milner, B., Zatorre, R., Meyer, E., & Evans, A. (1995). The neural substrates underlying word generation: A bilingual functional-imaging study. *Proceedings of the National Academy of Sciences of the United States of America, 92,* 2899–2903.

Klein, D., Zatorre, R. J., Milner, B., Meyer, E., & Evans, A. C. (1994). Left putaminal activation when speaking a second language: Evidence from PET. *NeuroReport, 5,* 2295–2297.

Klein, W. (1995). Language acquisition at different ages. In D. Magnusson (Ed.), *The lifespan development of individuals: Behavioral, neurobiological, and psychosocial perspectives. A synthesis* (pp. 244–264). Cambridge: Cambridge University Press.

Knudsen, E. I. (2004). Sensitive periods in the development of the brain and behavior. *Journal of Cognitive Neuroscience, 16,* 1412–1425.

Kroll, J. F., & Tokowicz, N. (2005). Models of bilingual representation and processing: Looking back and to the future. In J. F. Kroll & A. M. B. de Groot (Eds.), *Handbook of bilingualism: Psycholinguistic approaches* (pp. 531–553). New York: Oxford University Press.

Kuhl, P. K. (2000). A new view of language acquisition. *Proceedings of the National Academy of Sciences of the United States of America, 97,* 11,850–11,857.

Larsen-Freeman, D. (2005). Second language acquisition and the issue of fossilization: There is no end, and there is no state. In Z.-H. Han & T. Odlin, *Studies of fossilization in second language acquisition* (pp. 189–200). Clevedon, UK: Multilingual Matters.

Laufer, B. (2003). The influence of L2 on L1 collocational knowledge and on L1 lexical diversity in free written expression. In V. Cook (Ed.), *Effects of the second language on the first* (pp. 19–31). Clevedon, UK: Multilingual Matters.

Lee, N. (2004). The neurobiology of procedural memory. In J. H. Schumann, S. E. Crowell, N. E. Jones, N. Lee, S. A. Schuchert, & L. A. Wood, *The neurobiology of learning: Perspectives from second language acquisition* (pp. 43–73). Mahwah, NJ: Erlbaum.

Lenneberg, E. H. (1967). *Biological foundations of language*. New York: Wiley.

Li, S.-C., Lindenberger, U., & Sikström, S. (2001). Aging cognition: From neuromodulation to representation. *Trends in Cognitive Sciences*, 5, 479–486.

Long, M. H. (1990). Maturational constraints on language development. *Studies in Second Language Acquisition*, 12, 251–285.

Lupien, S., De Leon, M., De Santi, S., Convit, A., Tarshish, C., Nair, N. P. V., et al. (1998). Longitudinal increase in cortisol during human aging predicts hippocampal atrophy and memory deficits. *Nature Neuroscience*, 1, 69–73.

Lupien, S., Lecours, A. R., Lussier, I., Schwartz, G., Nair, N. P. V., & Meaney, M. J. (1994). Basal cortisol levels and cognitive deficits in human aging. *Journal of Neuroscience*, 14, 2893–2903.

Mack, M., Bott, S., & Boronat, C. B. (1995). Mother, I'd rather do it myself, maybe: An analysis of voice-onset time produced by early French-English bilinguals. *IDEAL*, 8, 23–55.

MacWhinney, B. (2005a). A unified model of language acquisition. In J. F. Kroll & A. M. B. de Groot (Eds.), *Handbook of bilingualism: Psycholinguistic approaches* (pp. 49–67). New York: Oxford University Press.

MacWhinney, B. (2005b). Emergent fossilization. In Z.-H. Han & T. Odlin, *Studies of fossilization in second language acquisition* (pp. 134–156). Clevedon, UK: Multilingual Matters.

Marinova-Todd, S. (2003). *Comprehensive analysis of ultimate attainment in adult second language acquisition*. Unpublished doctoral dissertation, Harvard University.

Marinova-Todd, S. H., Marshall, D. B., & Snow, C. E. (2000). Three misconceptions about age and L2 learning. *TESOL Quarterly*, 34, 9–34.

McCandliss, B. D., Fiez, J. A., Protopapas, A., Conway, M., & McClelland, J. L. (2002). Success and failure in teaching the [r]-[l] contrast to Japanese adults: Predictions of a Hebbian model of plasticity and stabilization in spoken language perception. *Cognitive, Affective and Behavioral Neuroscience, 2,* 89–108.

McClelland, J. L., Fiez, J. A., & McCandliss, B. D. (2002). Teaching the /r/-/l/ discrimination to Japanese adults: Behavioral and neural aspects. *Physiology & Behavior, 77,* 657–662.

McLaughlin, J., Osterhout, L., & Kim, A. (2004). Neural correlates of second-language word learning: Minimal instruction produces rapid change. *Nature Neuroscience, 7,* 703–704.

Michael, E. B., & Gollan, T. H. (2005). Being and becoming bilingual: Individual differences and consequences for language production. In J. F. Kroll & A. M. B. de Groot (Eds.), *Handbook of bilingualism: Psycholinguistic approaches* (pp. 497–515). New York: Oxford University Press.

Miller, A. K., Alston, R. L., & Corsellis, J. A. (1980). Variation with age in the volumes of grey and white matter in the cerebral hemispheres of man: Measurements with an image analyser. *Neuropathology and Applied Neurobiology, 6,* 119–132.

Moro, A., Tettamanti, M., Perani, D., Donati, C., Cappa, S. F., & Fazio, F. (2001). Syntax and the brain: Disentangling grammar by selective anomalies. *NeuroImage, 13,* 110–118.

Moyer, A. (1999). Ultimate attainment in L2 phonology: The critical factors of age, motivation and instruction. *Studies in Second Language Acquisition, 21,* 81–108.

Moyer, A. (2004). *Age, accent and experience in second language acquisition.* Clevedon, UK: Multilingual Matters.

Mueller, J. L., Hahne, A., Fujii, Y., & Friederici, A. D. (2005). Native and nonnative speakers' processing of a miniature version of Japanese as revealed by ERPs. *Journal of Cognitive Neuroscience, 17,* 1229–1244.

Newman, A. J., Pancheva, R., Ozawa, K., Neville, H. J., & Ullman, M. T. (2001). An event-related fMRI study of syntactic and semantic violations. *Journal of Psycholinguistic Research, 30,* 339–364.

Ojima, S., Nakata, H., & Kakigi, R. (2005). An ERP study of second language learning after childhood: Effects of proficiency. *Journal of Cognitive Neuroscience, 17,* 1212–1228.

Osterhout, L., McLaughlin, J., Kim, A., & Inoue, K. (2004). Sentences in the brain: Event-related potentials as real-time reflections of sentence comprehension and language learning. In M. Carreiras & C. Clifton, Jr. (Eds.), *The on-line study of sentence comprehension: Eyetracking, ERP, and beyond.* Philadelphia: Psychology Press.

Pallier, C., Dehaene, S., Poline, J.-B., LeBihan, D., Argenti, A.-M., Dupoux, E., et al. (2003). Brain imaging of language plasticity in adopted adults: Can a second language replace the first? *Cerebral Cortex, 13,* 155–161.

Papadopoulou, D., & Clahsen, H. (2004). Parsing strategies in L1 and L2 sentence processing: A study of relative clause attachment in Greek. *Studies in Second Language Acquisition, 25,* 501–528.

Paradis, M. (2004). *A neurolinguistic theory of bilingualism.* Amsterdam: Benjamins.

Paradis, M. (2005). Introduction to Part IV: Aspects and implications of bilingualism. In J. F. Kroll & A. M. B. de Groot (Eds.), *Handbook of bilingualism: Psycholinguistic approaches* (pp. 411–415). New York: Oxford University Press.

Park, D. C. (2000). The basic mechanisms accounting for age-related decline in cognitive function. In D. C. Park & N. Schwarz (Eds.), *Cognitive aging: A primer* (pp. 3–21). Philadelphia: Psychology Press.

Park, D. C., & Gutchess, A. H. (2005). Long-term memory and aging: A cognitive neuroscience perspective. In R. Cabeza, L. Nyberg, & D. Park (Eds.), *Cognitive neuroscience of aging: Linking cognitive and cerebral aging* (pp. 218–245). New York: Oxford University Press.

Patkowski, M. S. (1990). Age and accent in a second language: A reply to James Emil Flege. *Applied Linguistics, 11,* 73–89.

Perani, D., Dehaene, S., Grassi, F., Cohen, L., Cappa, S. F., Dupoux, E., et al. (1996). Brain processing of native and foreign languages. *NeuroReport, 7,* 2439–2444.

Perani, D., Paulesu, E., Sebastian-Galles, N., Dupoux, E., Dehaene, S., Bettinardi, V., et al. (1998). The bilingual brain: Proficiency and age of acquisition of the second language. *Brain, 121,* 1841–1852.

Pfefferbaum, A., Mathalon, D. H., Sullivan, E. V., Rawles, J. M., Zipursky, R. B., & Lim, K. O. (1994). A quantitative magnetic resonance imaging study of changes in brain morphology from infancy to late adulthood. *Archives of Neurology, 51,* 874–887.

Pienemann, M., Di Biase, B., Kawaguchi, S., & Hakansson, G. (2005). Processing constraints on L1 transfer. In J. F. Kroll & A. M. B. de Groot (Eds.), *Handbook of bilingualism: Psycholinguistic approaches* (pp. 128–153). New York: Oxford University Press.

Piller, I. (2002). Passing for a native speaker: Identity and success in second language learning. *Journal of Sociolinguistics, 6,* 179–206.

Pinker, S. (1994). *The language instinct: How the mind creates language.* New York: Morrow.

Pinker, S., & Ullman, M. (2002). The past and future of the past tense. *Trends in Cognitive Sciences, 6,* 456–463.

Proverbio, A. M., Cok, B., & Zani, A. (2002). Electrophysiological measures of language processing in bilinguals. *Journal of Cognitive Neuroscience, 14,* 994–1017.

Raz, N. (2000). Aging of the brain and its impact on cognitive performance: Integration of structural and functional findings. In F. I. M. Craik & T. A. Salthouse (Eds.), *The handbook of aging and cognition* (2nd ed., pp. 1–90). Mahwah, NJ: Erlbaum.

Raz, N. (2005). The aging brain observed *in vivo*: Differential changes and their modifiers. In R. Cabeza, L. Nyberg, & D. Park (Eds.), *Cognitive neuroscience of aging: Linking cognitive and cerebral aging* (pp. 19–57). New York: Oxford University Press.

Raz, N., Rodrigue, K. M., Kennedy, K. M., Head, D., Gunning-Dixon, F. M., & Acker, J. D. (2003). Differential aging of the human striatum: Longitudinal evidence. *American Journal of Neuroradiology, 24*, 1849–1856.

Resnick, S. M., & Maki, P. M. (2001). Effects of hormone replacement therapy on cognitive and brain aging. *Annals of the New York Academy of Sciences, 949*, 203–214.

Reuter-Lorenz, P. A. (2000). Cognitive neuropsychology of the aging brain. In D. C. Park & N. Schwarz (Eds.), *Cognitive aging: A primer* (pp. 93–114). Philadelphia: Psychology Press.

Rogers, W. A. (2000). Attention and aging. In D. C. Park & N. Schwarz (Eds.), *Cognitive aging: A primer* (pp. 57–73). Philadelphia: Psychology Press.

Sabourin, L. (2003). *Grammatical gender and second language processing: An ERP study*. PhD dissertation, University of Groningen, The Netherlands.

Salthouse, T. A. (1996). Constraints on theories of cognitive aging. *Psychonomic Bulletin & Review, 3*, 287–299.

Schaie, K. W. (1994). Developmental designs revisited. In S. H. Cohen & H. W. Reese (Eds.), *Life-span developmental psychology: Methodological considerations* (pp. 45–64). Hillsdale, NJ: Erlbaum.

Schumann, J. H. (1997). *The neurobiology of affect in language*. Oxford: Blackwell.

Schumann, J. H. (2001). Appraisal psychology, neurobiology, and language. *Annual Review of Applied Linguistics, 21*, 23–42.

Schumann, J. H., Crowell, S. E., Jones, N. E., Lee, N., Schuchert, S. A., & Wood, L. A. (2004). *The neurobiology of learning: Perspectives from second language acquisition*. Mahwah, NJ: Erlbaum.

Scovel, T. (1988). *A time to speak: A psycholinguistic inquiry into the critical period for human speech*. Rowley, MA: Newbury House.

Segalowitz, N., & Hulstijn, J. (2005). Automaticity in bilingualism and second language learning. In J. F. Kroll & A. M. B. de Groot (Eds.), *Handbook of bilingualism: Psycholinguistic approaches* (pp. 371–388). New York: Oxford University Press.

Seliger, H. W. (1978). Implications of a multiple critical periods hypothesis for second language learning. In W. Ritchie (Ed.), *Second language acquisition research: Issues and implications* (pp. 11–19). New York: Academic Press.

Selinker, L. (1972). Interlanguage. *International Review of Applied Linguistics, 10*, 209–231.

Singleton, D. (2005). The Critical Period Hypothesis: A coat of many colours. *International Review of Applied Linguistics, 43*, 269–286.

Singleton, D., & Ryan, L. (2004). *Language acquisition: The age factor* (2nd ed.). Clevedon, UK: Multilingual Matters.

Skehan, P. (1989). *Individual differences in second-language learning.* London: Edward Arnold.

Stafford, C. A. (2005). *Bilingualism, cognitive capacity and age: A computer-based study in L3 processing.* Unpublished doctoral dissertation, Georgetown University.

Stowe, L. A., & Sabourin, L. (2005). Imaging the processing of a second language: Effects of maturation and proficiency on the neural processes involved. *International Review of Applied Linguistics, 43*, 329–354.

Takagi, N. (2002). The limits of training Japanese listeners to identify /r/ and /l/: Eight case studies. *Journal of the Acoustical Society of America, 111*, 2887–2896.

Teichman, M., Dupoux, E., Kouider, S., Brugieres, J.-P., Boissé, M.-F., Baudic, S., et al. (2005). The Role of the striatum in rule application: The model of Huntington's disease at early stage. *Brain, 128*, 1155–1167.

Ullman, M. T. (2001). The neural basis of lexicon and grammar in first and second language: The declarative/procedural model. *Bilingualism: Learning and Cognition, 4*, 105–122.

Ullman, M. T. (in press). Is Broca's area part of a basal ganglia thalamocortical circuit? *Cortex.*

Van Hell, J. G., & Dijkstra, T. (2002). Foreign language knowledge can influence native language performance in exclusively native contexts. *Psychonomic Bulletin & Review, 9*, 780–789.

Ventureyra, V. A. (2005). *A la recherche de la langue perdue: Étude de l'attrition de la première langue chez des Coréens adoptés en France.* Unpublished doctoral dissertation, Ecole des Hautes Etudes en Sciences Sociales, Paris.

Volkow, N. D., Wang, G.-J., Fowler, J. S., Ding, Y.-S., Gur, R., Gatley, S. J., et al. (1998). Parallel loss of pre and postsynaptic dopamine markers in normal aging. *Annals of Neurology, 44*, 143–147.

Wartenburger, I., Heekeren, H. R., Abutalebi, J., Cappa, S. F., Villringer, A., & Perani, D. P. (2003). Early setting of grammatical processing in the bilingual brain. *Neuron, 37*, 159–170.

Weinert, F. E., & Perner, J. (1996). Cognitive development. In D. Magnusson (Ed.), *The lifespan development of individuals: Behavioral, neurobiological, and psychosocial perspectives: A synthesis* (pp. 207–222). Cambridge: Cambridge University Press.

White, L. (2003). Fossilization in steady state L2 grammars: Persistent problems with inflectional morphology. *Bilingualism: Language and Cognition, 6*, 128–141.

Xue, G., Dong, Q., Jin, Z., Zhang, L., & Wang, Y. (2004). An fMRI study with semantic access in low proficiency second language learners. *NeuroReport, 15*, 791–796.

L2 Acquisition, Age, and Generativist Reasoning. Commentary on Birdsong

Peter Coopmans
Utrecht University

Birdsong's highly informative overview of the theoretical issues and facts on age and second language acquisition (L2A) has made me realize again how deceptively appealing it is to put two and two together in this overall discussion of age and acquisition. Let me explain what I mean by this temptation. If one takes the view that the goal of linguistic theory is to seek an explanation for the problem of language acquisition, for some linguists, myself included, part of the solution will be found in postulating Universal Grammar (UG) as a theory of an innate language faculty. Such a nativist approach provides a plausible answer to the successful, rapid, and pretty much uniform attainment of L1. In the majority of cases of adult L2A, we see anything but successful, rapid, or uniform language attainment: "The outcome ... is demonstrably different" (Birdsong, this volume, p. 9). If an innate language faculty plays a decisive role in the success of acquisition, it is tempting to explain the lack of success by pointing to a hypothesized absence of that same faculty. If not total absence, then at least limited availability. If one adopts an essentially biolinguistic approach to the study of language, it is all too easy to feel confident that some form of explanation for a maturation-based critical period will ultimately make sense. Indeed, the maturational factor

---

Correspondence concerning this article should be addressed to Peter Coopmans, Utrecht Institute of Linguistics OTS, Utrecht University, Trans 10 3512 JK, Utrecht, The Netherlands. Internet: coopmans@let.uu.nl

in L2 acquisition has received a variety of explanatory mechanisms (cf. Birdsong, 1999; Singleton & Ryan, 2004).

It seems to me that this is the way many of us who have not specifically studied age and rates of development or attainment in L2 have maintained this fairly traditional and somewhat naive outlook on the differences between L1 and L2. No doubt, we were initially persuaded that the findings and views put forward in such contributions as those by Lenneberg (1967), Johnson and Newport (1989), and Pinker (1994) will immediately do as an explanation. The "window of opportunity" gets shut, and maturation is the key to why it gets closed. At such moments, it is good to be critically reminded by researchers like Birdsong that "a superficial difference in ends [here: nativelike performance] does not necessarily imply an underlying difference in means," and likewise that "similarity of ends/products [does not] necessarily imply similar means/processes" (p. 10).

That such warnings are necessary is, in my view, partly due to a sometimes observed misconception of UG. UG is a theory of the innate language faculty that allows natives to overcome the poverty-of-the-stimulus problem—the fact that our linguistic knowledge is underdetermined by the primary linguistic data. UG is not a theory of language acquisition. It is a central component in a complex architecture of cognitive modules that make native language acquisition successful, unique perhaps, but still just one of many components. In L2 performance UG's role might be confounded by a great variety of factors of cognitive development (cf. Singleton & Ryan, 2004), and Birdsong points out some, such as effects of a processing nature, quantitative differences as artifacts of the nature of bilingualism, and cognitive aging (pp. 28–31).

Birdsong repeats his argument that the notion of a critical period is "a poor fit for [L2A] age effect data" (Birdsong, 2005, p. 109). In quite a number of studies, the observed age effects are not limited to a temporally bounded period, but they persist over the entire span of the Age of Acquisition (AoA). The relationship is a linear function, suggesting that AoA can predict success even for postpubertal L2 learners. Birdsong's work on the temporal and the geometric features of the age function, his conceptual discussions of the timing of maturational effects, and his reinterpretation of the outcomes when the data are broken up force one to reconsider the traditional, somewhat naive look at the Critical Period Hypothesis (CPH).

Yet, despite the impressive range of studies that Birdsong lists with high incidences of nativelikeness as counterexamples to the CPH, my main interest as a grammarian in the properties of UG keeps my attention focused on the question of whether there is a qualitative difference between the learning mechanisms underlying L1 and L2; for example, what happens in those cases in which nativelike proficiency is *not* found in advanced L2 learners (i.e., when adult L2 end-state grammars are found to differ from native-speaker grammars)? Schwartz (1990) has pointed out that this does not necessarily mean that the two grammars are epistemologically nonequivalent. It is possible that adult L2 learners might not have nativelike L2 grammars, but that these grammars are nevertheless still constrained in the same way as native-speaker grammars. Here we can draw the analogy with work in L1 acquisition (L1A), where we can show that children's developing grammars have not reached the end state, yet are still very much UG-constrained.

So, if Birdsong succeeds in convincing us that there is no discontinuity to be found in the age gradient, we might suspect that there is no change in learning mechanisms, but we would still have to show that the developing L2 system is determined by something like UG, in the way it is conceived of in L1A. Given this perspective, I see no reason, like White (1989), to approach L2A in any other way than L1A, by stressing the logic of the poverty-of-the-stimulus argumentation, and to investigate the strongest possible instances for which a truly explanatory grammatical analysis can be given. In the words of Dekydtspotter, Sprouse, and Thyre (1999/2000, p. 266):

> it is important to realize that the strongest empirical arguments for UG involvement in L2 acquisition would in principle rely not on the mere compatibility of interlanguage data with the constraints of UG, but rather on the

demonstration that at least some aspects of interlanguage knowledge are underdetermined by the input in domain-specific ways—that is, they require UG.

Such a poverty-of-the-stimulus problem is posed by the interrogative *combien* for English L2 learners of French. It can appear together with or separated from its nominal complement, and the resulting continuous and discontinuous constituents differ in interpretation.

(1) a. Combien de livres est-ce que tous les étudiants lisent?
how many of books is-it that all the students read
How many books do all the students read?

b. Combien est-ce que tous les étudiants lisent de livres?
how many is-it that all the students read of books

Example (1a), like its English equivalent, can either be answered collectively or distributively. In a situation in which one student reads books A, B, and C and another student reads books B, C, and D, the collective answer would be "two" (because there are two books read by all of the students). The distributive reading would lead to the answer "three," asking for the number of books per student. Example (1b) can only be understood on the distributive reading. Example (1a) is an ambiguous question; (1b) is not. The unavailability of the collective reading in (1b) results from the interaction of various syntactic and interface constraints (Dekydtspotter, Sprouse, Swanson, & Thyre, 1999; Obenauer, 1984/1985).

The English L2 learner of French somehow has to come to know this fact, but nothing in the L2 input will lead him to this awareness. Because the equivalent form of (1b) does not exist in English, he cannot rely on his L1. There is nothing in the L2 input or in the L1 to prevent him from assuming—incorrectly—that the discontinuous pattern in (1b) is simply a rewrite variant of the continuous pattern in (1a). After all, the reading for (1b) forms a proper subset of the readings allowed by (1a).

Dekydtspotter et al. (1999) argued that if L2 learners demonstrate knowledge of this property of French, this must result from the L2 hypothesis space being severely constrained, in a similar fashion to L1A. Results from a judgment task show that beginning-intermediate group L2 learners failed to make a distinction between the continuous and discontinuous questions in the collective condition, whereas like the native controls, advanced L2 learners did make such a distinction. They consistently rejected the collective reading for the discontinuous question more frequently than for the continuous question. Dekydtspotter et al. (1999, p. 170) concluded that this knowledge "could not feasibly be acquired without a [UG-based] restricted relation between levels of syntactic and conceptual structure representations."

A comparable example is the phenomenon of object scrambling, extensively studied by Unsworth (2005). In Dutch and German, but not in English, the direct object can be moved away from the verb, across an adverb, or across negation. This is shown in (2), with the object—here an indefinite noun phrase—italicized.

(2) a. Het meisje heeft twee keer *een aap* gekieteld.
the girl has two times a monkey tickled
The girl tickled a(ny) monkey twice.

b. Het meisje heeft *een aap* twee keer gekieteld.
the girl has a monkey two times tickled
The girl tickled a (certain) monkey twice.

These different positions correspond with a difference in meaning. When the object appears to the right of the adverb in a nonscrambled position, the indefinite object receives a nonspecific interpretation, whereas when it occurs to the left of the adverb, in a scrambled position, it gets a specific interpretation. Example (2b) singles out a certain monkey that the girl tickled on two different occasions, whereas the nonscrambled indefinite object in (2a) can refer to just any monkey. The net result is that (2a) allows for multiple interpretations (one or two monkeys involved): the scrambled version in (2b) only one.

The English sentence *The girl tickled a monkey twice* has both meanings and might be compared to the unscrambled form in (2a). Learning the interpretive constraints on scrambled (indefinite) objects once again constitutes a poverty-of-the-stimulus problem, for the native learner as well as the English L2 learner of Dutch. Like in the *combien* case, there is nothing in the input to inform L2 learners that (2b) is not simply a rewrite variant of (2a). Successful acquisition of these syntax-semantics interface properties must be enabled by UG intervention. Unsworth (2005) meticulously showed how English adult learners of Dutch are able to acquire nativelike knowledge of these properties of object scrambling. They demonstrated knowledge of the specific interpretation of the scrambled indefinite object and appeared to share the native preferences in their responses on the nonscrambled version.

What is more, Unsworth (2005) took the phenomenon of scrambling as an ideal testing ground for Schwartz's (1992) thesis that to determine the exact role of UG in postmaturational L2A, one should really compare the developmental patterns of adult and child L2 learners with the same L1. Two factors are important in this respect: age and L1 transfer. L2 children and L2 adults are, by definition, similar in that both groups have knowledge of another language and this allows for the potential of L1 transfer. However, these two groups differ with respect to age, because the L2 children are younger than L2 adults, and in this sense more similar to L1 children, insofar as their language-learning capacities can still be assumed to be constrained by UG. Unsworth investigated whether the three groups (L1 children, L2 adults, and L2 children) pass through the same developmental sequences in their acquisition of direct object scrambling in Dutch. She found truly remarkable similarities in patterns of development between the L2 children and L2 adults, of course with an initial L1 transfer stage that differentiates them from L1 children.

This, I think, is the best we can do from a nativist grammarian's perspective. Finding empirical evidence for a language faculty that not only allows us to get around a poverty-of-the-stimulus problem but also helps in shedding light on why we can find similarities in patterns and rates of development in L1A and L2A, both early and late. This is where I see the link with the interesting work and multitude of ideas in Birdsong's overview.

References

Birdsong, D. (1999). Introduction: Whys and why nots of the Critical Period Hypothesis. In D. Birdsong (Ed.), *Second language acquisition and the Critical Period Hypothesis* (pp. 1–22). Mahwah, NJ: Erlbaum.

Birdsong, D. (2005). Interpreting age effects in second language acquisition. In J. F. Kroll & A. M. B. de Groot (Eds.), *Handbook of bilingualism: Psycholinguistic approaches* (pp. 109–127). New York: Oxford University Press.

Dekydtspotter, L., Sprouse, R. A., Swanson, K. A., & Thyre, R. (1999). Semantics, pragmatics and second language acquisition: The case of *combien* extractions. In A. Greenhill, H. Littlefield, & C. Tano (Eds.), *Proceedings of the 23rd Annual Boston University Conference on Language Development* (pp. 162–171). Somerville, MA: Cascadilla Press.

Dekydtspotter, L., Sprouse, R. A., & Thyre, R. (1999/2000). The interpretation of quantification at a distance in English-French interlanguage: Domain specificity and second-language acquisition. *Language Acquisition, 8*, 265–320.

Johnson, J. S., & Newport, E. L. (1989). Critical period effects in second language learning: The influence of maturational state on the acquisition of English as a second language. *Cognitive Psychology, 21*, 60–99.

Lenneberg, E. H. (1967). *Biological foundations of language*. New York: Wiley.

Obenauer, H.-G. (1984/1985). On the identification of empty categories. *The Linguistic Review, 4*, 153–202.

Pinker, S. (1994). *The language instinct: How the mind creates language*. New York: Morrow.

Schwartz, B. D. (1990). Un-motivating the motivation for the Fundamental Difference Hypothesis. In P. Rounds (Ed.), *Variability in second language acquisition* (pp. 667–684). Eugene: University of Oregon Press.

Schwartz, B. D. (1992). Testing between UG-based and problem-solving models of L2A: Developmental sequence data. *Language Acquisition, 2*, 1–19.

Singleton, D., & Ryan, L. (2004). *Language acquisition: The age factor*. Clevedon, UK: Multilingual Matters.

Unsworth, S. (2005). *Child L2, adult L2, child L1: Differences and similarities. A study on the acquisition of direct object scrambling in Dutch.* Doctoral dissertation, University of Utrecht, The Netherlands.

White, L. (1989). *Universal Grammar and second language acquisition*. Amsterdam: Benjamins.

Development of the Human Cortex and the Concept of "Critical" or "Sensitive" Periods

H. B. M. Uylings
Royal Netherlands Academy of Sciences
VU University Medical Center

This review describes the prenatal and postnatal development of the human cortex. Neurogenesis, neuronal migration, dendrite maturation, synaptogenesis, and white matter development are discussed. In addition, the concept of "critical" or "sensitive" periods is discussed as well as genetic and environmental influences (Nature-Nurture). The effects of irradiation, alcohol, smoking, and prenatal maternal influenza and stress on brain functions and language performance are reviewed. The periods of plasticity are reviewed for stereoscopic vision, brain lesions, social neglect, and first and second language acquisition.

This article illustrates that "critical" or "sensitive" periods exist for those functions for which axonal rewiring across a long distance in the brain is necessary. This is virtually impossible after completion of the developmental competition between different axonal systems.

I acknowledge the long-standing collaboration with the group of Dr. Ivica Kostović (Croatian Institute for Brain Research, Zagreb, Croatia), and the stimulating discussion with Dr. Kai Kaila (University of Helsinki, Finland) about the issue of genetic and environmental influences on human brain at the IBRO-FENS Summerschool "Development and Plasticity of the Human Cerebral Cortex" at Zadar (September 2005). This discussion led to the composition of Figure 8 for this review. I thank Ms. W. T. P. Verweij for her secretarial assistance and Mr. H. Stoffels and Mr. G. van der Meulen for their drawings and figures.

Correspondence concerning this article should be addressed to H. B. M. Uylings, Netherlands Institute for Neurosciences, Meibergdreef 33, 1105 AZ Amsterdam, The Netherlands. Internet: H.Uylings@nin.knaw.nl

Furthermore, genotyping allows the maximum of structural and functional properties, but the eventual outcome is under the influence of the environment in positive or negative directions. Under dramatic circumstances, environmental factors during the first years of life can be quite decisive for the development of language, social, and other intellectual functions.

Progress in brain development allows development of cognitive capabilities and, in its turn, the exercising of cognitive abilities shapes further brain development. Some capabilities (such as binocular vision) can only be developed within rather limited sensitive or "critical" developmental periods. For other capabilities, such as "experience" learning in a so-called enriched environment, such restricted developmental time windows do not exist (e.g., Uylings, 2001) because of the plasticity potential of the underlying brain circuitry elements necessary for these abilities.

Various cognitive functions are involved in mastering a second language (L2) perfectly (e.g., Stowe & Sabourin, 2005). At least for speaking a L2 without an accent (Flege, MacKay, & Meador, 1999) and nativelike mastering of the grammar of L2 (e.g., Johnson & Newport, 1989, 1991), a sensitive period appears to be present (DeKeyser & Larson-Hall, 2005), although this is not obviously the case for semantic processes (see Birdsong, this volume).

The development of the cerebrum is fascinating because of the multitude and complexity of the processes involved (Figure 1). At different time courses, a multitude of neurons (e.g., 25×10^9 for cerebral neocortex and 150×10^9 in the whole brain) are generated and migrate across large distances, through other fibers and cell groups. These neurons differentiate mainly when they have reached their final location or destination and are connected with particular neurons. In this way, neuronal systems and cortical area specifications arise that are necessary for the execution of many functions (O'Leary, 1989; Rakic, 1988, 2005; Sur & Rubenstein, 2005). It is the general rules of development that are

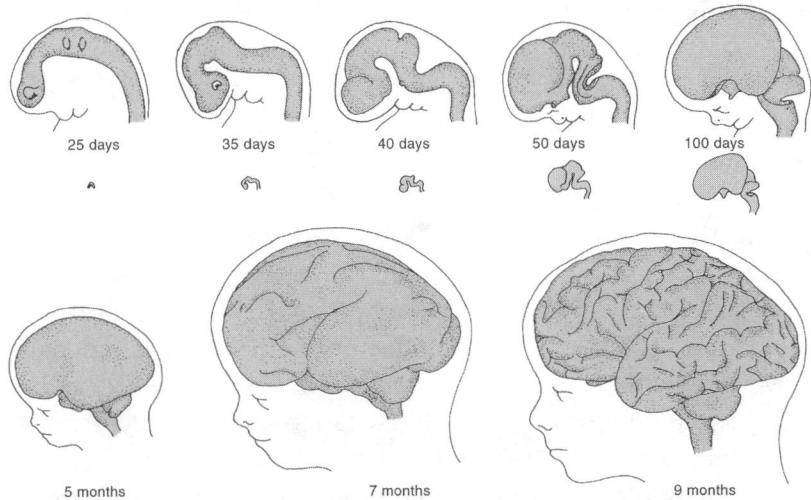

Figure 1. The macroscopic development of the human brain before birth (modified from Hochstetter, 1919). The drawings are all in relative proportion, while the first five stages have also been enlarged to show the morphological alterations in the neural tube (adapted from Uylings, 2001).

genetically programmed rather than each individual connection, which would be impossible because of the incredibly large number of synapses (e.g., approximately 10^{14} in the neocortex alone).

In this article, I will review the structural human cortical development, which might be correlated with particular phases of human cognitive development: the period of cortical neurogenesis, the development of cortical dendrites, synaptogenesis, and chemical/neurotransmitter maturation. In addition, some aspects related to the concept of critical or sensitive period are discussed.

Neuronal Development

Neurogenesis

In animal studies, neurogenesis in adulthood has been found to be restricted to the olfactory bulb and a part of the hippocampus (i.e., fascia dentate; e.g., Kempermann, 2006; Taupin, 2005) and to be related to plasticity in some hippocampus-dependent learning (Kempermann; Shors, Townsend, Zhao, Kozorovitskiy, & Gould, 2002). Genesis of human cortical neurons, however, appears to be restricted mainly to the prenatal period between approximately 6 and 18 weeks of gestation (Rakic, 1995; Rakic & Sidman, 1968). During this period, about 225,000 cortical neurons per minute on average are generated (Uylings, 2001). In the literature, there is no discussion that, after birth, cortical neurons are generated in the dentate gyrus and for the olfactory bulb. There is a debate, however, whether this also occurs for neurons in primate nonhuman associational cortices (Gould & Gross, 2002; Gould, Reeves, Graziano, & Gross, 1999; Rakic, 2002a, 2002b). In human cortices, Shankle et al. (1998, 1999) reported a postnatal doubling of the number of neurons in the areas studied—among others, Broca's area. However, the quantitative methods applied in the latter articles are disputed (e.g., Korr & Schmitz, 1999). In addition, our stereological study[1] of the total neuron number in the human Broca's area after birth shows that a doubling of the total number of neurons is quite unlikely (Uylings et al., 2005). This study suggested no alteration in the total number of neurons in Broca's area after birth. Larsen et al. (in press) did not find a postnatal increase in the total number of neurons in the whole human cerebral cortex either. We can thus conclude that the present data indicate no significant postnatal increase in the total number of neurons in the human neocortex.

Migration of Neocortical Neurons

The period of major neocortical neuron migration lasts until about 26 weeks of gestation (e.g., Marin-Padilla, 1992). This is also the time when the thalamocortical fibers enter the frontal cortex (Kostović & Goldman-Rakic, 1983). The radial migration of pyramidal neurons from the ventricular zone is guided by radial glia fibers (Figure 2). By interaction of the migrating cortical neuron with radial glial fibers, an inside-out pattern of migration

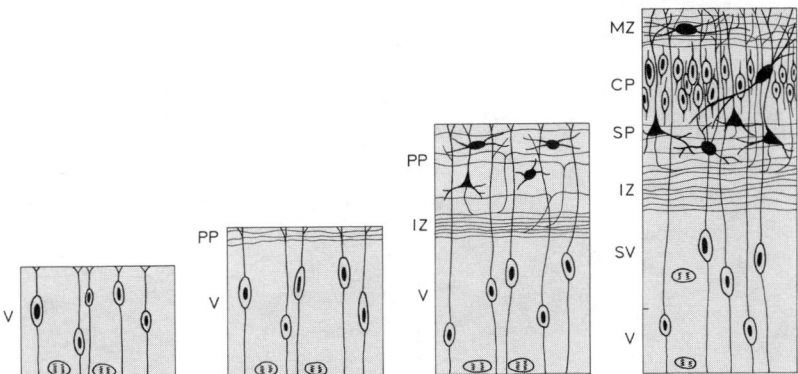

Figure 2. Four different stages in mammalian fetal lamination in the cerebral wall around the cerebral vesicles. Cell division occurs at the inner side of the wall. The first cellular layer in the outer zone of the wall is the preplate (PP). Cortical plate neurons migrate radially along radial glial fibers that connect the inner side with the outer side of the cerebral wall. These cortical plate (CP) neurons subdivide the preplate in a subplate (SP) and marginal zone (MZ). Cellular division occurs in the ventricular zone (V) and in later phases also in the subventricular zone (SV). IZ: intermediate zone. (Adapted from Uylings et al., 1994.)

occurs. This means that the neurons that are the last to be generated migrate to the peripheral part of the cortical plate just below the marginal zone and pass along the previously generated neurons. In rodents, the majority of nonpyramidal (inter-) neurons migrate tangentially from the ganglionic eminence (Figure 3; Parnavelas, 2000; Marin & Rubenstein, 2001). In primates, however, the largest part of the nonpyramidal neurons arrives after a radial migration (Letinic, Zoncu, & Rakic, 2002). The tangential migration is guided by glial and axonal fibers.

Maturation of Dendrites

After the arrival of thalamocortical fibers in the cerebral cortex at around 26 weeks of gestation (Kostović & Goldman-Rakic, 1983), we see a somewhat accelerated growth of the pyramidal dendrites (Mrzljak, Uylings, Van Eden & Judáš, 1990; Mrzljak,

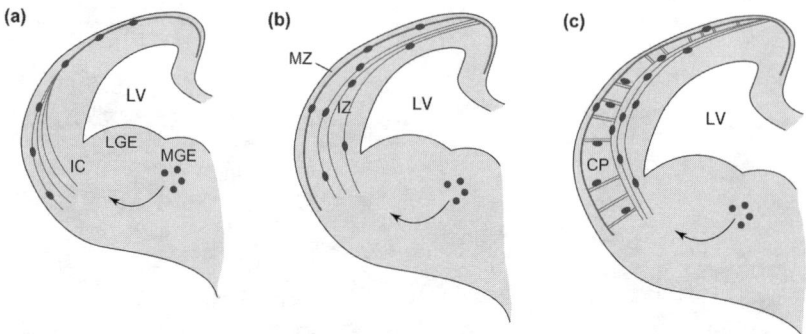

Figure 3. Cross-sections through the cerebrum at three different stages to display schematically how nonpyramidal neurons migrate tangentially to the neocortex from the medial ganglionic eminence (MGE). The tangential migration is along axons and glial fibers: first in the preplate and marginal zone, MZ (a), and later in subplate and the intermediate zone, IZ (b). From the intermediate zone and the subplate nonpyramidal neurons eventually migrate radially into the cortical plate (CP) along radially arranged axons (c). IC: internal capsule; LV: lateral ventricle; LGE: lateral ganglionic eminence. (Adapted from Parnavelas The origin and migration of cortical neurones; new vistas, *Trends in Neurosciences*, *23*, 126–131, 2000, with kind permission from Elsevier Science.)

Uylings, Kostović & Van Eden, 1992); see Figure 4. The major outgrowth, however, starts after birth and continues during the first 2–3 years (Koenderink & Uylings, 1995; Koenderink, Uylings, & Mrzljak, 1994); see Figures 5 and 6. This phase of the fastest dendritic outgrowth parallels the phase of fast cerebral cortex expansion, which also lasts until about 4 years of age. Giedd et al. (1999) reported that certain human cortical areas, such as the frontal cortex and parietal cortex, keep increasing until the age of about 12–13 years. From that age onward, the volume of these cortices should diminish. However, there is not yet agreement on this decrease of gray matter volume in human cortices (Paus, 2005). Zilles (1978) has already demonstrated for the tree shrew that different cortical regions follow their own pattern of development. Each cortex has a particular age at which it reaches a maximum value, after which some cortices decline. Such a decline is not detectable in the volume of the whole cerebral cortex or whole brain of either humans or tree shrews. We might

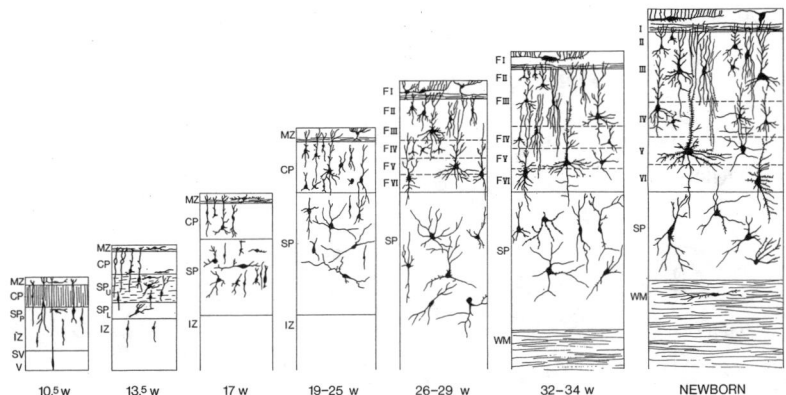

Figure 4. The prenatal development of neurons and cortical layers in the human frontal cortex. SP: subplate zone; SP_P = first part of SP; Sp_U = upper part of SP; SP_L = lower part of subplate. MZ: marginal zone; CP: cortical plate; IZ: intermediate zone; SV: subventricular zone; V: ventricular zone; FI = fetal layer I., FII = fetal layer II, and so forth. (Adapted from Mrzljak, L., Uylings, H.B.M., Kostovic, I., & Van Eden, C.G. The prenatal development of neurons in the human prefrontal cortex. I. A qualitative Golgi study. *Journal of Comparative Neurolology*, 271, 355–386, 1988, with kind permission from Wiley-Liss.)

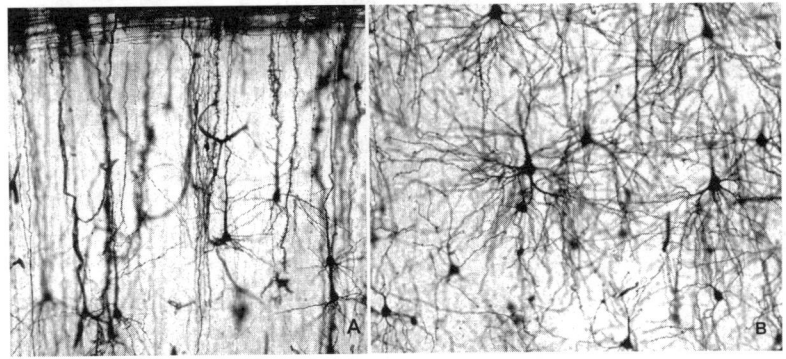

Figure 5. Photograph of pyramidal neurons and their dendrites in the human frontal cortex layer III in (a) newborn and (b) at 5 years of age. Golgi staining.

expect that a decline in a particular cortical volume is paralleled by a decline in dendritic trees present in this cortex. However, we did not observe a clear developmental decline in dendritic size in our cross-sectional study of the dorsolateral prefrontal cortex

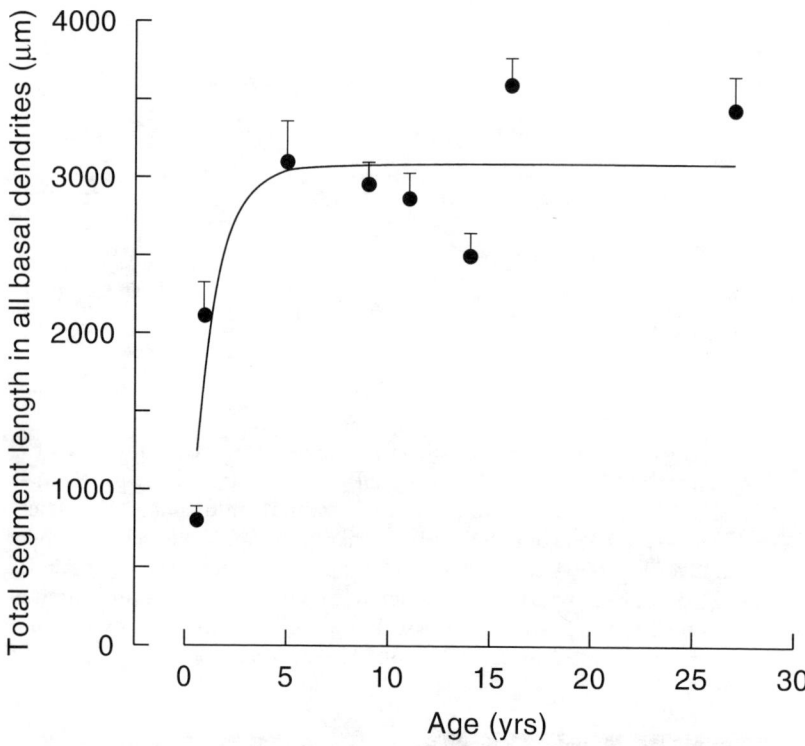

Figure 6. The postnatal development of basal dendrites of layer V pyramidal neurons in the human frontal cortex. (Adapted from Koenderink & Uylings, 1995.)

(Figure 6; Koenderink et al.; Koenderink & Uylings; Petanjek, Judáš, Kostović, & Uylings, in preparation) and further research is needed to establish the presence of a decline in the human association cortex and its underlying factors.

After 5 years of age, a large interindividual variation in dendritic tree size becomes apparent. On the basis of the animal studies on environmental enrichment, for example (Greenough, Withers, & Wallace, 1990; Uylings, Kuypers, Diamond, & Veltman, 1978) we might hypothesize that this variation is due to genetic and environmental effects.

Cerebral White Matter Development

In a large-cohort magnetic resonance imaging (MRI) study, Giedd et al. (1999) found a continuous increase in overall volume of human cerebral white matter during the studied age-period from 4 to 21 years. Also, a region-specific growth has been detected for cerebral white matter. For example, during this period, the posterior part of the corpus callosum changed continuously, but not the anterior part (e.g., Giedd, 1999). These MRI studies (Barnea-Goraly et al., 2005; Giedd; Giedd et al., 1999) expanded the former postmortem studies (e.g., Flechsig, 1920; Kaes, 1907; Yakovlev & Lecours, 1967), which showed an area-specific pattern; that is, the myelin development below and in the primary motor and sensory areas are ahead of those of the associational frontal, parietal, and temporal areas and continues into adulthood.

Increase in white matter means an increase in myelin volume and indicates a better isolation of the transport of different electric signals. It cannot indicate, however, whether the number of axonal fibers is higher. In human brains, the development of axonal systems cannot be examined in the way that it can be examined in animals, where exuberant growth (i.e., overgrowth followed by pruning) has been demonstrated—for instance, for callosal fibers in the cat (e.g., Innocenti & Price, 2005) and for cortical pyramidal axons growing into the brainstem and spinal cord in the rat (O'Leary, 1992; O'Leary et al., 1990; Innocenti & Price). Pruning is considered to be the result of the competition between different developing fiber systems for the receptive neurons.

Synaptogenesis

On the basis of the dendritic maturation, we could expect the major synaptogenesis to occur postnatally. This has been established in the human brain by Huttenlocher and colleagues (Huttenlocher & Dabholkar, 1997; Huttenlocher, de Courten,

Garey, & Van der Loos, 1982). They have also reported a temporal overgrowth in synaptic numbers, which peaked around 2–4 years and fell afterward, reaching a mature figure around 10–15 years of age at about two thirds of the peak value. This reduction in synaptic number, detected in cross-sectional studies, is quite extensive and not paralleled by the dendritic maturation (Figure 6). After the age of 15 years we can still expect that some new synapses will be formed by stimulation (Holtmaat et al., 2005; Trachtenberg et al., 2002; Zito & Svoboda, 2002)—for example, by so-called environmental enrichment.

In addition to the number of synapses, we should also take into consideration the developmental changes in synaptic types (symmetric and asymmetric synapses, which are inhibitory and excitatory, respectively).

Chemical/Transmitter Maturation

Although the outgrowth of cortical neurons is at its mature level between 2 and 4 years of age, the chemical, monoaminergic, and cholinergic transmitter maturation continues until the end of puberty (Goldman-Rakic & Brown, 1982; Herlenius & Lagercrantz, 2004; Kalsbeek, De Bruin, Feenstra, & Uylings, 1990; Kostović, 1990; Lewis & Sesack, 1997); for example, the mature level for nonpyramidal neuropeptide Y (NPY) cortical neurons has been observed from 8 years of age (Delalle, Evers, Kostović, & Uylings, 1997; Uylings & Delalle, 1997) and from 12 to 15 years for the calcium-binding proteins calbindin and parvalbumin (Uylings, Delalle, Petanjek, & Koenderink, 2002), whereas for acetylcholinesterase, the mature pattern is reached from about 18 years of age onward (Kostović). Also for dopamine, noradrenaline, and serotonin, a mature pattern might be expected at the end of puberty/early adulthood (Goldman-Rakic & Brown; Kalsbeek et al.; Lewis & Sesack). As a consequence, we also expect a late maturation of several transmitter receptors on the cortical neurons around the end of puberty and early adulthood. Transmitters together with their specific receptors affect neuronal excitation or inhibition. This depends on the type of transmitter and the type of receptor (e.g., Cooper, Bloom, & Roth, 2002).

Metabolic Rate

Chugani et al. (1987) did not detect any region-specific developmental pattern in the local glucose metabolic rate of the cerebral cortex. This local metabolic rate was highest between 4 and 9 years of age, declined somewhat during puberty, and after the age of 16 years, it reached a lower adult level in the frontal, parietal, temporal, and occipital cortex. This developmental pattern probably reflects the neurochemical and hormonal maturation (e.g., Cameron, 2004).

Critical Periods in Cortical Development

Various definitions of the term "critical period" are used in the literature. One definition is the period in which a subject is particularly susceptible to damage (i.e., "period of susceptibility"). Another definition of critical period is the restricted period in which recovery or a flexible response occurs (i.e., a "period of plasticity" or "sensitive period").

Critical "Period of Susceptibility"

The brain is susceptible to environmentally inflicted insults, such as irradiation, alcohol use, smoking, drug abuse (e.g., cocaine), and stress, especially during the period of major neurogenesis of cerebral cortical neurons (6–18 weeks of gestation), migration, and early path-finding of growing axons. The effects of irradiation and drugs are age dependent and dose dependent. Small doses at ages of major neurogenesis have a dramatically greater effect than higher doses in adulthood.

Irradiation

In the developing rat an irradiation dose of 1.5 Gy on day 12 of gestation (i.e., in the first period of major neocortical neurogenesis) led to massive cell death and cerebral malformation. However, on postnatal day 7 (i.e., after the major neocortical neurogenesis and neuronal migration), a dose of 4 Gy did not induce acute cell death (Hicks & D'Amato, 1978). Studies on the effect of the atom bombs in Hiroshima and Nagasaki also showed that the period of major neocortical neurogenesis (i.e., 6–18 weeks of gestation) is the most sensitive period as far as deleterious effects of irradiation are concerned. Children who were exposed *in utero* to a dose higher than 0.5 Gy between 8 and 15 weeks of gestation suffered from a significantly higher incidence of severe mental retardation and epileptic seizures than children who were exposed to the same doses during a later period (i.e., between 16 and 25 weeks of gestation). MRI investigation showed that especially the children exposed to a dose between 0.5 and 1.5 Gy between 8 and 15 weeks of gestation had a smaller brain (e.g., smaller cortex and corpus callosum) and ectopic cells around the lateral ventricle (Shigematsu, Ito, Kamada, Akiyama, & Sasaki, 1995; UNSCEAR, 1986).

Alcohol

Drinking one or two glasses of an alcoholic beverage a day in adulthood is thought to be healthy. However, during pregnancy, two glasses a day is too much and leads to lower mental capacities in the child (Streissguth, Barr, & Sampson, 1990). Alcohol not only affects cerebral neurogenesis but also the correct wiring of neuronal connections (Miller, 1987) and causes widespread neuronal death in development (e.g., Olney, 2002). All three trimesters of pregnancy have therefore been implicated as vulnerable periods for alcohol. Concomitant use of other substances will aggravate the effects of alcohol (Malanga & Kosofsky, 2004); therefore, it is essential for pregnant women to temporarily refrain from drinking alcohol for the well-being of the fetus. Children exposed *in utero* to alcohol will suffer a reduction in both mental and social capabilities, which, in turn, will cause difficulties at school (Malanga & Kosofsky).

Smoking

Smoking also affects prenatal development (Slotkin, 1998). Chronic maternal smoking during pregnancy leads to behavioral and cognitive impairment in children and to an increased risk of low birth weight and attention-deficit hyperactivity disorder (Kline, Stein, & Susser, 1989; Weitzman, Gortmaker, & Sobol, 1992).

Cocaine

Prenatal exposure to cocaine can lead to impaired fetal growth, smaller head size at birth, and altered behavior in orientation and alertness. At ages 3 and 5 years, a significantly poorer receptive and expressive language performance has been reported, occurring independently of potential confounders like prenatal alcohol and tobacco exposure (Bandstra, Vogel, Morrow, Xue, & Anthony, 2004; Morrow et al., 2004; Nulman et al., 2001). So far as is known, early intervention and preschool programs appear to be beneficial for prenatally cocaine-exposed children (Frank et al., 2005). However, the prenatal cocaine effects are less severe than those caused by alcohol (Malanga & Kosofsky, 2004).

Maternal Influenza and Stress

Studies on schizophrenia show that exposure to environmental insults, such as influenza, malnutrition, and stress, during the second trimester of pregnancy leads to an increased incidence of schizophrenia (Huttunen & Niskanen, 1978; Susser et al., 1996).

Prenatal maternal stress, particularly during early gestation, can also lower general intellectual and language abilities in human toddlers (Laplante et al., 2004).

Medical Drugs

Fortunately, large malformations due to drug use during the main period of neurogenesis seldom occur and are to be considered tragic accidents. Examples are an absent cerebral cortex and hypothalamus (anencephaly), spina bifida, caused by valproate, a drug against epilepsy; or large malformations of legs and arms, caused by the sleep drug thalidomide (Softenon). Functional teratological effects are, however, less rare. For instance, clonidine, an α_2-receptor agonist used as an antihypertensive reduces noradrenaline release and interferes with growth hormone secretion; also evident are a disturbance of the learning ability (by, e.g., lead from the environment), mental retardation (by mercury), increased aggression (by, e.g., progestativa taken during pregnancy or particular types of artificial food [color] additive), increased depression (by diethylstilbestrol [DES] during pregnancy), and so forth. For a more detailed review on this subject of functional teratology, I refer to the works by Olney (2002) and Olney, Young, Wozniak, Jevtovic-Todorovic, and Ikonomidou (2004).

In summary, the most "susceptible period" is the period of major neurogenesis, migration, and early path-finding of developing axons to make the appropriate connections. Environmentally inflicted insults occurring in this period generally lead to a significant, persistent deficit in the number of neurons and in miswiring of axons.

Critical "Period of Plasticity" / Sensitive Period

Some capabilities (such as binocular vision) can only be developed within rather restricted sensitive or "critical" developmental periods. For other capabilities, such as "experience" learning in a so-called enriched environment, such restricted developmental time windows do not exist (e.g., Uylings, 2001) because of the plasticity potential of the underlying brain circuitry elements necessary to develop these abilities.

Plasticity to develop new functions such as binocular vision requires, among others things, new elongation of axons over a long distance. After the developmental competition has been completed between different axonal systems for establishing connections with target neurons (e.g., O'Leary, 1992), new 'long distance' axonal elongation is unfeasible to create new connections and reopen competition with existing fiber systems (Knudsen, 2004). As described earlier, brain development is a multiphasic process: Different phases (recently reviewed by Hensch, 2004) are required to establish functions during critical or sensitive periods. This leads to different developmental patterns of critical periods for different functions, after which no plasticity is left for the pertinent functions (see Figures 7A and 7B).

"Experience" learning requires local dendritic and synaptic alterations. This kind of plasticity does not stop in adulthood. This does not imply that the capacity for plasticity remains unchanged during life. In practice, a decline with age has been observed in the paradigm of the so-called environmental enrichment (Figure 7C). On the other hand, specially designed strategies might be necessary to obtain (better) plastic changes in adult learning (e.g., Bergan, Ro, Ro, & Knudsen, 2005).

Examples of critical periods for different functions are summarized in the following subsections.

Stereoscopic Vision

Critical periods have been detected for shape and stereoscopic vision. Originally, congenital cataracts (cataracts are opacities of the lens that interfere with vision) were surgically removed between the ages of 10 and 20 years. After studies had demonstrated a permanent impairment in form vision at these ages, these operations appeared to be more beneficial if they were

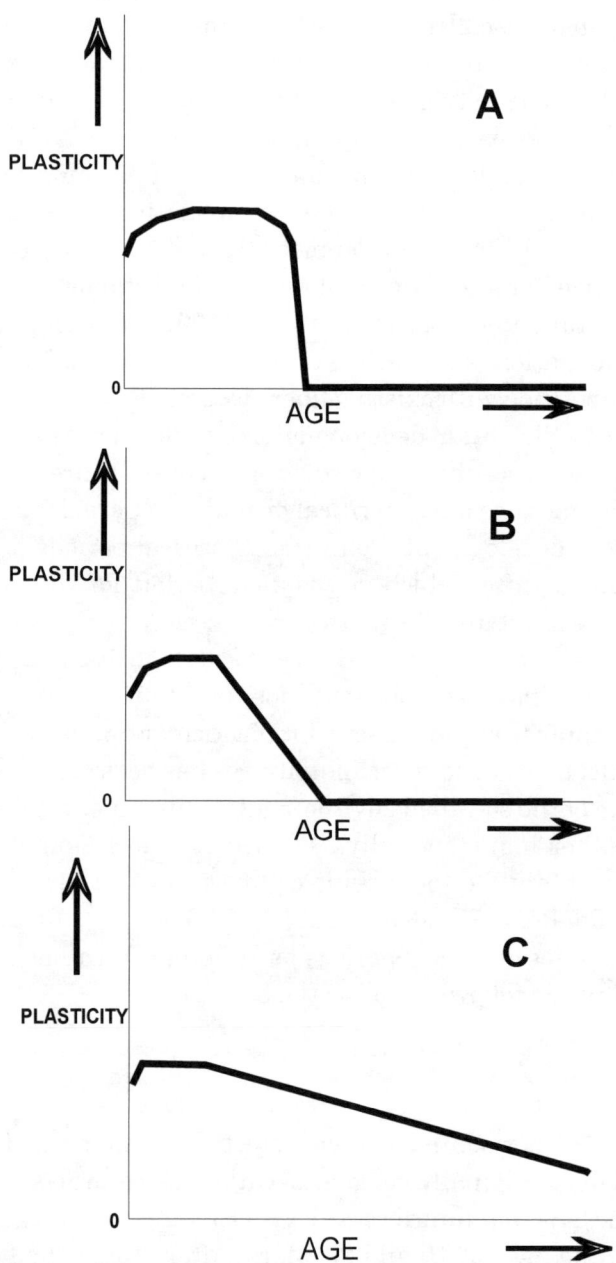

Figure 7. (A) and (B) different patterns of "critical" or "sensitive" periods. (C) Persisting plasticity in adulthood, although to a lesser extent.

performed earlier, before the age of 8 years. Stereoscopic vision develops during the period in which the ocular dominance columns of the primary visual cortex (Huttenlocher & Dabholkar, 1997) and the corticocortical connections (Kandel, Jessell, & Sanes, 2000) between the visual and parietal cortex are formed (Hyvärinen, 1982). When stereoscopic vision does not develop in this period of approximately age 8, due to an imbalance in the wiring of the pertinent circuits, it will probably remain permanently impaired. For a recent review on possible mechanisms, I refer to Hensch (2005).

Gonadal Hormones

There are also critical periods for persisting effects of testicular hormones on the developing nervous system (e.g., see Gorski, 2000, for a review). The masculine characteristics of brain structure and function are only imposed during the development of the brain by these hormones. These hormones interfere with programmed cell death in nuclei of the hypothalamus, spinal cord, and visual cortex, which leads to a sexually dimorphic number of neurons in these nuclei (see, e.g., Gorski; Hofman & Swaab, 1989; Nuñez, Sodhi, & Juraska, 2002). In addition to these influences on neuronal cell number, which are restricted to development, gonadal hormones can also modify particular structural changes such as spines and dendrites in the hippocampus during the estrous cycle in adulthood (McEwen, 2004; Woolley & McEwen, 1992).

Thyroid Hormones

Thyroid hormone deficiency due to a diet with insufficient iodine leads to severe mental retardation and neurological deficits, which can be largely prevented by adding iodine to the diet soon after birth (Kooistra et al., 1994). Because prenatal thyroid deficiency affects prenatal brain development from about

15 weeks of gestation (Thorpe-Beeston, Nicolaides, Felton, Butler, & McGregor, 1991), complete recovery is not possible.

Phenylalaline

Phenylketonuria is a syndrome of severe mental retardation due to elevated levels of phenylalaline (Yudkoff, 1994). Without treatment, the majority of the afflicted individuals have an IQ of less than 50 and have to be institutionalized for their entire life (Heminger, 1999). If this inborn error is diagnosed soon after birth, early treatment by restriction of the dietary intake of phenylalaline can generally prevent the mental retardation. Some mild cognitive deficits might remain if the brain damage started prenatally (Heminger).

Brain Lesions

A lesion in the human brain caused by a perinatal infarct can be restored in the first years of life, as shown by imaging at different ages (Kostović et al., 1989). Studies by Rasmussen and Milner (1977) showed that early unilateral lesions in left cortical language areas did not have a great deal of effect because the language functions under study shifted to the contralateral side. In all likelihood, such lesions are "compensated for" more readily at the youngest ages of postnatal development, due to a transient exuberance of neurons and connections in normal development. This view seems to be in line with the Kennard principle: The earlier the damage, the less severe the functional loss (see discussion in Kolb, 1999). However, studies performed in the past couple of decades show that this view is too simplistic. Both rat and human studies (e.g., Kolb; Kolb & Whishaw, 1990, chaps. 25 and 26) indicated that earlier is not always better, due to the existence of a critical period of plasticity. Before and after such a period, the effect of the same kind of lesions is more severe. In addition, the type of lesion and, especially, the location of the lesion in the brain are of importance for the existence of a particular critical period of plasticity. Lesions in the left hemisphere before the age of 1 year impair both verbal and performance IQ, whereas after 1 year of age, left-hemisphere lesions do not affect either kind of IQ compared to controls. For a recent review, see Tranel, Damasio, Denburg, and Bechara (2005). Similar kinds of lesion in the right hemisphere before and after the age of 1 year effect only a significantly lower performance IQ (Riva & Cazzaniga, 1986; Woods & Teuber, 1973). Also, in rat studies, a circumscribed critical period was detected during which frontal lesions did not have an effect on frontal functions such as spatial delayed alternation in later life, whereas earlier and later lesions in adulthood did have such an effect (Kolb, Forgie, Gibb, Gorny, & Rowntree, 1998; Kolb & Tomie, 1988).

In addition, Anderson, Bechara, Damasio, Tranel, and Damasio (1999) described that very early lesions in the rostromedial and orbitomedial frontal cortex before 16 months of age have more severe effects than similar lesions in adulthood. Their report on two cases indicates a functional worsening compared to the effects of similar lesions in adulthood. The two cases with lesions sustained around the age of 1 year showed not only difficulties with social behavior but also a lack of awareness of proper moral reasoning and of the consequences of social misbehavior.

Parental Depression / Social Isolation or Social Neglect

Critical periods for socialization or social neglect have been reported. Harlow (1958) showed that in monkeys, isolation/maternal deprivation for 6 months during the first 18 months leads to permanent deficits in social behavior. This did not happen when older monkeys underwent a 6-month period of isolation. The classic studies of Spitz (1946) in children who lacked proper maternal/parental care in the first year of their life showed that they were withdrawn, that their intellectual performance was below standard, and that they were more susceptible to infections, which is indicative of a weakened immune system (see

Ader, Felten, & Cohen, 1991; Petitto, Cruess, Repetto, & Evans, 2004). Global neglect (i.e., minimal exposure to language, touch, and social interaction during the first 5 years of life) leads to a permanently smaller head circumference, smaller brain size, and impaired ability to learn language and to develop normal social behavior (e.g., Chugani et al., 2001; Glaser, 2000; Perry, 2002). When these neglected children are removed from their deprived environment and brought up in foster care during the first 4 years of life, recovery occured: The earlier the environmental improvement, the more recovery (Perry). After 4 years of age, recovery is insignificant, as the major development of the brain has finished. For language acquisition, early communication is essential (Trevarthen, 2001).

Recent studies (Hay et al., 2001; Kim-Cohen, Moffitt, Taylor, Pawlby, & Caspi, 2005) have even indicated that maternal postnatal depression during the first 3 months can lead to a considerably lower IQ in boys (group mean value of IQ = 86), due to a shortage of maternal care. The effect on girls was noticeably less. For maternal depression and child development, see also Bernard-Bonnin et al. (2004).

First and Second Language Acquisition

Early postnatal childhood experience can decisively interfere with first language (L1) acquisition, including the L1 acquisition of American Sign Language (e.g., Cohen, 2002; DeKeyser & Larson-Hall, 2005; Grimshaw, Adelstein, Bryden, & MacKinnon, 1998; Newton, 2002; Sakai, 2005). Feral children who have grown up in the wild (e.g., with wolves) or in isolated conditions during the first 4–6 years of their life during which time they are not exposed to human interaction and language, appear unable to speak complete sentences or learn Sign Language (e.g., Victor & Genie in Newton). These reports showed that although the genetic contribution is essential, environmental conditions can be decisive for the functional outcome.

For second language (L2) acquisition, the age of acquisition is of critical importance for nativelike grammar and pronunciation (e.g., DeKeyser & Larson-Hall, 2005; Flege et al., 1999; Johnson & Newport, 1989, 1991). Until the age of about 7 years, the L2 can be learned to a level that is grammatically virtually indistinguishable from that of native speakers (Johnson & Newport, 1989, 1991). From 8 to 10 years onward, however, it becomes increasingly difficult to master the grammar of a L2 completely. This is the conclusion drawn from a study that compared the final level of grammatically correct English of immigrants who entered the United States at varying ages with that of native speakers of American English (Johnson & Newport, 1989). The use of the term "critical" period in the strict sense of the definition is debatable for grammar mastering, because the Johnson and Newport (1989) data showed that acquisition of grammar persists in adulthood, although not at the nativelike level. For semantic processing, a critical period might not exist (Neville & Bavelier, 1998; Stowe & Sabourin, 2005). For further assessments of the evidence, see Birdsong's article in this volume, which is entirely devoted to assessing the evidence for or against a sensitive period in L2 in various linguistic domains.

So-Called Environmental Enrichment

On the other hand, the period during which the brain responds favorably to an environmental life condition that persists to be novel and challenges the experience learning is not restricted. Even in adulthood, the rat brain reacts to the stimulation of environmental enrichment with an increase in spines and a lengthening of dendrites (e.g., Uylings et al., 1978). The extent of the flexible reaction is, however, somewhat smaller than during early postnatal development and becomes smaller with advancing age (Greenough et al., 1990). This is to be expected, as the size of dendrites and the maturation of synapses have generally reached a stable level in adulthood. The environmental enrichment effects are detected in a number of cortical areas, but not in all, and in male rats, they are different than those in female rats. The greatest differences are found in the hippocampal dendrites of female rats and in the occipital cortical dendrites of male rats (Juraska, 1991). It is to be expected that, in adulthood, learning also induces morphological alterations in synapses, spines, and dendrites, in addition to additional neurogenesis of interneurons in the dentate gyrus (e.g., Farmer et al., 2004).

Environmental enrichment effects on human brains can be derived from a full cross-fostering study on adoptive children, for instance (Chapron & Duyme, 1989; Duyme, Dumaret, & Tomkiewicz, 1999). In these studies, environmental life condition enrichment, in addition to the contribution of genetic constitution, appears to increase intellectual capabilities. On the other hand, environmental impoverishment in adulthood leads to a decrease in synapses, spine numbers, and dendrites in the same brain regions that react favorably to environmental enrichment (e.g., Greenough et al., 1990).

Conclusion

Critical periods of flexibility coincide with periods of transient exuberance in the number of neurons, axonal projections, and synapses. After the periods of selective reduction in neuron number, axonal connections, and synapses in development, flexible responses in these brain areas are virtually impossible or reduced. Critical periods exist for functions that require plastic rewiring of axons over long distances in the brain. Adult axonal rewiring or axonal extension over a long distance hardly ever occurs when the developmental competition between different axonal systems has finished. This does not mean that the brain cannot change further in adulthood; for example, environmental enrichment remains capable of challenging the brain, even in adulthood.

The review given in this article shows that genotype allows the maximum of structural and functional properties, but the eventual outcome is under the influence of environment in

Genes & Environment

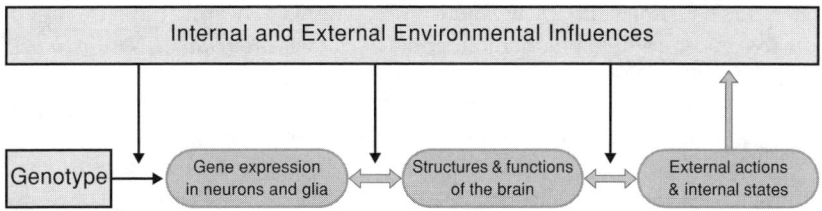

Figure 8. Diagram to show the interaction sites of genotype and environment leading to structure and function of the brain (phenotype).

positive or negative directions. Figure 8 specifies that environmental influences can modulate different aspects of development. Under dramatic circumstances, environmental factors during the first years of life can even be quite decisive for language, social, and other intellectual functions.

Note

[1] Stereology is a methodology to achieve quantitative descriptions of the geometry and number of three-dimensional structures from measurements that are made on (nearly) two-dimensional images.

References

Ader, R., Felten, D. L., & Cohen, N. (Eds.) (1991). *Psychoneuroimmunology*. New York: Academic Press.

Anderson, S. W., Bechara, A., Damasio, H., Tranel, D., & Damasio, A. R. (1999). Impairment of social and moral behavior related to early damage in human prefrontal cortex. *Nature Neuroscience, 2*, 1032–1037.

Bandstra, E. S., Vogel, A. L., Morrow, C. E., Xue, L., & Anthony, J. C. (2004). Severity of prenatal cocaine exposure and child language functioning through age seven years: A longitudinal latent growth curve analysis. *Substance Use Misuse, 39*, 25–59.

Barnea-Goraly, N., Menon, V., Eckert, M., Tamm, L., Bammer, R., Karchemskiy, A., et al. (2005). White matter development during childhood and adolescence: A cross-sectional diffusion tensor imaging study. *Cerebral Cortex*, *15*, 1848–1854.

Bergan, J. F., Ro, P., Ro, D., & Knudsen, E. I. (2005). Hunting increases adaptive auditory map plasticity in adult barn owls. *Journal of Neuroscience*, *25*, 9816–9820.

Bernard-Bonnin, A. C., & Psychosocial Paediatrics Committee (2004). Maternal depression and child development. Canadian Paediatric Society Position Statement PP 2004-03. *Paediatric Child Health*, *9*, 575–583.

Cameron, J. L. (2004). Interrelationships between hormones, behavior, and affect during adolescence. Understanding hormonal, physical, and brain changes occurring in association with pubertal activation of the reproductive axis. *Annals of the New York Academy of Sciences*, *1021*, 110–123.

Chapron, C., & Duyme, M. (1989). Assessment of effects of socio-economic status of IQ in a full cross-fostering study. *Nature*, *340*, 552–554.

Chugani, H. T., Behen, M. E., Muzik, O., Juhasz, C., Nagy, F., & Chugani, D. C. (2001). Local brain functional activity following early deprivation: A study of post-institutionalized Romanian orphans. *NeuroImage*, *14*, 1290–1301.

Chugani, H. T., Phelps, M. E., & Mazziotta, J. C. (1987). Positron emission tomography study of human brain functional development. *Annals of Neurology*, *22*, 487–497.

Cohen, D. (2002). *How the child's mind develops*. Hove, UK: Routledge.

Cooper, J. R., Bloom, F. E., & Roth, R. H. (2002). *The biochemical basis of neuropharmacology*. New York: Oxford University Press.

DeKeyser, R., & Larson-Hall, J. (2005). What does the critical period really mean? In J. F. Kroll & A. M. B. de Groot (Eds.), *Handbook of bilingualism: Psycholinguistic approaches* (pp. 89–108). Oxford: Oxford University Press.

Delalle, I., Evers, P., Kostović, I., & Uylings, H. B. M. (1997). Laminar distribution of neuropeptide-Y immunoreactive neuron in human prefrontal cortex during development. *Journal of Comparative Neurology*, *379*, 523–540.

Duyme, M., Dumaret, A. C., & Tomkiewicz, S. (1999). How can we boost IQs of "dull children"?: A late adoption study. *Proceedings of the National Academy of Sciences of the United States of America*, *96*, 8790–8794.

Farmer, J., Zhao, X., Van Praag, H., Wodtke, K., Gage, F. H., & Christie, B. R. (2004). Effects of voluntary exercise on synaptic plasticity and gene expression in dentate gyrus of adult male Sprague-Dawley rats in vivo. *Neuroscience*, *124*, 71–79.

Flechsig, P. (1920). *Anatomie des menschlichen Gehirns und Rückenmarks auf myelogenetischer Grundlage*. Leipzig: Georg Thieme.

Flege, J. E., MacKay, I. R., & Meador, D. (1999). Native Italian speakers' perception and production of English vowels. *Journal of the Acoustical Society of America, 106,* 2973–2987.

Frank, D. A., Rose-Jacobs, R., Beeghly, M., Wilbur, M. A., Bellinger, D., & Cabral, H. (2005). Level of prenatal cocaine exposure and 48 month IQ: Importance of preschool enrichment. *Neurotoxicology & Teratology, 27,* 15–28.

Giedd, J. N. (1999). Development of the human corpus callosum during childhood and adolescence: A longitudinal MRI study. *Progress in Neuropsychopharmacology & Biological Psychiatry, 23,* 571–588.

Giedd, J. N., Blumenthal, J., Jeffries, N. O., Castellanos, F. X., Liu, H., Zijdenbos, A., et al. (1999). Brain development during childhood and adolescence: A longitudinal MRI study. *Nature Neuroscience, 2,* 861–863.

Glaser, D. (2000). Child abuse and neglect and the brain: A review. *Journal of Child Psychology and Psychiatry, 41,* 97–116.

Goldman-Rakic, P. S., & Brown, R. M. (1982). Postnatal development of monoamine content and synthesis in the cerebral cortex of rhesus monkeys. *Developmental Brain Research, 4,* 339–349.

Gorski, R. A. (2000). Sexual differentiation of the nervous system. In E. R. Kandel, J. H. Schwartz, & T. M. Jessell (Eds.), *Principles of neural science* (pp. 1131–1148). New York: McGraw-Hill.

Gould, E., & Gross, C. G. (2002). Neurogenesis in adult mammals: Some progress and problems. *Journal of Neuroscience, 22,* 619–623.

Gould, E., Reeves, A. J., Graziano, M. S., & Gross, C. G. (1999). Neurogenesis in the neocortex of adult primates. *Science, 286,* 548–552.

Greenough, W. T., Withers, G. S., & Wallace, C. S. (1990). Morphological changes in the nervous system arising from behavioral experience: What is evidence that they are involved in learning and memory? In L. R. Squire & E. Lindenlaub (Eds.), *The biology of memory* (pp. 159–192). Stuttgart: Schattauer.

Grimshaw, G. M., Adelstein, A., Bryden, M. P., & MacKinnon, G. E. (1998). First-language acquisition in adolescence: evidence for a critical period for verbal language development. *Brain and Language, 63,* 237–255.

Harlow, H. F. (1958). The nature of love. *American Psychologist, 13,* 673–685.

Hay, D. F., Pawlby, S., Sharp, D., Asten, P., Mills, A., & Kumar, R. (2001). Intellectual problems shown by 11-year-old children whose mothers had postnatal depression. *Journal of Child Psychology & Psychiatry, 42,* 871–889.

Heminger, G. R. (1999). Special challenges in the investigation of the neurobiology of mental illness. In D. S. Charney, E. J. Nestler, & B. S. Bunney (Eds.), *Neurobiology of mental illness* (pp. 89–99). New York: Oxford University Press.

Hensch, T. K. (2004). Critical period regulation. *Annual Review of Neuroscience, 27*, 549–579.

Hensch, T. K. (2005). Critical period plasticity in local cortical circuits. *Nature Reviews Neuroscience, 6*, 877–888.

Herlenius, E., & Lagercrantz, H. (2004). Development of neurotransmitter systems during critical periods. *Experimental Neurology, 190*, S8–S21.

Hicks, S. P., & D'Amato, C. J. (1978). Effects of ionizing radiation on developing brain and behavior. In G. Gottlieb (Ed.), *Studies on the development of behavior and the nervous system* (pp. 36–72). New York: Academic Press.

Hochstetter, F. (1919). *Beiträge zur Entwicklungsgeschichte des menschlichen Gehirns*. Wien: I. Deuticke.

Hofman, M. A., & Swaab, D. F. (1989). The sexual dimorphic nucleus of the preoptic area in the human brain: A comparative morphometric study. *Journal of Anatomy (London), 164*, 55–72.

Holtmaat, A. J., Trachtenberg, J. T., Wilbrecht, L., Shepherd, G. M., Zhang, X., Knott, G. W., et al. (2005). Transient and persistent dendritic spines in the neocortex in vivo. *Neuron, 45*, 279–291.

Huttenlocher, P. R., & Dabholkar, A. S. (1997). Developmental anatomy of prefrontal cortex. In N. A. Krasnegor, G. R. Lyon, & P. S. Goldman-Rakic (Eds.), *Development of the prefrontal cortex: Evolution, neurobiology, and behavior* (pp. 69–83). Baltimore: Paul H. Brookes.

Huttenlocher, P. R., de Courten, C., Garey, L. G., & Van der Loos, H. (1982). Synaptogenesis in human visual cortex: Evidence for synapse elimination during normal development. *Neuroscience Letters, 33*, 247–252.

Huttunen, M. O., & Niskanen, P. (1978). Prenatal loss of father and psychiatric disorders. *Archives of General Psychiatry, 35*, 429–431.

Hyvärinen, J. (1982). *The parietal cortex of monkey and man*. Berlin: Springer-Verlag.

Innocenti, G. M., & Price, D. J. (2005). Exuberance in the development of cortical networks. *Nature Reviews Neuroscience, 6*, 955–965.

Johnson, J. S., & Newport, E. L. (1989). Critical period effects in second language learning: The influence of maturational state on the acquisition of English as a second language. *Cognitive Psychology, 21*, 60–99.

Johnson, J. S., & Newport, E. L. (1991). Critical period effects on universal properties of language: The status of subjacency in the acquisition of a second language. *Cognition, 39*, 215–258.

Juraska, J. M. (1991). Sex differences in "cognition" regions of the rat brain. *Psychoendocrinology, 16*, 105–119.

Kaes, T. (1907). *Die Grosshirnrinde des Menschen in ihren Massen und in ihrem Fasergehalt. Ein Gehirnanatomischer Atlas*. Jena: Gustav Fischer.

Kalsbeek, A., De Bruin, J. P. C., Feenstra, M. G. P., & Uylings, H. B. M. (1990). Age-dependent effects of lesioning the mesocortical dopamine system upon prefrontal cortex morphometry and PFC-related behaviors. *Progress in Brain Research, 85,* 257–283.

Kandel, E. R., Jessell, T. M., & Sanes, J. R. (2000). Sensory experience and the fine-tuning of synaptic connections. In E. R. Kandel, J. H. Schwartz, & T. M. Jessell (Eds.), *Principles of neural science* (pp. 1115–1130). New York: McGraw-Hill.

Kempermann, G. (2006). *Adult neurogenesis: Stem cells and neuronal development in the adult brain.* New York: Oxford University Press.

Kim-Cohen, J., Moffitt, T. E., Taylor, A., Pawlby, S. J., & Caspi, A. (2005). Maternal depression and children's antisocial behavior: Nature and nurture effects. *Archives of General Psychiatry, 62,* 173–181.

Kline, J., Stein, Z., & Susser, M. (1989). *Conception to birth: Epidemiology of prenatal development.* New York: Oxford University Press.

Knudsen, E. I. (2004). Sensitive periods in the development of the brain and behavior. *Journal of Cognitive Neuroscience, 16,* 1412–1425.

Koenderink, M. J. Th., & Uylings, H. B. M. (1995). Postnatal maturation of layer V pyramidal neurons in the human prefrontal cortex: A quantitative Golgi analysis. *Brain Research, 678,* 233–243.

Koenderink, M. J. Th., Uylings, H. B. M., & Mrzljak, L. (1994). Postnatal maturation of the layer III pyramidal neurons in the human prefrontal cortex: A quantitative Golgi analysis. *Brain Research, 653,* 173–182.

Kolb, B. (1999). Brain development, plasticity and behavior. *American Psychologist, 44,* 1203–1212.

Kolb, B., Forgie, M., Gibb, R., Gorny, G., & Rowntree, S. (1998). Age, experience and the changing brain. *Neuroscience Biobehavioral Reviews, 22,* 143–159.

Kolb, B., & Tomie, J. A. (1988). Recovery from early cortical damage in rats. IV. Effects of hemidecortication at 1, 5 or 10 days of age on cerebral anatomy and behavior. *Behavioral Brain Research, 28,* 259–274.

Kolb, B., & Whishaw, I. Q. (1990). *Fundamentals of human neuropsychology.* San Francisco: W. H. Freeman.

Kooistra, L., Laane, C., Vulsma, T., Schellekens, J. M. H., Van der Meere, J. J., & Kalverboer, A. F. (1994). Motor and cognitive development in children with congenital hypothyroidism: A long-term evaluation of the effects of neonatal treatment. *Journal of Pediatrics, 124,* 903–909.

Korr, H., & Schmitz, C. (1999). Facts and fictions regarding postnatal neurogenesis in the developing human cerebral cortex. *Journal of Theoretical Biology, 200,* 291–297.

Kostović, I. (1990). Structural and histochemical reorganization of the human prefrontal cortex during perinatal and postnatal life. *Progress in Brain Research, 85,* 223–240.

Kostović, I., & Goldman-Rakic, P. S. (1983). Transient cholinesterase staining in the mediodorsal nucleus of the thalamus and its connections in the developing human and monkey brain. *Journal of Comparative Neurology, 219,* 431–447.

Kostović, I., Lukinovic, N., Judáš, M., Bogdanovic, N., Mrzljak, L., Zecevic, N., et al. (1989). Structural basis of the developmental plasticity in the human cerebral cortex: The role of the transient subplate zone. *Metabolic Brain Disease, 4,* 17–23.

Laplante, D. P., Barr, R. G., Brunet, A., Galbaud du Fort, G., Meaney, M. L., Sauchier, J. F., et al. (2004). Stress during pregnancy affects general intellectual and language functioning in human toddlers. *Pediatric Research, 56,* 400–410.

Larsen, C. C., Larsen, K. B., Bogdanovic, N., Laursen, H., Græm, N., Samuelsen, G. B., et al. (2006). Total number of cells in the human newborn telencephalic wall. *Neuroscience,* in press.

Letinic, K., Zoncu, R., & Rakic, P. (2002). Origin of GABAergic neurons in the human neocortex. *Nature, 417,* 645–649.

Lewis, D. A., & Sesack, S. R. (1997). Dopamine systems in the primate brain. In A. Bjorklund, T. Hökfelt, & F. E. Bloom (Eds.), *Handbook of clinical neuroanatomy: Vol. 13. The primate nervous system, part 1* (pp. 263–375). Amsterdam: Elsevier.

Malanga, C. J., & Kosofsky, B. E. (2004). Effect of drugs of abuse on brain development. In D. S. Charney & E. J. Nestler (Eds.), *Neurobiology of mental illness* (pp. 720–739). New York: Oxford University Press.

Marin, O., & Rubenstein, J. L. R. (2001). A long remarkable journey: Tangential migration in the telencephalon. *Nature Reviews Neuroscience, 2,* 780–790.

Marin-Padilla, M. (1992). Ontogenesis of the pyramidal cell of the mammalian neocortex and developmental cytoarchitectonics: A unifying theory. *Journal of Comparative Neurology, 321,* 223–240.

McEwen, B. S. (2004). Structural and functional plasticity in the hippocampal formation: Stress, adaptation, and disease. In D. S. Charney & E. J. Nestler (Eds.), *Neurobiology of mental Illness* (2nd ed., pp. 558–583). New York: Oxford University Press.

Miller, M. W. (1987). Effect of prenatal exposure to alcohol on the distribution and time of corticospinal neurons in the rat. *Journal of Comparative Neurology, 257,* 372–382.

Morrow, C. E., Vogel, A. L., Anthony, J. C., Ofir, A. Y., Dausa, A. T., & Bandstra, E. S. (2004). Expressive and receptive language functioning in preschool children with prenatal cocaine exposure. *Journal of Pediatrics, 29*(7), 543–554.

Mrzljak, L., Uylings, H. B. M., Kostović, I., & Van Eden, C. G. (1988). The prenatal development of neurons in the human prefrontal cortex: I. A qualitative Golgi study. *Journal of Comparative Neurology, 271,* 355–386.

Mrzljak, L., Uylings, H. B. M., Kostović, I., & Van Eden, C. G. (1992). The prenatal development of neurons in the human prefrontal cortex: A quantitative Golgi study. *Journal of Comparative Neurology, 316,* 485–496.

Mrzljak, L., Uylings, H. B. M., Van Eden, C. G., & Judáš, M. (1990). Neuronal development in human prefrontal cortex in prenatal and postnatal stages. *Progress in Brain Research, 85,* 185–222.

Neville, H. J., & Bavelier, D. (1998). Neural organization and plasticity of language. *Current Opinions in Neurobiology, 8,* 485–496.

Newton, M. (2002). *Savage girls and wild boys: A history of feral children.* London: Faber & Faber.

Nulman, I., Rovet, J., Greenbaum, R., Loebstein, M., Wolpin, J., Pace-Asciak, P., et al. (2001). Neurodevelopment of adopted children exposed in utero to cocaine: The Toronto Adoption Study. *Clinical Investigation in Medicine, 24,* 129–137.

Nuñez, J. L., Sodhi, J., & Juraska, J. M. (2002). Ovarian hormones after postnatal day 20 reduce neuron number in the rat primary visual cortex. *Journal of Neurobiology, 52,* 312–321.

O'Leary, D. D. M. (1989). Do cortical areas emerge from a protocortex? *Trends in Neuroscience, 12,* 400–406.

O'Leary, D. D. M. (1992). Development of connectional diversity and specificity in the mammalian brain by the pruning of collateral projections. *Current Opinions in Neurobiology, 2,* 70–77.

O'Leary, D. D. M., Bicknese, A. R., De Carlos, J. A., Heffner, C. D., Koester, S. E., Kutka, L. J., et al. (1990). Target selection by cortical axons: alternative mechanisms to establish axonal connections in the developing brain. In *Cold Spring Harbor Symposia on Quantitative Biology: Vol LV. The brain* (pp. 453–468). Plainview, NY: Cold Spring Harbor Laboratory Press.

Olney, J. W. (2002). New insights and new issues in developmental neurotoxicology. *Neurotoxicology, 23,* 659–668.

Olney, J. W., Young, C., Wozniak, D. F., Jevtovic-Todorovic, V., & Ikonomidou, C. (2004). Do pediatric drugs cause developing neurons to commit suicide? *Trends in Pharmacological Sciences, 25,* 135–139.

Parnavelas, J. G. (2000). The origin and migration of cortical neurones: New vistas. *Trends in Neuroscience, 23*, 126–131.

Paus, T. (2005). Mapping brain maturation and cognitive development during adolescence. *Trends in Cognitive Sciences, 9*, 60–68.

Perry, B. D. (2002). Childhood experience and the expression of genetic potential: What childhood neglect tells us about Nature and Nurture. *Brain and Mind, 3*, 79–100.

Petanjek, Z., Judaš, M., Kostović, I., & Uylings, H. B. M. (in preparation). *Lifespan alterations in human prefrontal cortex basal dendrites: A layer specific pattern.*

Petitto, J. M., Cruess, D. G., Repetto, M. J., & Evans, D. L. (2004). Neuroimmunology. In D. S. Charney & E. J. Nestler (Eds.), *Neurobiology of mental illness* (pp. 171–179). New York: Oxford University Press.

Rakic, P. (1988). Specification of cerebral cortical areas. *Science, 241*, 170–176.

Rakic, P. (1995). A small step for the cell, a giant leap for mankind: A hypothesis of neocortical expansion during evolution. *Trends in Neuroscience, 18*, 383–388.

Rakic, P. (2002a). Adult neurogenesis in mammals: An evaluation of the evidence. *Nature Reviews Neuroscience, 3*, 65–71.

Rakic, P. (2002b). Adult neurogenesis in mammals: An identity crisis. *Journal of Neuroscience, 22*, 614–618.

Rakic, P. (2005). Less is more: Progenitor death and cortical size. *Nature Neuroscience, 8*, 981–982.

Rakic, P., & Sidman, R. L. (1968). Supravital DNA synthesis in the developing human and mouse brain. *Journal of Neuropathology & Experimental Neurology, 27*, 246–276.

Rasmussen, T., & Milner, B. (1977). The role of early left-brain injury in determining lateralization of cerebral speech functions. *Annals of the New York Academy of Sciences, 299*, 355–369.

Riva, D., & Cazzaniga, L. (1986). Late effects of unilateral brain lesions sustained before and after age one. *Neuropsychologica, 24*, 423–428.

Sakai, K. (2005). Language acquisition and brain development. *Science, 310*, 815–819.

Shankle, W. R., Landing, B. H., Rafii, M. S., Schiano, A., Chen, J. M., & Hara, J. (1998). Evidence for a postnatal doubling of neuron number in the developing human cerebral cortex between 15 months and six years. *Journal of Theoretical Biology, 191*, 115–140.

Shankle, W. R., Rafii, M. S., Landing, B. H., & Fallon, J. H. (1999). Approximate doubling of numbers of neurons in postnatal human cerebral cortex and in 35 specific cytoarchitectural areas from birth to 72 months. *Pediatric Developmental Pathology, 2*, 244–259.

Shigematsu, I., Ito, C., Kamada, N., Akiyama, M., & Sasaki, H. (1995). *Effects of A-bomb radiation on the human body.* Chur: Harwood.

Shors, T. J., Townsend, D. A., Zhao, M., Kozorovitskiy, Y., & Gould, E. (2002). Neurogenesis may relate to some but not all types of hippocampal-dependent learning. *Hippocampus, 12,* 578–584.

Slotkin, T. A. (1998). Fetal nicotine or cocaine exposure: Which one is worse? *Journal of Pharmacology and Experimental Therapeutics, 285,* 931–945.

Spitz, R. A. (1946). Hospitalism: A follow-up report on investigation described in volume 1, 1945. *Psychoanalitical Study of Children, 2,* 113–117.

Stowe, L. A., & Sabourin, L. (2005). Imaging the processing of a second language: Effects of maturation and proficiency on the neural processes involved. *International Review of Applied Linguistics in Language Teaching, 43,* 329–353.

Streissguth, A. P., Barr, H. M., & Sampson, P. D. (1990). Moderate prenatal alcohol exposure: Effects on child IQ and learning problems at age $7^{1}/_{2}$ years. *Alcohol, Clinical and Experimental Research, 14,* 662–669.

Sur, M., & Rubenstein, J. L. R. (2005). Patterning and plasticity of the cerebral cortex. *Science, 310,* 805–810.

Susser, E., Neugebauer, R., Hoek, H., Brown, A., Lin, S., Labovitz, D., et al. (1996). Schizophrenia after prenatal famine: Further evidence. *Archives of General Psychiatry, 53,* 25–31.

Taupin, P. (2005). Adult neurogenesis in the mammalian central nervous system: Functionality and potential clinical interest. *Medical Science Monitor, 11,* 247–252.

Thorpe-Beeston, J. G., Nicolaides, K. H., Felton, C. V., Butler, J., & McGregor, A. M. (1991). Maturation of the secretion of thyroid hormone and thyroid-stimulating hormone in the fetus. *New England Journal of Medicine, 324,* 532–536.

Trachtenberg, J. T., Chen, B. E., Knott, G. W., Feng, G., Sanes, J. R., Welker, E., et al. (2002). Long-term in vivo imaging of experience-dependent synaptic plasticity in adult cortex. *Nature, 420,* 788–794.

Tranel, D., Damasio, H., Denburg, N. L., & Bechara, A. (2005). Does gender play a role in functional asymmetry of ventromedial prefrontal cortex? *Brain, 128,* 2872–2881.

Trevarthen, C. (2001). The neurobiology of early communication: Intersubjective regulations in human brain development. In A. F. Kalverboer & A. Gramsbergen (Eds.), *Handbook of brain and behaviour in human development* (pp. 841–881). Amsterdam: Kluwer Academic.

UNSCEAR (United Nations Scientific Committee on Effects of Atomic Radiation) (1986). *Genetic and somatic effects of ionizing radiation.* Report to the General Assembly. New York: United Nations.

Uylings, H. B. M. (2001). The human cerebral cortex in development. In A. F. Kalverboer & A. Gramsbergen (Eds.), *Handbook of brain and behaviour in human development* (pp. 63–80). Amsterdam: Kluwer Academic.

Uylings, H. B. M., & Delalle, I. (1997). Morphology of neuropeptide Y-immunoreactive neurons and fibers in human prefrontal cortex during prenatal and postnatal development. *Journal of Comparative Neurology, 379*, 523–540.

Uylings, H. B. M., Delalle, I., Petanjek, Z., & Koenderink, M. J. Th. (2002). Structural and immunocytochemical differentiation of neurons in prenatal and postnatal human prefrontal cortex. *Neuroembryology, 1*, 176–186.

Uylings, H. B. M., Kuypers, K., Diamond, M. C., & Veltman, W. A. M. (1978). Effects of differential environments on plasticity of dendrites of cortical pyramidal neurons in adult rats. *Experimental Neurology, 62*, 658–677.

Uylings, H. B. M., Malofeeva, L. I., Bogolepova, I. N., Jacobsen, A. M., Amunts, K., & Zilles, K. (2005). No postnatal doubling of number of neurons in human Broca's area (BA 44 and 45)? A stereological study. *Neuroscience, 136*, 715–728.

Uylings, H. B. M., Van Pelt, J., Parnavelas, J. G., & Ruiz-Marcos, A. (1994). Geometrical and topological characteristics in the dendritic development of cortical pyramidal and non-pyramidal neurons. In J. Van Pelt, M. A. Corner, H. B. M. Uylings, & F. H. Lopes da Silva (Eds.), *The self-organizing brain: From growth cones to functional networks* (pp. 109–123). Amsterdam: Elsevier.

Weitzman, M., Gortmaker, S., & Sobol, A. (1992). Maternal smoking and behavior problems of children. *Pediatrics, 90*, 342–349.

Woods, B. T., & Teuber, H. L. (1973). Early onset of complementary specialization of cerebral hemispheres in man. *Transactions of the American Neurological Association, 98*, 113–117.

Woolley, C. S., & McEwen, B. S. (1992). Estradiol mediates fluctuation in hippocampal synapse density during the estrous cycle in the adult rat. *Journal of Neuroscience, 12*, 2549–2554.

Yakovlev, P. I., & Lecours, A. R. (1967). The myelogenetic cycles of regional maturation of the brain. In A. Minkowski (Ed.), *Regional development of the brain in early life* (pp. 3–70). Oxford: Blackwell Scientific.

Yudkoff, M. (1994). Disorders of amino acid metabolism. In G. J. Siegel, B. W. Agranoff, R. W. Albers, & P. B. Molinoff (Eds.), *Basic neurochemistry* (pp. 824–826). New York: Raven Press.

Zilles, K. (1978). Ontogenesis of the visual system. *Advances in Anatomy, Embryology and Cell Biology, 54*, 1–138.

Zito, K., & Svoboda, K. (2002). Activity-dependent synaptogenesis in the adult mammalian cortex. *Neuron, 35*, 1015–1017.

What We Cannot Learn From Neuroanatomy About Language Learning and Language Processing. Commentary on Uylings

Peter Hagoort
F. C. Donders Centre for Cognitive Neuroimaging,
Radboud University Nijmegen

Cognitive neuroscience aims at specifications of human cognition as instantiated in the human brain. This requires that information from multiple levels be taken into consideration and that information at one level be used to guide the investigations at another level. With regard to the human language faculty, linguistics specifies structural descriptions of the relevant knowledge types such as phonological, syntactic, and semantic structures. Psycholinguistics specifies the processing architectures of different language functions such as speaking, reading, and listening. These processing architectures explain how, in real time, the different sources of linguistic knowledge in long-term memory are recruited and exploited in mapping sound or orthography onto meaning (in listening and reading) or vice versa (in speaking and writing). Finally, (cognitive) neuroscience specifies how the processing architectures are instantiated in the human brain. In this context, an account of the neural architecture of language should specify which brain areas are recruited for different language functions, as well as how these areas communicate while these

---

I am grateful for the comments of Pascal Fries, Peter Indefrey, and Ivan Toni on an earlier version of my contribution.

Correspondence concerning this article should be addressed to Peter Hagoort, F. C. Donder's Centre for Cognitive Neuroimaging, P.O. Box 9101, 6500 HB Nijmegen, The Netherlands. Internet:Peter.Hagoort@fcdonders.ru.nl

functions are executed. Classically, neuropsychological data from patients with a lesion in the language-relevant cortex have been used in studies on the neural architecture of language. The advent of a whole generation of advanced neuroimaging techniques (positron-emission tomography [PET], functional magnetic resonance imaging [fMRI], magnetic-encephalography [MEG]) in the recent decades, however, replaced lesion studies as the major tool for investigating neural architecture issues.

As everyone involved in cognitive neuroscience of language realizes, it is far from trivial to relate to each other the levels of linguistic description, processing architecture, and neural hardware. This is partly due to the differences in the conceptual vocabulary in which accounts at the distinct levels are couched. However, we also suffer from the fact that we do not have a full grasp of the mapping relations between the different levels. For instance, is the mapping between specific components of the cognitive architecture and a set of anatomically defined brain areas one-to-one, one-to-many, or many-to-one (Mehler, Morton, & Jusczyk, 1984)? It would be even worse if there does not exist a transparent or lawful relation between cognitive architecture and neural architecture, as some have claimed (Fodor, 1975). In the latter case, the attempt to connect brain and behavior is a useless endeavor. I take the success of cognitive neuroscience in many domains of cognition as an indication that even in the absence of full knowledge about the exact nature of the mapping relation between cognitive and neural architectures, this relation is less opaque and more direct than proponents of the functional stance have claimed. Such is the general assumption underlying cognitive neuroscience research. Within this broad framework of cognitive neuroscience, one has to look in more detail what the constraints and contributions of different subdomains are in designing a full account of the neurocognition of language. In the remainder, I will focus on the constraints from neuroanatomy, a field that is discussed in this volume by Uylings.

A prime example of the contribution of neuroanatomy is the famous map by Brodmann (1869–1918). This map consists of 47 different areas, usually referred to by expressions such as BA 44 for Brodmann Area 44. The numbers of the Brodmann Areas were determined by the order in which Brodmann went through the brain, analyzing one area after the other. Brodmann's classification is based on the cytoarchitectonics of the brain, which refers to the structure, form, and position of the cells in the six layers of the cortex. Quantification was done by Brodmann on postmortem brains. These were sectioned into slices of 5–10 μm thickness that underwent Nissl staining and were then inspected under the microscope. In this way, the distribution of different cell types across cortical layers and brain areas could be determined. Even today, Brodmann's map, which was published in 1909, is seen as a hallmark in the history of neuroscience. It was published in Leipzig by the publisher Barth after almost 10 years of hard work under the title *Vergleichende Lokalisationslehre der Grosshirnrinde in ihren Prinzipien dargestellt auf Grund des Zellenbaues*. Brodmann's work reveals that the composition of the six cortical layers, in terms of cell types, varies across the brain. Also, cell numbers can vary. The primary visual cortex, for instance, has about twice as many neurons per cortical column as other brain areas (Amaral, 2000).

The classical view among neuroanatomists is that these architectural differences in brain structure are indicative of functional differences and, conversely, that functional differences demand differences in architecture (Bartels & Zeki, 2005; Brodmann, 1905; Vogt & Vogt, 1919; Von Economo & Koskinas, 1925). Following the classical view, through different ways of characterizing brain structure (i.e., cyto-, myelo-, and receptorarchitectonics; Zilles & Palomero-Gallagher, 2001), brain areas can be identified, for which differences in structural characteristics imply functional differences. From this view, it follows that one should look for the structural features that determine why a particular brain area can support, for instance, the processing of a first or second language.

In contrast to the classical view in neuroanatomy, more recent accounts have argued that from a computational perspective, different brain areas are very similar. For instance, Douglas and Martin (2004) argued that

> The same basic laminar and tangential organization of the excitatory neurons of the neocortex, the spiny neurons, is evident wherever it has been sought. The inhibitory neurons similarly show a characteristic morphology and patterns of connections throughout the cortex (...) all things considered, many crucial aspects of morphology, laminar distribution, and synaptic targets are very well conserved between areas and between species. (p. 439)

Functional differences between brain areas are in this perspective mainly due to variability of the input signals in forming functional specializations. Domain specificity of a particular piece of cortex might thus not so much be determined by heterogeneity of brain tissue, but by the way in which its functional characteristics are shaped by the input.

Recent neuroimaging studies provide support for this view. A number of remarkable forms of neural plasticity have been reported in recent years. I will here discuss one example. In a recent study by Amedi, Raz, Pianka, Malach, and Zohary (2003), the authors report that they found increased activation in the primary visual cortex (V1) during a verbal memory task in congenitally blind subjects. Moreover, the stronger the activation in V1, the better the memory performance. If the structural properties of V1 had been decisive for its functional capacities, then it would be hard to see how the same neurons that in seeing people support vision could be recruited in the blind for verbal memory. This demonstrates that the cytoarchitectonic constraints for specifications of cognitive function are rather loose. Presumably, the input and the patterns of connectivity between areas are a more relevant functional parameter than the differences in the composition of cortical layers.

With respect to the language function, comparative neuroanatomy further supports my point. In a recent study (Petrides, Cadoret, & Mackey, 2005), an area in the macaque monkey has been described that is cytoarchitectonically comparable to BA 44 in humans, which is part of Broca's area. BA 44 in humans has been found to be involved in syntactic processes, among other things. The homologous area in the macaque monkey subserves orofacial movements. One could make an argument that in the course of evolution, human language has developed out of a system of vocalizations that require the involvement of orofacial musculature. However, my point here is that whatever the evolutionary trajectory to language, an area in the macaque monkey with a very similar cytoarchitectonic structure as Broca's area has a clearly different function than in humans. This again supports the claim that the purely cytoarchitectonic constraints for specifications of cognitive functions do not seem to be very strong.

All of this does not imply that I deny the great importance of cytoarchitectonic structures for human cognition. Clearly, without these basic building blocks of the brain, cognition would not be possible. Without neurons, glia, and axons, the cognitive machinery would not work. However, the issue here is that these building blocks enter into processes of functional specialization. My claim is that the exact nature of these functional specializations is more easily inferred from an analysis of input and connectivity than by looking at the cytoarchitectonic characteristics.

What are the consequences for accounts of language learning and language processing? I suggest that important evidence can come from different patterns of brain activation for different aspects of language processing (e.g., syntax vs. semantics). At the same time, I do not think that at the moment there is much evidence for the claim that much hinges on whether a particular activation is found in, say, BA 44 rather than BA 45. An approach based on reversed inferences from structural anatomy to cognitive function does not seem well constrained enough in the light of our current knowledge. It is functional anatomy that counts, and that might provide stronger constraints than structural anatomy for specifications of the different forms of human cognition. This implies that we should look at patterns of associations and dissociations in measures of brain activity. These might provide more direct insights into the neural instantiations of cognitive functions than the classical anatomical measures.

In summary, a cognitive neuroscience approach of language takes information and constraints from different levels of analysis into consideration, in the service of a full account of the neurocognition of language. The assumption hereby is that different levels can be connected in a transparent way. At the same time, not all constraints have the same force. Here it is argued that the constraints provided by the classical anatomical measures (cytoarchitectonics and myeloarchitectonics) in our current understanding are only very loose constraints for detailed specifications of cognitive functions, including language learning and language processing. However, measures of the computational features of brain tissue might provide stronger constraints. For the time being, our best bet for understanding cognitive specialization is to focus on measures of functional rather than structural neuroanatomy.

References

Amaral, D. G. (2000). The anatomical organization of the central nervous system. In E. R. Kandel, J. H. Schwartz, & Th. M. Jessell (Eds.), *Principles of neural science* (4th ed., pp. 317–336). New York: McGraw-Hill.

Amedi, A., Raz, N., Pianka, P., Malach, R., & Zohary, E. (2003). Early "visual" cortex activation correlates with superior verbal memory performance in the blind. *Nature Neuroscience, 6,* 758–766.

Bartels, A., & Zeki, S. (2005). The chronoarchitecture of the cerebral cortex. *Philosophical Transactions of the Royal Society of London B, 360,* 733–750.

Brodmann, K. (1905). Beiträge zur histologischen Lokalisation der Grosshirnrinde. Dritte Mitteilung: Die Rindenfelder der niederen Affen. *Journal of Psychology and Neurology, 4,* 177–226.

Douglas, R. J., & Martin, K. A. (2004). Neuronal circuits of the neocortex. *Annual Review of Neuroscience, 27,* 419–451.

Fodor, J. D. (1975). *The language of thought.* New York: Thomas Y. Crowall.

Mehler, J., Morton, J., & Jusczyk, P. W. (1984). On reducing language to biology. *Cognitive Neuropsychology, 1,* 83–116.

Petrides, M., Cadoret, G., & Mackey, S. (2005). Orofacial somatomotor responses in the macaque monkey homologue of Broca's area. *Nature, 435*, 1235–1238.

Vogt, C., & Vogt, O. (1919). Allgemeinere Ergebnisse unserer Hirnforschung. Vierte Mitteilung. Die physiologische Bedeutung der architektonischen Rindenfelderung auf Grund neuer Rindenreizungen. *Journal für Psychologie und Neurologie, 25*, 279–462.

Von Economo, C., & Koskinas, G. N. (1925). *Die Cytoarchitektonik der Hirnrinde des erwachsenen Menschen.* Wien, Berlin: Springer-Verlag.

Zilles, K., & Palomero-Gallagher, N. (2001). Cyto-, myelo-, and receptor architectonics of the human parietal cortex. *Neuroimage, 14*, S8–S20.

Convergence, Degeneracy, and Control

David W. Green, Jenny Crinion, and Cathy J. Price
University College London

Understanding the neural representation and control of language in normal bilingual speakers provides insights into the factors that constrain the acquisition of another language, insights into the nature of language expertise, and an understanding of the brain as an adaptive system. We illustrate both functional and structural brain changes associated with acquiring other languages and discuss the value of neuroimaging data in identifying individual differences and different phenotypes. Understanding normal variety is vital too if we are to understand the consequences of brain damage in bilingual and polyglot speakers.

A natural language allows us to coordinate action. Language is a tool to achieve such coordination. From this perspective, utterances are instructions to build a particular mental representation (the one a speaker has in mind) and speech production is a process of building such instructions. In acquiring a language, we learn how to create instructions using a range of lexical, grammatical, and prosodic means.

David W. Green, Department of Psychology; Jenny Crinion and Cathy J. Price, Wellcome Department of Imaging Neuroscience

We thank the Wellcome Trust for support. We thank Bob Turner and his Japanese colleagues, Joe Devlin, and members of our immediate team Alice Grogan and Nilufa Ali. A number of MSc students, Kath Stockton, Christos Pliatsikas, Marie-Luise Mechias, and Sylvie Orabona, also contributed, and we thank them for their efforts.

Correspondence concerning this article should be addressed to David W. Green, University College London, Gower Street, London, WC1E 6BT. Internet: d.w.green@ucl.ac.uk

Languages differ in the way they attach to these different linguistic means. In English, for instance, "who did what to whom" is conveyed by word order. In other languages, word order is less constrained and such information is conveyed by word endings or by prosody in tone languages (see Baker, 2003). These different linguistic means or signals require different neural devices for their processing. The particular properties of the language will affect the demands on particular processes. For instance, a tonal language will place greater demands on the neural devices involved in prosody. On the simplest assumption, the acquisition of a second language will utilize existing devices and the particular properties of the language will affect the relative demands on these devices. So, for instance, acquiring a tone language will require using the system for intonation and prosody for purposes of lexical semantics. According to this proposal, a natural expectation is that the neural representation of a second language (L2) will converge with that of the native speakers of that language as proficiency improves (Green, 2003). Convergence predicts that the acquisition of another language will have both functional and structural consequences. We consider this issue in the first section and draw out the role of neuroimaging studies in enhancing our understanding of how bilingual speakers control their languages.

In the second section, we argue that any challenge to convergence that relies on showing that different neuronal mechanisms are involved in task performance in L2 speakers needs to establish the extent to which different mechanisms are in fact invoked by native speakers. For instance, words in English can be read in different ways and task and individual differences might lead one strategy to predominate over another. Neuroimaging data can help establish the range of phenotypes and their structural and functional bases.

Neuroimaging (in contrast to many current psychological models) points to the complexity of the circuits involved in language selection and control. In the third section, we stress the need to discover how these processes are orchestrated. We argue that it is also important to devise studies that take account of the evolutionary context of language so as to obtain a more coherent view of the functional role that different neural regions play. We conclude that a cognitive neuroscience approach to L2 acquisition is essential if we are to understand individual differences (phenotypes) in the ease of acquiring other languages or particular bilingual skills. It is vital also if we are to understand the intimate connections between language and the self and to theorize effectively about the consequences of brain damage in bilingual and polyglot speakers.

Convergence: Representation and Control

An L2 might be represented in different neural substrates or its neural representation might converge with the neural representations activated in native speakers. Proponents for the differential representation of an L2 might appeal to evidence regarding the dedicated nature of neural machinery for language acquisition in infants, especially in terms of grammar. Those in favor of converging representations can urge that particular substrates are better at processing certain types of information and will continue to do so (Green, 2003). We consider two kinds of consequence in the acquisition of another language: functional and structural. Structural consequences might arise as a result of functional demands. In functional terms, we focus on evidence concerning the neural regions activated in task performance.

Consider by way of illustration the acquisition of vocabulary in an L2. Grant that a set of regions mediates the representations of words and their meanings (e.g., Vandenberghe, Price, Wise, Josephs, & Frackowiak, 1996). Functional consequences will arise because words in different languages will access a common conceptual, sensory-motor system. Presenting words in one language will prime the semantic representation of its translation equivalent; that is, there will be a short-term neuronal adaptation. However, individuals must also be able to distinguish words of one language from those of another. Conceivably then, in some regions neuronal adaptation might be language-sensitive.

Other functional consequences of acquiring a second vocabulary are possible. One consequence arises because a given object can now be named in different languages and so individuals must select the relevant name thereby, potentially at least, giving rise to a problem of control.

On the supposition that functional demands induce structural changes, acquiring the vocabulary of another language means that regions involved in registering the phonologic-orthographic and motor mappings must adapt in order to cope with an increase in the number of words to be represented. Such a structural adaptation should be detectable. If it is the case that acquisition increases the demands on the systems involved in controlling language, there should also be evidence of structural changes in the regions associated with control. We consider evidence from our own work for such functional and structural consequences.

Functional Consequences: Neuronal Adaptation

Prior neuroimaging studies on bilinguals have examined the existence of a common system using word generation tasks and metalinguistic judgment tasks that involve a large number of other component processes. Other studies have used semantic judgment tasks using single words. For instance, Illes et al. (1999) required a group of fluent English/Spanish speakers to perform a semantic task (Is the presented word abstract or concrete?) and a nonsemantic task (Is the word in upper or lower case?) in both English and Spanish. They observed no language differences. Relative to the nonsemantic task, the semantic task on both English and Spanish words activated a common left inferior frontal region. Likewise, Pillai et al. (2003) found that Spanish/English bilinguals activated the same left frontal and temporal regions during a noun-verb matching task irrespective of whether the words were in English or Spanish. However, language differences did emerge in the Pillai et al. study during rhyming decision (Do two presented words rhyme or not?) with increased activation in right frontal and parietal regions when the words were presented in English.

To dissociate language-sensitive effects within shared semantic regions, we used a semantic priming technique in which a target word is preceded by a prime word that is closely related to it (e.g., the prime might be semantically related to the target or be a translation equivalent of it). In behavioral research, normal proficient bilinguals show priming effects in both first language (L1) and in L2, but the pattern of cross-language priming depends on the nature of task. When individuals are required to perform a lexical decision task, priming is greater from L1 to L2 than from L2 to L1. This pattern also obtains where the prime is in one script (an alphabetic one, for instance) and the target is in another script (a logographic one such as Chinese or Japanese; see Jiang & Forster, 2001). The lexical decision task is a metalinguistic task and arguably sensitive to strategic factors or to possible differences in the semantic organization of words in the L1 and L2 (see Finkbeiner, Forster, Nicol, & Nakamura, 2004). In contrast, when moderately fluent bilinguals perform a semantic categorization task, priming, at least for translation equivalents, is also robust from L2 to L1 (Grainger & Frenck-Mestre, 1998; Finkbeiner et al.). We sought to minimize any effects of language direction and thus used a property verification semantic task. If words in different languages access the same neural populations for the meaning and the referents of concrete words, then semantic priming should be the same in the between-language condition as in the within-language condition. The effect of priming should be to induce neuronal adaptation—decreased activation to a target primed by a translation equivalent or by a semantically related word.

Previous functional magnetic resonance tomography (fMRI) studies (Chee, Soon, & Lee, 2003; Nakamura, Dehaene, Jobert, Le Bihan, & Kouider, 2005) have used priming techniques to examine script-specific language responses. In Chee et al. (2003),

English-Mandarin speakers silently read English words preceded either by the same or different word in English or in Chinese. There was evidence of neuronal adaptation irrespective of language but no evidence of language-sensitive neuronal adaptation. In Nakamura et al. (2005), native-Japanese participants decided whether a target word referred to a natural object or to an artifact. Targets were preceded subliminally by the same or different words written in either the same script or in a different script. Neuronal adaptation was sensitive to the script, but it is unclear whether this effect arose at the visual or lexical level.

In our priming task (Crinion et al., submitted), individuals were presented with a property contrast (e.g., short legs vs. long legs) and had to decide which property best described the object referred to by a target word. The target word (e.g., BULL) was preceded by a prime (for 250 ms) that was either related to it semantically (i.e., from the same category or subcategory; e.g., cow) or that was unrelated to it (e.g., locust). In addition, the target and prime were either in the same language (either L1 or L2) or in a different language.[1] Words were presented visually with the prime in lowercase letters and the target in uppercase letters. To perform the task, individuals must recognize words, retrieve their meanings and a representation of their referents, discriminate the referent of the target, verify a property of it, and respond by pressing one response key or another.

We sought to identify regions that were sensitive to both the semantic relationship between the prime and the target and to the language of the stimuli (i.e., where semantic adaptation arises only when prime and target are in the same language but not otherwise).

We established generality and robustness using three different bilingual groups. One group of German-English bilinguals ($n = 11$) participated in a positron-emission tomography (PET) experiment. A second group of German-English bilinguals ($n = 13$) and a third group of Japanese-English bilinguals ($n = 9$) participated in fMRI experiments. All participants had moderate to high proficiency in English and, indeed, were equally accurate in both languages. As expected, relative to unrelated controls, individuals were faster to respond to targets when primed by a semantically related word irrespective of language.

The neuroimaging data confirm and amplify the behavioral data. As expected, a common semantic system was activated. A semantically related prime reduced activation in a region of the left anterior temporal region irrespective of the language of the prime or the target. This region was in a region of interest predicted on the basis of previous semantic priming effects in monolingual speakers (e.g., Mummery, Shallice, & Price, 1999).

The most robust language-selective response lay in a subcortical structure—the left caudate nucleus. Here a semantically related prime only reduced activation (i.e., elicited neuronal adaptation) if it was in the same language as the target (whether that language was German, English, or Japanese). Expressed differently, left caudate activation was higher for semantically related words in different languages. Such sensitivity to the language of input is compatible with this region's involvement in language switching (see Abutalebi, Miozzo, & Cappa, 2000) and in the present case would seem to imply a "cost" of switching languages. In the next subsection, we consider the costs of using an L2 in a picture-naming task.

Functional Consequences: Interference in Naming

Functional data from bilinguals need to be interpreted in the light of the problems of control. For instance, demonstrating that bilingual speakers show increased frontal activation in certain tasks (e.g., Abutalebi, Cappa, & Perani, 2001) does not imply that the neural representation for an L2 is more extensive than that for an L1. It is more likely to reflect increased effort in selecting among competing alternatives or in producing the less familiar name (Abutalebi & Green, submitted).

In a PET study, De Bleser et al. (2003) examined picture naming in proficient Dutch-French bilinguals using a design in which subjects named pictures in either Dutch or French in the same experimental session. They found increased activation in naming in the L2 for noncognate names in a left inferior temporal region associated with semantic retrieval (e.g., Vandenberghe et al., 1996). There was also increased activation in regions of the left inferior frontal gyrus associated with phonological retrieval or decisions and semantic retrieval in the face of competition (see Devlin, Matthews, & Rushworth, 2003).

We sought to extend our understanding of the production of noncognate names by nonnative speakers by using a range of tasks (e.g., semantic judgments, reading, and naming). Relative to native speakers, relatively proficient L2 speakers might be slower to access meaning and slower to retrieve and to generate the phonological representation of a word. Processes associated with an increase in the difficulty of access, retrieval, or generation might be contrasted with those specifically associated with resolving any competition between languages. Such a view suggests that it might be possible to identify separate regions in the inferior frontal cortex associated with naming objects versus reading words.

We contrasted the activation patterns of Greek-English bilinguals (of moderate to high proficiency in English, $N = 14$) with a group of native-English speakers ($N = 16$) matched for task performance in a 2×2 design with the factors of task (semantic decision and naming) and modality (pictures of familiar objects and their written names). For each task by modality condition, there was a specific baseline.

In the semantic task on words, subjects decided whether the top word of a triad was more closely related in meaning to the lower left or lower right word (Figure 1, 1a). Its baseline comprised three meaningless strings of symbols (Figure 1, 1b) and subjects judged whether the top stimulus looked identical to the meaningless stimulus on the lower left or lower right. In the condition where subjects reached a semantic decision on pictures, they judged whether the top picture was related more closely to the picture on the lower left or lower right (Figure 1, 2a). For the baseline task, they decided whether the meaningless object on the top was identical or not to the stimulus on the lower left or lower

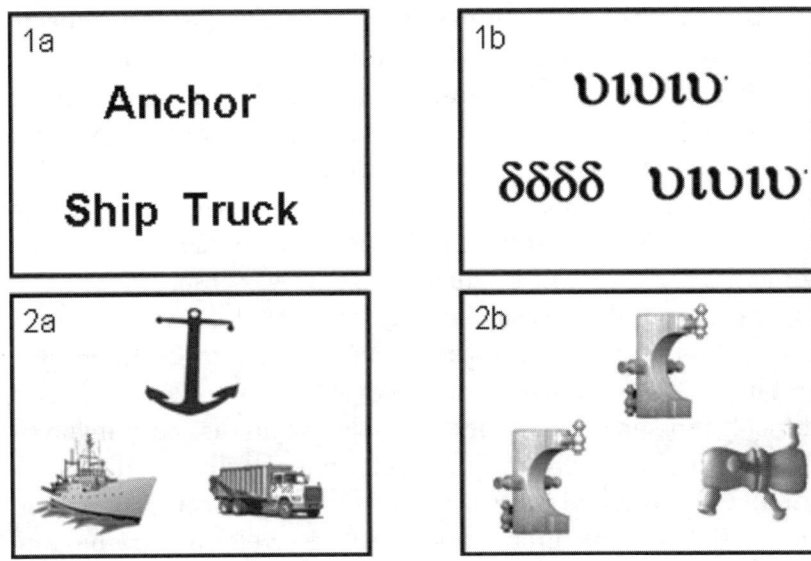

Figure 1. Examples of types of stimulus used in the fMRI naming study of Greek-English bilinguals.

right (Figure 1, part 2b). In the reading task, they read the three words aloud (Figure 1, part 1a) or said "one, two, three" to the set of meaningless strings (Figure 1, part 1b). In the naming task, subjects named all three pictures starting with the top stimulus, then the lower left stimulus, and, finally, the lower right stimulus (Figure 1, part 2a). Again, to control for visual input and motor output, the naming baseline task required subjects to say "one, two, three" while looking at the three meaningless objects (Figure 1, part 2b). Response accuracy and time were recorded for semantic or perceptual button presses and vocal responses (reading, naming aloud, and saying "one, two, three") were recorded and subsequently analyzed for accuracy.

As expected, during semantic decisions, both groups activated the left inferior and middle temporal regions that have been associated with semantic retrieval in many previous studies (e.g., Vandenberghe et al., 1996) and there were no group differences in activation for semantic relative to perceptual judgments on either words or pictures. Group differences did emerge, however, during the vocal conditions even when response accuracy was closely matched. Greek-English bilinguals showed increased activation in three left-lateralized frontal regions (premotor cortex, ventral inferior frontal cortex, and dorsal inferior frontal cortex).

In the premotor cortex, the effect is likely to reflect increased demands associated with articulation because increased premotor activation in Greek-English bilinguals was observed for naming, reading, and the vocal baseline (saying "one, two, three") but not during the semantic or perceptual matching tasks. In the ventral inferior frontal region, the effect is likely to reflect increased difficulty in semantic retrieval, as this region is more involved in semantic than phonological retrieval (Devlin et al., 2003). Finally, in the dorsal inferior frontal cortex, the increased activation in the bilingual group was only observed for picture-naming in comparison to reading. The latter result is consistent with the idea that naming elicits between-language competition. One test of this idea would be to examine the effect on this dorsal region of the left inferior frontal cortex after the name of the picture had been primed in the other language (see Kroll, Wodniecka, Bobb, & Green, 2005, for behavioral research using this competitor priming technique).

Structural Consequences: Vocabulary Acquisition

Acquiring another language involves acquiring additional vocabulary (word forms and morphemes) that must link to the conceptual, functional, and sensory representations of their referents and interface with combinatorial syntactic and phonological processes. To the extent that a common substrate is involved, there might be adaptive changes in brain structure relative to monolingual speakers. We can address this question by examining differences in brain structure (morphometry) on a voxel-by-voxel basis across the whole brain. The technique, known as voxel-based morphometry (VBM), compares the density or volume of gray or white matter in stereotactically normalized structural

MRI brain images. Prior research provides evidence of plastic changes as a function of expertise in various tasks (e.g., Maguire et al., 2000; Draganski et al., 2004).

Mechelli et al. (2004) examined the possibility of gray matter differences in the brains of bilinguals compared to monolingual controls. They found increased gray matter for bilinguals in the inferior parietal lobes and in a sample of Italian-English speakers; gray matter density in the same regions was found to correlate with second language proficiency. This same region is sensitive to vocabulary growth in children and young adults: As English vocabulary increases, there is a concomitant increase in gray matter density in children and young adults (Lee et al., submitted) irrespective of verbal or nonverbal IQ or verbal fluency abilities.

In current work, we are examining the effect of acquiring multiple languages. Our opportunistic sample comprises 87 bilingual and multilingual individuals (53 bilinguals, 21 trilinguals, 8 quadrilinguals, and 5 individuals with five or more languages). We search the whole brain for language correlated effects, rather than focusing on any specific cortical region of interest. Preliminary analyses indicate an area in the left parietal cortex that shows a significant effect of the number of languages spoken. The peak of the effect is anterior to the area shown in the Mechelli et al. (2004) work. Further comparisons (comparing between two languages, and three/four languages and three/four languages vs. five/six/seven languages) suggest that the acquisition of multiple languages leads to an expansion of a gray matter, rather than greater gray matter density and we are currently exploring its basis.

Structural Consequences: Control

Work by Bialystok (e.g., Bialystok, Craik, Klein, & Viswanathan, 2004) suggested that the acquisition of another language increases skills associated with selective attention in the young and protects against loss of skills in selective attention in the elderly. It is natural to wonder what structural changes this implies. If there are structural changes, then this suggests that acquisition can have long-term effects because such structural changes should make the acquisition of other languages easier. However, in the normal case (i.e., where an individual develops no specialized skill such as becoming a simultaneous interpreter), it might not be the case that the regions associated with control become more extensive as the number of languages acquired increases. Conceivably, changes in the way lexical information is represented reduce control demands. For instance, connections within the language might become stronger than those between the languages (see, e.g., Paradis, 2004; Abutalebi & Green, submitted) and the acquisition of a further language might establish such a linkage from the start.

In the first instance, evidence for specific effects might be better sought by examining the consequences of acquiring specific and unique skills associated with becoming bilingual. We examined structural effects (using VBM analyses of structural MRIs) in the brains of six simultaneous interpreters. Our preliminary analyses suggest that the acquisition of specific skills is indeed associated with changes in a number of subcortical regions.

We found gray matter density to be higher for our sample of six simultaneous interpreters compared to other monolingual, bilingual, and multilingual speakers in three regions: bilateral putamen, the inferior (and superior) colliculi, and the bilateral dorso-medial thalami. We had expected an effect in the putamen on the basis of our prior imaging work on translation in normal bilinguals (Price, Green, & Von Studnitz, 1999) but had not predicted the latter findings. Also, perhaps contrary to expectations, we found no effects in the caudate. Based on Behrens et al. (2003), it is probable that the regions of the thalamus connect to both the temporal lobes and the prefrontal cortex. Further, these regions connect strongly to the inferior colliculi, which are involved in auditory processing (the superior colliculi are involved in visual processing). It is conceivable then that these regions form part of the same circuit involved in the storage and manipulation of linguistic input.

In summary so far, current research indicates that the neural representation of an L2 converges with that of an L1 and that this has both functional and structural consequences. Of course, current structural studies need to be supplemented by longitudinal studies to examine plastic changes as a language or language skill is acquired. Our data point to the relevance of such studies, especially if we are to disentangle effects of acquisition from pre-existing differences. The notion that the neural representation of an L2 converges with that of an L1 does not deny that there might be language-specific neuronal sites within these regions. Studies using cortical stimulation mapping (e.g., Lucas, McKhann, & Ojemann, 2004; Ojemann & Whitaker, 1978) identify sites at which naming is selectively disrupted for one language but not for another. Such selective disruption might reflect differences in the relative computational demands for producing a particular word.

Degeneracy and Individual Differences

On the face of it, convergence is refuted by evidence that nonnative speakers activate different substrates compared to native speakers of that language. For instance, given that the same anterior regions are involved in grammatical operations for comprehension and production within an L1 (Indefrey, Hellwig, Herzog, Seitz, & Hagoort, 2004), evidence that grammatical operations in the L2 are handled by a distinct posterior system (see Ullman, 2001) would seem to refute the claim of convergence. Our aim is not to review the evidence on this point, nor its computational basis (see Green, 2003), but to consider the logic of any such challenge. A challenge of this type hinges on the presupposition that native speakers invariably process their language using the same method and a single substrate. However, native speakers might use different methods and different substrates for achieving a common end: There might be functional degeneracy.

Degeneracy (see Edelman & Gally, 2001; Price & Friston, 2002) refers to the ability of structurally different elements to perform a similar function or achieve the same outcome. Functional degeneracy operates when more than one system is sufficient for the same function. Friston and Price (2003) gave the example of waving goodbye. The function "waving goodbye" is degenerate because waving with either the right hand or with the left hand will suffice. Redundancy (e.g., waving with both hands) contrasts with degeneracy and can only be expressed given a degenerate set (i.e., here, a set of two hands). Degeneracy is ubiquitous in biological systems (see Edelman & Gally, Table 1, p. 13,764; e.g., in the immune response, populations of antibodies are degenerate—they bind equally well to the same antigen; in human communication there are many different ways to transmit the same message). Indeed, degeneracy is prerequisite for natural selection because selection pressures can only operate on dissimilar organisms (Edelman & Gally, p. 13,766). In general, different gene networks contribute to each phenotypic feature, and because selection has no way to assign responsibility to particular gene loci, degenerate systems will be maintained and favored (see also Tononi, Sporns, & Edelman, 1999). The presence of a degenerate structure or mechanism does not mean that each mechanism is functionally equivalent. Each variant offers novel properties. A degenerate system can therefore generate different outputs in different contexts. Therefore, it is adaptable in the context of different and perhaps novel environmental demands.

The concept of degeneracy allows us to take account of intersubject variability in a neuroimaging study (see Price & Friston, 2002, p. 417): A distinct pattern of activation in one subject might reflect not random error, but a degenerate system for performing the task. The notion of degeneracy therefore provides a way to address the important issue of individual differences.

What kinds of signature exist for the presence of degenerate systems in normal individuals? One signature is that where there are two degenerate systems, activation in one system is inversely related to activity in the other. Normal variability in performing the same task offers a strong guide as to the presence of distinct systems (see Price & Crinion, 2005). A second signature is distinct patterns of activation under different stimulus constraints (e.g., under noise, time pressure, or mental load). In an innovative study on bilinguals, Chee, Soon, Lee, and Pallier (2004; see Perani, 2005 for a review) contrasted two types of English-Chinese bilingual in Singapore, both with excellent L1 attainment but who differed in L2 attainment despite strong pressures to achieve high proficiency in both languages. They measured brain activation as a function of phonological working memory load using foreign language words (bisyllabic French words and pseudowords). Both groups showed activation changes as a function of load. A region in the left parietal lobe, linked to phonological storage, increased with load in both groups. Despite common performance levels in the two groups, the patterns of activation with increasing load differed in other regions. For the equal-proficiency group, there was a larger increase in the left insula and the left inferior frontal gyrus. For the unequal-proficiency group, there was greater activation in the anterior cingulate gyrus and reduced activation in areas of the parietal cortex (frontal medial and posterior). These data suggest that bilinguals showing equal proficiency in both languages were better at engaging a system suited to vocabulary acquisition, such as phonological working memory (see Baddeley, Gathercole, & Papagno, 1998; Papagno & Vallar, 1995). The findings for the unequal-proficiency group are consistent with the idea that they need to allocate greater attention to the task.

A third possible signature comes from brain lesion data. So, for example, the apparent use of the right hemisphere poststroke in monolinguals might reflect the use of a preexisting degenerate system (Crinion & Price, 2005). Such an inference (see Price & Crinion, 2005) might be explored by examining whether performance is further impaired or is improved by a temporary lesion induced in the right hemisphere by transcranial magnetic stimulation (e.g., Thiel et al., 2005). In bilingual stroke patients, if the predominant pattern is of impairment, then we can infer that there is a degenerate system. Poststroke, such a system becomes functionally useful. (The presence of such a degenerate system does not imply a greater incidence of aphasia in bilingual individuals with right hemisphere stroke; see Fabbro, 1999.)

Degeneracy is important in terms of understanding the neural effects of acquiring a given L2 in the context of different native languages. For instance, is it the case that different neural systems are involved in reading English by individuals whose first script is logographic (e.g., Chinese; see Wang, Koda, & Perfetti, 2003, for a relevant behavioral study and Tan et al., 2003, for a functional imaging study)? In order to address this question, we need to know the range of normal variation in native speakers and we need to determine if nonnative speakers show the same or different response when required, for example, to read aloud pronounceable nonwords.

Similar considerations apply when we consider the processing of regular and irregular verb forms in English. The regular past tense in English can be made by adding the affix /-d/ to the stem of the verb [bless-blessed]. The irregular past tense associated with a closed set of about 160 verbs requires the rote learning of each instance (e.g., make-made). On the basis of priming data from normal, native speakers of English, Tyler, Stamatakis, Post, Randall, and Marslen-Wilson (2005) argued that whereas irregular forms (teach-taught) are handled by a (bilateral) temporal lobe system storing whole forms, regular verbs (bless-blessed) require the additional involvement of left frontal systems supporting processes involved in assembly and disassembly (i.e., processes that identify the stem and affix, e.g., bless plus -ed). Whereas the past tense of irregular verbs must be learned as rote forms, it is possible that the past tense of regular words could be retrieved rather than computed (see Ullman, 2001). Any finding that nonnative speakers use retrieval rather than computation does not by itself refute convergence. We need first to establish the nature of individual differences in native speakers where the precise method used for processing regulars might depend on the task and/or on individual differences in the ease of performing a given computation.

Different phenotypes might arise not only as a result of the nature of the first acquired language but also as a result of preexisting structural or connectivity differences. A distinctive cytoarchitecture in right and left Broca's area seems to be implicated in the amazing facility of Emil Krebs (1867–1930), who spoke more than 60 languages (Amunts, Schleicher, & Zilles, 2004). Work by Golestani, Paus, and Zatorre (2002) suggested that individuals who acquire a difficult phonetic contrast more easily had more white matter in parietal regions, especially on the left. They speculated that this might reflect greater myelination, allowing more efficient processing. Conceivably, small structural or connectivity differences triggered the different patterns of processing revealed in the study by Chee et al. (2003). If research should identify examples of the impact of prior structural or connectivity differences in the acquisition process, then there will be practical consequences for the design of L2 teaching.

Prospects

Neuroimaging data can help us discover how individuals become skilled users because they help us identify the network of control that mediates the use of language. Current psychological models of speech production in bilingual speakers often assume a single locus for language selection whether at the level of the goal, lexical node, or the stage of response (e.g., Costa, 2005; La Heij, 2005; Roelofs, 2003) and ignore the network.

At the most general level, as Sperry (1986) argued, language in the brain is asymmetrically organized and normal use might involve suppression of homologous areas in the right hemisphere so that their distinct contributions can be orchestrated effectively (e.g., syntactically organized sequences generated in the left and prosody and tone added in the right). Normal use then might involve the suppression of degenerate systems that could potentially carry out the task.

Even a cursory analysis of the task of speech production in bilinguals points to numerous processes in language selection and control.[2] A particular type of preverbal specification might be required (de Bot & Schreuder, 1993); the goal to speak in one language rather than another must be maintained and competing tasks (e.g., translating the input or repeating it) managed (Green, 1986, 1998a). It is likely that these different levels of control are mapped onto distinct components of a fronto-parietal-subcortical circuit (see Abutalebi et al., 2000; Abutalebi & Green, submitted; Price et al., 1999). Given its unique connectivity to a wide range of sensory, motor, and associative systems, the prefrontal cortex can exert "top-down" influences on a wide range of processes. It interconnects strongly with the parietal cortex and this circuit is implicated in the selection of competing responses.

Subcortical regions, in addition to cortical regions, are essential to the acquisition of perceptual-motor mappings (Callan et al., 2003) and to cortical control. A left basal ganglia-left prefrontal cortex circuit might subserve language planning and suppress competing responses (e.g., Friederici, Kotz, Werheid, Hein, & Von Cramon, 2003; Longworth, Keenan, Barker, Marslen-Wilson, & Tyler, 2005).

We have focused primarily on the question of the nature of the neural systems involved. We also need to understand the dynamic orchestration of these neural systems and how that changes with proficiency. Does convergence in terms of the neural representation of a language imply that the dynamics of speaking an L2 will be identical to that in native speakers of that language? In other words, does the processing profile in understanding a sentence or producing an utterance converge with that of the nature of speakers? If there is neural convergence, then, as we have already argued, there is the possibility of competitive effects at output and at input (see Dijkstra & Van Heuven, 2002).

Mapping conceptual intentions onto speech might also invoke competing grammatical operations. Indeed, the very act of acquiring another language might affect the grammatical options used in the person's native language. Convergence implies that the bilingual must adopt means to regulate two systems (e.g., Green, 1998a; Grosjean, 1998), but it leaves open the question of the extent to which systems under control evolve to become equivalent to those of the native system as proficiency increases (e.g., Paradis, 2004, chap. 7) and the factors (motivational, neural, and social) that might limit or constrain such equivalence.

Identifying Higher Order Functions

Granted that different regions coordinate to achieve task performance, it is still puzzling that a particular region might activate in seemingly different tasks (see Stowe, Haverkort, & Zwarts, 2005). There are a number of possible responses to this state of affairs (see Price & Friston, 2005). Conceivably, within a region, there is a specialization of function that is outside of the resolving power of current methods. Alternatively, the region might mediate some unknown higher order function. The second does not preclude the first in that a higher order function might require specialized circuits for particular domains (e.g., language). In order to explore the possibility that there is some higher order function, we need to examine a range of different tasks and to have some principle of selection. We offer one discovery heuristic based on the view that language is a tool for coordinated action.

On evolutionary grounds we might think there is a close connection between language and action even if specialized circuits evolve. Wolpert (2003) remarked that the primary function of brains is to control movement. We move, of course, to approach things that we want and to avoid things that we do not want. However, we also act to shape the world to create what we want and to avoid what we do not want. We can consider possible future states of affairs and how to bring these about. We use our hands to shape the world and we use our hands to construct tools to better shape the world as we want it (see Johnson-Frey, 2003, for a review of the neural underpinnings of humans' tool use). Language is one of the tools that we use to achieve coordinated action to achieve goals.[3]

Whatever the nexus of changes that led to the emergence of natural language (see Arbib, 2005) and its subsequent cultural development, in acquiring language we become skilled in using a tool that allows us to coordinate actions. The cognitive operations involved in retrieving meaning and constructing sentences are analogous to the operations that we would use in achieving some change in some actual or imagined world.

One discovery heuristic for identifying higher order functions is this: contrast patterns of activation on verbal and nonverbal tasks that involve actions. For instance, what is the relationship between generating a sentence specifying a sequence of actions on particular objects and imagining or performing that sequence of actions?

The precise nature of our actions depends on the tool. Clearly, language production involves control over oral-facial muscles as part of the production of vocal gestures. Moving, or envisaging moving, a hand involves a distinct system (although there is a connection; see Gentilucci, 2003), so the interest lies in the nature of any patterns of common activity linking regions associated with the representation of objects, their position in space, and the semantics of changing one state into another. Our actions depend not only on the tool and its properties (e.g., consider the difference between tennis and badminton) but also our proficiency in using it. By contrasting activation patterns in different domains, we can identify higher order functions mediated by particular neural regions and so begin to probe different acquisition trajectories and different bilingual phenotypes in a more coherent fashion.

Language as a Neuroprosthesis

There is an important relationship between tool use and the human body that might be relevant to our understanding about the effects of language acquisition. One of the phenomenological consequences of proficient tool use is that the tool becomes an extension of the body. Prior to that point, use is subjectively unnatural. We can think of language as a neuroprosthetic device—it helps shape how we sense and perceive the world. The use of a language becomes part of the self although the extent to which it does so—and becomes natural—depends on our proficiency.

The autobiographies of bilingual writers point to the importance of language to the sense of self ("this language is beginning to invent another me"; see Green, 1998b; Pavlenko, 2001). Language also functions as a means for the internalization of cultural values. Marian and Kaushanskaya (2004) confirmed earlier work showing how the language in which memories are to be recalled affected the content of what is recalled and the manner of its description. Recalling memories in Russian as opposed to English increased the use of group pronouns ("we," "our") over personal pronouns ("I," "me") consistent with the more collectivist nature of Russian culture. The switching of cultural frames of reference might also be expected to shift personality differences in a direction consistent with the cultural values (see Ramirez-Esparza, Gosling, Benet-Martínez, Potter, & Pennebaker, 2006, for results consistent with this notion).

Acquisition then involves a potential shift in the way in which individuals engage in the world. Neuroimaging studies of bilinguals might be expected to contribute to a deeper understanding of ourselves as languaging beings.

Conclusions

Acquisition of another language induces both functional and structural brain changes. Functional neuroimaging methods offer a way to understand individual differences in the process of acquisition and in the manner in which proficiency is expressed both in terms of the nature of the neural representations involved and in their control. Understanding such variety is vital if we are to understand the nature of recovery patterns in bilingual and polyglot speakers. We need, for instance, to know the extent to which recovery is mediated by a normal degenerate system or solely by spared tissue.

Notes

[1] There was one further condition in which the prime was the translation equivalent of the target word. Lexical primes yielded faster reaction times than semantically related and unrelated primes but did not differ in accuracy. The largest effect of lexical priming was in the calcarine sulcus, which might indicate less visual imagery when the prime and target refer to the same object than when they refer to two different objects.

[2] Such a view is compatible with the recent framework proposed by Hagoort (2005) with the proviso that aspects of control permeate all of the functional components distinguished and not just the one labeled "control." Hagoort distinguished three functional components of language processing: a Memory component, a Unification component, and a Control component. The Memory component comprises the types of language information including retrieval operations. The Unification component refers to the processes involved in representing multiword utterances and the Control component refers to the use of language in a given situation and covers the selection of a language and turn-taking.

[3] Certainly recursiveness, considered the key feature of natural language (Hauser, Chomsky, & Fitch, 2002), is also a property of composite tool construction (Ambrose, 2001). Composite tools, such as an axe, are structured compositions of different units (a shaft, a stone insert, and binding materials). Variation in basic units and alternative compositions create functionally different tools (e.g., a scraper, a spear). The formal analogy is to grammatical language where different structured sound sequences convey differ phrases. Perhaps such a relation is unsurprising and is a correlate of a change in the capacity of the brain, as a result of an increase in connectivity (see Treves, 2005), to generate infinite sequences of elements from a finite set of such elements.

References

Abutalebi, J., Cappa, S. F., & Perani, D. (2001). The bilingual brain as revealed by functional neuroimaging. *Bilingualism: Language and Cognition, 4,* 179–190.

Abutalebi, J., & Green, D. W. (submitted). Bilingual language production: The neurocognition of language representation and control.

Abutalebi, J., Miozzo, A., & Cappa, S. F. (2000). Do subcortical structures control language selection in bilinguals? Evidence from pathological language mixing. *Neurocase, 6,* 101–106.

Ambrose, S. H. (2001). Paleolithic technology and human evolution. *Science*, *291*, 1748–1753.
Amunts, K., Schleicher, A., & Zilles, K. (2004). Outstanding language competence and cytoarchitecture in Broca's speech region. *Brain and Language*, *89*, 346–353.
Arbib, M. A. (2005). From monkey-like action recognition to human language: An evolutionary framework for neurolinguistics. *Behavioural and Brain Sciences*, *28*, 105–124.
Baddeley, A. D., Gathercole, S. E., & Papagno, C. (1998). The phonological loop as a language learning device. *Psychological Review*, *105*, 158–173.
Baker, M. C. (2003). Linguistic differences and language design. *Trends in Cognitive Sciences*, *7*, 349–353.
Behrens, T. E. J., Johansen-Berg, H., Woolrich, M. W., Smith, S. S., Wheeler-Kingshott, C. A. M., Boulby, P. A., et al. (2003). Non-invasive mapping of connections between the human thalamus and cortex using diffusion imaging. *Nature Neuroscience*, *6*, 750–757.
Bialystok, E., Craik, F. I. M., Klein, R., & Viswanathan, M. (2004). Bilingualism, aging, and cognitive control: Evidence from the Simon task. *Psychology and Aging*, *19*, 290–303.
Callan, D. E., Tajima, K., Callan, A. M., Kubo, R., Masaki, S., & Akahane-Yamada, R. (2003). Learning-induced plasticity associated with improved identification performance after training of a difficult second-language phonetic contrast. *NeuroImage*, *19*, 113–124.
Chee, M. W. L., Soon, C. S., & Lee, H. W. (2003). Common and segregated neuronal networks for different languages revealed using functional magnetic resonance adaptation. *Journal of Cognitive Neuroscience*, *15*, 85–97.
Chee, M. W. L., Soon, C. S., Lee, H. W., & Pallier, C. (2004). Left insula activation: A marker for language attainment in bilinguals. *Proceedings of the National Academy of Sciences of the United States of America*, *101*, 15,265–15,270.
Costa, A. (2005). Lexical access in bilingual speech production. In J. F. Kroll & A. M. B. de Groot (Eds.), *Handbook of bilingualism: Psycholinguistic approaches* (pp. 308–325). Oxford: Oxford University Press.
Crinion, J., & Price, C. J. (2005). Right anterior superior temporal activation predicts auditory sentence comprehension following aphasic stroke. *Brain*, *128*, 2858–2871.
Crinion, J., Turner, R., Grogan, A., Usui, K., Noppeney, U., Devlin, J. T., Aso, T., Urayama, S., Fukuyama, H., Stockton, K., Usui, K., Green, D. W., & Price, C. J. (2006). Language control in the bilingual brain. *Science*, *312*, 1537–1540.

De Bleser, R., Dupont, P., Postler, J., Bormans, G., Speelman, D., Mortelmans, L., et al. (2003). The organization of the bilingual lexicon: A PET study. *Journal of Neurolinguistics, 16,* 439–456.

De Bot, K., & Schreuder, R. (1993). Word production and the bilingual lexicon. In R. Schreuder & B. Weltense (Eds.), *The bilingual lexicon* (pp. 191–214). Amsterdam: Benjamins.

Devlin, J. T., Matthews, P. M., & Rushworth, M. F. S. (2003). Semantic processing in the left inferior prefrontal cortex: A combined functional magnetic resonance imaging and transcranial magnetic stimulation study. *Journal of Cognitive Neuroscience, 15,* 71–84.

Dijkstra, T., & Van Heuven, W. J. B. (2002). The architecture of the bilingual word recognition system: From identification to decision. *Bilingualism: Language and Cognition, 5,* 175–197.

Draganksi, B., Gaser, C., Busch, V., Schuierer, G., Bogdahn, U., & May, A. (2004). Changes in grey matter induced by training. *Nature, 427,* 311–312.

Edelman, G. M., & Gally, J. A. (2001). Degeneracy and complexity in biological systems. *Proceedings of the National Academy of Sciences of the United States of America, 98,* 13,763–13,768.

Fabbro, F. (1999). *The neurolinguistics of bilingualism: An introduction.* Hove, UK: Psychology Press.

Finkbeiner, M., Forster, K., Nicol, J., & Nakamura, K. (2004). The role of polysemy in masked semantic and translation priming. *Journal of Memory and Language, 51,* 1–22.

Friederici, A. D., Kotz, S. A., Werheid, K., Hein, G., & Von Cramon, D. Y. (2003). Syntactic comprehension in Parkinson's disease: Investigating early automatic and late integrational processes using event-related brain potentials. *Neuropsychology, 17,* 133–142.

Friston, K. J., & Price, C. J. (2003). Degeneracy and redundancy in cognitive anatomy. *Trends in Cognitive Science, 7,* 151–152.

Gentilucci, M. (2003). Grasp observation influences speech production. *European Journal of Neuroscience, 17,* 179–184.

Golestani, N., Paus, T., & Zatorre, R. J. (2002). Anatomical correlates of learning novel speech sounds. *Neuron, 35,* 997–1010.

Grainger, J., & Frenck-Mestre, C. (1998). Masked priming by translation equivalents in bilinguals. *Language and Cognitive Processes, 13,* 601–623.

Green, D. W. (1986). Control, activation and resource: A framework and a model for the control of speech in bilinguals. *Brain and Language, 27,* 210–223.

Green, D. W. (1998a). Mental control of the bilingual lexico-semantic system. *Bilingualism: Language and Cognition, 1,* 67–81.

Green, D. W. (1998b). Bilingualism and thought. *Psychologica Belgica, 38*, 253–278.

Green, D. W. (2003). The neural basis of the lexicon and the grammar in L2 acquisition: The convergence hypothesis. In R. van Hout, A. Hulk, F. Kuiken, & R. Towell (Eds.), *The interface between syntax and the lexicon in second language acquisition* (pp. 197–218). Amsterdam: Benjamins.

Grosjean, F. (1998). Studying bilinguals: Methodological and conceptual issues. *Bilingualism: Language and Cognition, 1*, 131–149.

Hagoort, P. (2005). On Broca, brain, and binding: A new framework. *Trends in Cognitive Science, 9*, 416–423.

Hauser, M. D., Chomsky, N., & Fitch, W. T. (2002). The faculty of language: What is it, who has it, and how did it evolve? *Science, 298*, 1569–1579.

Illes, J., Francis, W. S., Desmond, J. E., Gabrieli, J. D. E., Glover, G. H., Poldrack, R., et al. (1999). Convergent cortical representation of semantic processing in bilinguals. *Brain and Language, 70*, 347–363.

Indefrey, P., Hellwig, F., Herzog, H., Seitz, R. J., & Hagoort, P. (2004). Neural responses to the production and comprehension of syntax in identical utterances. *Brain and Language, 89*, 312–319.

Jiang, N., & Forster, K. I. (2001). Cross-language priming asymmetries in lexical decision and episodic recognition. *Journal of Memory and Language, 44*, 32–51.

Johnson-Frey, S. H. (2003). What's so special about human tool use? *Neuron, 39*, 201–204.

Kroll, J. F.,Wodniecka, Z., Bobb, S., & Green, D. W. (2005, March). *Modulating cross-language activation in a primed language-switching paradigm.* Paper presented at the 5th International Symposium on Bilingualism, Barcelona.

La Heij, W. (2005). Monolingual and bilingual lexical access in speech production: Issues and models. In J. F. Kroll & A. M. B. de Groot (Eds.), *Handbook of bilingualism: Psycholinguistic approaches* (pp. 289–307). Oxford: Oxford University Press.

Lee, H. L., Shakeshaft, C., Stewart, L. H., Brennan, A., Glensman, J., Pitcher, K., et al. (submitted). Adolescent inferior parietal cortex increases with vocabulary.

Longworth, C. E., Keenan, R. A., Barker, R. A., Marslen-Wilson, W. D., & Tyler, L. K. (2005). The basal ganglia and rule-governed language use: Evidence from vascular and degenerative conditions. *Brain, 128*, 584–596.

Lucas, T. H., McKhann, G. M. M., & Ojemann, G. A. (2004). Functional separation of languages in the bilingual brain: A comparison of electrical stimulation language mapping in 25 bilingual patients and 117 monolingual control patients. *Journal of Neurosurgery, 101,* 449–457.

Maguire, E. A., Gadian, D. G., Johnsrude, I. S., Good, C. D., Ashburner, J., Frackowiack, R. S. J., et al. (2000). Navigation related structural changes in the hippocampi of taxi drivers. *Proceedings of the National Academy of Sciences of the United States of America, 97,* 4398–4403.

Marian, V., & Kaushanskaya, M. (2004). Self-construal and emotion in bicultural bilinguals. *Journal of Memory and Language, 51,* 190–201.

Mechelli, A., Crinion, J. T., Noppeney, U., O'Doherty, J., Ashburner, J., Frackowiack, R. S., & Price, C. J. (2004). Structural plasticity in the bilingual brain. *Nature, 431,* 757.

Mummery, C. J., Shallice, T., & Price, C. J. (1999). Dual-process model in semantic priming: A functional imaging perspective. *NeuroImage, 9,* 516–525.

Nakamura, K., Dehaene, S., Jobert, A., Le Bihan, D., & Kouider, S. (2005). Subliminal convergence of Kanji and Kana words: Further evidence for functional parcellation of the posterior temporal cortex in visual word perception. *Journal of Cognitive Neuroscience, 17,* 954–968.

Ojemann, G. A., & Whitaker, H. A. (1978). The bilingual brain. *Archives of Neurology, 35,* 409–412.

Papagno, C., & Vallar, G. (1995). Verbal short-term memory and vocabulary learning in polyglots. *Quarterly Journal of Experimental Psychology, 48,* 98–107.

Paradis, M. (2004). An integrated neurolinguistic perspective on bilingualism. In M. Paradis (Ed.), *A neurolinguistic theory of bilingualism* (pp. 187–231). New York: Benjamins.

Pavlenko, A. (2001). "In the world of the tradition, I was unimagined": Negotiation of identities in cross-cultural autobiographies. *International Journal of Bilingualism, 5,* 317–344.

Perani, D. (2005). The neural basis of language talent in bilinguals. *Trends in Cognitive Science, 9,* 211–213.

Pillai, J., Araque, J., Allison, J., Sethuraman, S., Loring, D., Thiruvaiyaru, D., et al. (2003). Functional MRI study of semantic and phonological language processing in bilingual subjects: Preliminary findings. *NeuroImage, 19,* 565–576.

Price, C. J., & Crinion, J. (2005). The latest on functional imaging studies of aphasic stroke. *Current Opinion in Neurology, 18,* 429–434.

Price, C. J., & Friston, K. J. (2002). Degeneracy and cognitive anatomy. *Trends in Cognitive Sciences, 6,* 416–421.

Price, C. J., & Friston, K .I. (2005). Functional ontologies for cognition: The systematic definition of structure and function. *Cognitive Neuropsychology, 22,* 262–275.

Price, C. J., Green, D. W., & von Studnitz, R. (1999). A functional imaging study of translation and language switching. *Brain, 122,* 2221–2235.

Ramirez-Esparza, N., Gosling, S. D., Benet-Martínez, V., Potter, J. P., & Pennebaker, J. W. (2006). Do bilinguals have two personalities? A special case of cultural frame-switching. *Journal of Research in Personality, 40,* 99–120.

Roelofs, A. (2003). Goal-referenced selection of verbal action: Modelling attentional control in the Stroop task. *Psychological Review, 110,* 88–125.

Sperry, R. (1986). Consciousness, personal identity and the divided brain. In F. Leppore, M. Petito, & H. H. Jasper (Eds.), *Two hemispheres—One brain: Functions of the corpus callosum* (pp. 3–20). New York: Alan R. Liss.

Stowe, L. A., Haverkort, M., & Zwarts, F. (2005). Rethinking the neurological basis of language. *Lingua, 115,* 997–1042.

Tan, L. H., Spinks, J. A., Feng, C.-M., Siok, W. T., Perfetti, C. A., Xiong, J., et al. (2003). Neural systems of second language reading are shaped by native language. *Human Brain Mapping, 18,* 158–166.

Thiel, A., Habedank, B., Winhuisen, L., Herholz, K., Kessler, J., Haupt, W. F., et al. (2005). Essential language function of the right hemisphere in brain tumour patients. *Annals of Neurology, 57,* 128–131.

Tononi, G., Sporns, O., & Edelman, G. M. (1999). Measures of degeneracy and redundancy in biological networks. *Proceedings of the National Academy of Sciences of the United States of America, 96,* 3257–3262.

Treves, A. (2005). Frontal latching networks: A possible neural basis for infinite recursion. *Cognitive Neuropsychology, 22,* 276–291.

Tyler, L. K., Stamatakis, E. A., Post, B., Randall, B., & Marslen-Wilson, W. (2005). Temporal and frontal systems in speech comprehension: An fMRI study of past tense processing. *Neuropsychologia, 43,* 1963–1974.

Ullman, M. T. (2001). The neural basis of lexicon and grammar in first and second language: The declarative/procedural model. *Bilingualism: Language and Cognition, 4,* 105–122.

Vandenberghe, R., Price, C. J., Wise, R., Josephs, O., and Frackowiak, R. S. J. (1996). Functional anatomy of a common semantic system for words and pictures. *Nature, 383,* 254–256.

Wang, M., Koda, K., & Perfetti, C. A. (2003). Alphabetic and nonalphabetic L1 effects in English word identification: A comparison of Korean and English Chinese L2 learners. *Cognition, 87,* 120–149.

Wolpert, L. (2003). Causal belief and the origins of technology. *Philosophical Transactions of the Royal Society London A, 361,* 1709–1719.

The Plastic Bilingual Brain: Synaptic Pruning or Growth? Commentary on Green, et al.

Kees de Bot
University of Groningen

The main issue the authors focused on in this contribution is to what extent the acquisition and use of a second/third/fourth language leads to functional and structural changes in the brain. Related to this is the reverse question: To what extent do functional and structural characteristics of individual brains play a role on the development of an additional language? These two problems are discussed from various angles and the data provided show the complexity of the relation between second language development (SLD) and the functioning of the brain. Here I want to focus on one aspect that is discussed under various headings: the impact of SLD on structures in the brain.

In their contribution, Green et al. referred to a number of studies that have provided data showing a relation between the structure of the human brain and certain learning tasks. The best known study is probably Maguire et al.'s (2003) study on differences in brain structures in experienced and novice taxi drivers in London. This cross-sectional study showed that the amount of experience in driving a cab correlated with gray matter density and the size of certain brain areas. The brains of bilinguals have been studied in a voxel-based morphology study by Mechelli et al. (2004). They compared a group of monolinguals with a group of early bilinguals (age of onset of L2 acquisition < 5 years) and

---

Correspondence concerning this article should be addressed to Kees de Bot, Department of Applied Linguistics, University of Groningen, P.O. Box 716, 9700 AS Groningen, The Netherlands. Internet: c.l.j.de.bot@rug.nl

a group of late bilinguals (age of onset of L2 acquisition between 10 and 15 years). In addition, the former group had been using the language regularly since acquisition, whereas the latter group used the language regularly during the last 5 years only. No information on type or amount of use, or possible effects of structural differences between the first language (L1) and L2 is given in the report. The outcomes of this first study showed greater gray matter density in the inferior parietal cortex in the bilinguals than in the monolinguals. This effect was greater for the early than for the late bilinguals. In a second study, Mechelli et al. (2004) looked at Italian learners of English with a range of age of onsets (2–34 years) and a range in proficiency levels in L2. A very high positive correlation was found between gray matter density and level of proficiency, whereas a similarly high negative correlation was found between gray matter density and age of onset of acquisition of the L2. This suggests that there is an increase of density with language learning and a decline with the increase of age of onset of acquisition.

Draganski et al. (2004) looked at the relation between the development of juggling skills and learning-induced plasticity in the brain. They compared a group that was learning to juggle with a nonjuggling group. The subjects' brains were scanned before the juggling training program started for the juggling group, 3 months after the beginning of the training program, and 3 months later. There were no differences between the two groups in terms of gray matter density for the first scan, but there was a significant bilateral expansion in gray matter in the mid-temporal area and in the left posterior intraparietal sulcus when the second scan was compared to the first one. This difference decreased again during the period after the second scan. During this period, there was no juggling activity in either group. More plasticity was found in the visual areas than in the motor areas, which might have to do with the specific task demands of the three-ball cascade juggling routines practiced. So, in contrast to the Mechelli et al. (2004) study and the Maguire et al. (2003) study, this was a longitudinal study including information on the initial conditions in the brains of both the control and the experimental group. Green et al. briefly discussed some of their own findings with bilinguals and multilinguals and mention that "preliminary analyses indicate an area in the left parietal context that shows a significant effect of the number of languages spoken" (p. 109). On the basis of the literature discussed and their own findings, they concluded that SLD leads to structural changes in the brain, in particular an expansion of gray matter or an increase of gray matter in specific areas.

Then, in the section on degeneracy and individual differences, they seemed to take a slightly different position on this matter when they discuss the amazing Emil Krebs, who reportedly spoke more than 60 languages (Amunts, Schleicher, & Zilles, 2004) and the work by Golestani, Paus, and Zatorre (2002), which suggested that individuals who acquire a difficult phonetic contrast more easily had more white matter in parietal regions, especially on the left. So rather than an increase of white matter due to learning, this suggested the preexistence of structures that enable or facilitate specific aspects of SLD.

The Mechelli et al. (2004) findings with respect to the effect of age of onset of acquisition are particularly interesting here. They can be explained in two ways: Either the density is higher because the L2 has been used more over the years, or the density is higher in younger years and is maintained over time. So we now have two positions:

1. SLD leads to structural changes in the brain.

2. Specific characteristics of the brain facilitate language learning.

There might be another option to be considered. In recent years, a considerable body of research has been done on synaptic pruning in development. Chechik, Meilijson, and Ruppin (1997) referred to this as follows:

> Beginning at early stages of the fetus development, synaptic density rises at a constant rate, until a peak level is

> attained (at age 2–3 years of age in humans). Then, after a relatively short period of stable synaptic density, an elimination process begins: synapses are being constantly removed yielding a marked decrease in synaptic density. This process proceeds until puberty, when synaptic density stabilizes at adult levels which are maintained until old age. (p. 1759)

In addition, there is evidence that synaptic elimination correlates with experience-dependent activity (Roe, Pallas, Hahm, & Sur, 1990; Stryker, 1986). Monolingual children use only one language and, accordingly, only a part of the system that might be there. Monolingualism is then defined as the nonuse of multiple languages for which the brain is equipped, as evidenced by studies on early bilingualism and its positive effects on cognitive functioning (Bialystok, 2005). This means that a decline of pruning due to nonuse could be the explanation for differences in gray matter: Bilingualism does not lead to expansion of gray matter, but monolingualism leads to extensive synaptic pruning. Of course, this is very speculative at the moment but definitely not completely off the mark. What is needed to test different hypotheses is longitudinal data in which relevant aspects, such as levels of proficiency, structural distance between languages, amount and type of contact with the languages involved, and age of onset of acquisition, are taken into account in the setup of the study. It is very likely that in the end, cases will be found that support different positions. It is very unlikely that people will become excellent multilinguals on a large scale because of specific structural characteristics of their brains, but as the Emil Krebs case and other cases of extraordinary multilingual skills reported on in the older aphasia literature (Paradis, 1983) show, there are individuals who will have both exceptional language skills and deviant brain structures. Given what we know about the plasticity of the brain at the moment, it is likely that learning might have an impact on brain structures, although it is unclear how plastic the brain is and to what extent specific teaching or learning methods might enhance plasticity or make optimal use of it.

There might also be a relation between these findings and the discussion on critical periods/ultimate attainment (see Birdsong, this volume): Individual differences in brain plasticity might explain differences in SLD. This means that more plasticity might lead to the recruitment of more brain cells for language learning and use. Also, the timing of synaptic pruning and the extent to which this can be influenced through learning might be related to different critical periods that have been discussed in the literature. In the Dutch educational setting in which foreign language learning has traditionally been important since the middle of the 19th century, language aptitude is seen as a crucial issue in language learning. There is even a very specific word for a knack of languages: "Talenknobbel," literally a cranial protrusion that results from the outgrowth of parts of the brain that are involved in language learning. This goes back to the old days of phrenology when structures at the outside surface of the skull were assumed to reflect structures in the brain. Of course, the whole idea of a "Talenknobbel" was discredited with the demise of phrenology, but it is interesting that through the research on the relation between SLD and brain structures, this old myth is reemerging based on very modern and advanced neuroimaging technology.

References

Amunts, K., Schleicher, A., & Zilles, K. (2004). Outstanding language competence and cytoarchitecture in Broca's speech region. *Brain & Language*, *89*, 346–353.

Bialystok, E. (2005). Consequences of bilingualism for cognitive development. In J. Kroll & A. M. B. de Groot (Eds.), *Handbook of Bilingualism: Psycholinguistic approaches* (pp. 417–432). Oxford: Oxford University Press.

Chechik, G., Meilijson, I., & Ruppin, E. (1997). Synaptic pruning in development: A computational account. *Neural Computation*, *10*, 1759–1777.

Draganski, B., Gaser, C., Busch, V., Schuierer, G., Bogdahn, U., & May, A. (2004). Changes in gray matter induced by training. *Nature*, *427*, 311–312.

Golestani, M., Paus, T., & Zatorre, R. J. (2002). Anatomical correlates of learning novel speech sounds. *Neuron, 35,* 997–1010.

Maguire, E., Spiers, H., Good, C., Hartley, T., Frackowiak, R., & Burgess, N. (2003). Navigation expertise and the human hippocampus: A structural brain imaging analysis. *Hippocampus, 13,* 250–259.

Mechelli, A., Grinion, J., Noppeney, U., O'Doherty, J., Ashburner, J., Frackowiak, R., et al. (2004). Structural plasticity in the bilingual brain. *Nature, 432,* 757.

Paradis, M. (1983). Bilingualism and aphasia then and now. In M. Paradis (Ed.), *Readings on aphasia in bilinguals and polyglots* (pp. 802–815). Montreal: Didier.

Roe, A., Pallas, S., Hahm, J., & Sur, M. (1990). A map of visual space induced in primary auditory cortex. *Science, 250,* 818–820.

Stryker, M. (1986). Binocular impulse blockade prevents the formation of ocular dominance columns in cat visual cortex. *Neuroscience, 6,* 2117–2133.

Executive Control in Bilingual Language Processing

A. Rodriguez-Fornells
University of Barcelona

R. De Diego Balaguer
*INSERM U421-IM3 / Paris XII
and
École Normale Supérieure*

T. F. Münte
Otto von Guericke University

A. Rodriguez-Fornells, Institució Catalana de Recerca i Estudis Avançats (ICREA) & Faculty of Psychology; R. de Diego Balaguer, Equipe Avenir–INSERM U421-IM3/Paris XII and Département d'Études Cognitives; T. F. Münte, Department of Neuropsychology.

We would like to acknowledge the contributions and helpful comments of the following colleagues (in alphabetical order): B. Britti, A. Costa, T. Cunillera, M. J. Del Rio, C. Escera, J. Festman, T. Gomila, H. J. Heinze, B. Jansma, A. van der Lugt, T. Lutz, J. Möller, E. Moreno, W. Nager, T. Nösselt, and M. Rotte. We would especially like to thank Luis Becerra and Carmen Moreno for their help in recruiting the German-Spanish bilingual samples and all of the bilingual speakers who participated in the studies. The data presented in this article have appeared in the following references: Rodriguez-Fornells, Rotte, et al. (2002), Rodriguez-Fornells, Schmitt, et al. (2002), Rodriguez-Fornells et al. (2005), De Diego Balaguer et al. (2005), Rodriguez-Fornells, Lutz, and Münte (submitted), and Rodriguez-Fornells, Del Rio, Escera, and Münte (unpublished data). This work has been supported by the DFG program (to TFM) and MCYT grant (BSO2002-01211 to ARF) and a predoctoral grant from the Catalan government (2000FI 00069 to RDB).

Correspondence concerning this article should be addressed to Antoni Rodriguez-Fornells, Dept. Psicologia Bàsica. Facultat de Psicologia, Universitat de Barcelona, Passeig de la Vall d'Hebron 171, Barcelona 08035, Spain. Internet: antoni.rodriguez@icrea.es

Little is known in cognitive neuroscience about the brain mechanisms and brain representations involved in bilingual language processing. On the basis of previous studies on switching and bilingualism, it has been proposed that executive functions are engaged in the control and regulation of the languages in use. Here, we review the existing evidence regarding the implication of executive functions in bilingual processing using event-related brain potentials and functional magnetic resonance imaging. Several brain potential experiments have shown an increased negativity at frontocentral areas in bilinguals, probably related to the activation of medial prefrontal regions, for different tasks, languages, and populations. Enhanced cognitive control is required in bilinguals, which also involves the recruitment of the left dorso-lateral prefrontal cortex. The degree of activation of this mechanism is also discussed considering the similarity of languages in use at the lexical, grammatical, and phonological levels. We propose that the prefrontal cortex probably mediates cognitive control in bilingual speakers through the interplay between a top-down selection-suppression mechanism and a local inhibitory mechanism in charge of changing the degree of selection-suppression of the different lexicons.

An unresolved issue in bilingualism is how different languages are represented in the brain and which cognitive mechanisms are required to regulate their use. Although the level of performance achieved by an adult learner might be less than optimal, the cognitive mechanisms involved in language acquisition enable us to learn a second language (L2) throughout our life span. In learning (and sustaining) a second (or third) language, the brain has to build on a neural network that enables the segregation of the new language from the native one, the creation of its corresponding activation and inhibitory links at the lexical, morphological, and syntactic level, and, finally, the development of the ability to correctly select a word and its syntactic properties in the target language (i.e., the language currently in use). The creation and crystallization of a full new lexicon can be considered a highly interesting natural experiment. The present article will deal with the cognitive mechanisms responsible for the regulation of two or more languages after they have been (partially) learned. We will argue at the end of the article that knowledge gained in the field is also important for the understanding of the acquisition of a new language.

Another goal of this article is to highlight how the cognitive neuroscience approach might prove helpful in understanding how an L2 is mastered and regulated in the brain. Building on a previous idea (Münte, Rodriguez-Fornells, & Kutas, 1999), we argue that brain imaging should not only be used to test predictions derived from psycholinguistic models, but that patterns of activations might be used to derive brain-inspired hypotheses about processing differences. We believe that this approach will be very fruitful in the future, as is attested to by the recent studies (e.g., de Diego Balaguer et al., in press) or brain-inspired models of language functions (Hagoort, 2005). The same idea has been raised recently by Poldrack and Wagner (2004): "knowledge of functional localization (i.e., the location of brain activations) can inform cognitive theories through the approach of reverse inference, wherein activation in a particular brain region (or regions) is taken as a marker of engagement of a particular cognitive process" (p. 177). In this article, we seek to apply this approach to the study of executive functions in bilinguals while recognizing the inherent limitations of this approach in terms of causality (see Poldrack & Wagner). Naturally, a data-driven brain-inspired interpretation of our data and a more adequate picture will emerge if this approach is applied to a set of studies addressing bilingual processing with varying paradigms.

Neural Implementation of Various Languages in the Brain

The bilingual's ability when speaking to select a word in a target language without too much interference is puzzling in light of recent neuroimaging studies that demonstrate overlapping neuroanatomical representation of languages at the macroscopic level (Chee, Caplan, et al., 1999, 2000; Chee, Hon, Lee, & Soon, 2001; Chee, Soon, & Lee, 2003; Chee, Tan, & Thiel, 1999b; Hasegawa, Carpenter, & Just, 2002; Hernandez, Dapretto, Mazziotta, & Bookheimer, 2001; Hernandez, Martinez, & Kohnert, 2000; Illes et al., 1999; Klein, Milner, Zatorre, Zhao, & Nikelski, 1999; Perani et al., 1998; Price, Green, & von Studnitz, 1999; for a recent review, see Franceschini, Zappatore, & Nitsch, 2003; Perani & Abutalebi, 2005). A critical issue with regard to this data is whether the spatial resolution of functional magnetic resonance (fMRI) (or positron-mission tomography [PET]) would allow us to demonstrate a separation of language representations in bilinguals.

In fact, noninvasive brain imaging studies do not agree completely with intraoperative electrocortical stimulation mapping and intraoperative optical imaging in bilingual populations. Early observations by Penfield and Roberts (1959) suggested that bilinguals have common language areas and pathways in the brain. However, Ojemann and Whitaker (1978) described two individuals who had brain regions that were shared by both of their languages along with other regions unique to each language. Similarly, Lucas, McKhann, and Ojemann (2004) reported that L1 and L2 have distinct and shared cortical representations. In 95% of the bilinguals (22 patients), stimulation produced deficits when naming pictures in only one of the languages and 43% of the patients showed specific language sites producing deficits in L1 or L2. Several restricted perisylvian areas were found in which L2 could not be mapped when compared to L1 representation in a large group of monolinguals. These areas were the inferior frontal and precentral gyrus, posterior temporal gyrus (extending to the supramarginal gyrus), and the posterior middle temporal gyrus. This pattern suggests that several language-related areas are exclusively dedicated to the processing of the native language. The acquisition and representation of an L2 might rely on neural assemblies, which are partially shared with L1 and partially specific (e.g., posterior temporal, parietal, precentral, and prefrontal regions). It should be noted, however, that a case study of a Spanish (L1)/English (L2) bilingual using optical imaging (Pouratian et al., 2000) showed the inverse pattern to what would be expected from Lucas's analysis (see Roux & Trémoulet, 2002, for similar results, but see also Walker, Quinones-Hinojosa, & Berger, 2004). Although it is therefore important to consider the limitations of electrical and optical intraoperative brain mapping techniques,[1] these studies are clearly in favor of a partial functional separation of bilingual lexicons in the brain, as has been suggested by a minority of neuroimaging experiments (e.g., Dehaene et al., 1997; Kim, Relkin, Lee, & Hirsch, 1997). Such a view could also explain the clinical observations in aphasic bilinguals showing selective impairment or differential recovery of one language (Fabbro, 2001; Paradis, 1995).

Although definitive statements regarding language representation in bilinguals are premature at this stage, the brain regions sustaining the representation and processing of different languages appear to be partially overlapping and partially segregated. This leads to the question as to how access to the different languages is controlled and interference between the languages is prevented. Penfield and Roberts (1959) have argued for the existence of a "language switch" mechanism (see also Macnamara & Kushnir, 1971). As we show in the next sections, this "switch" appears to be instantiated by the brain's executive system.

The Role of Executive Functions in Bilinguals

Switching and mixing languages are frequent in many bilingual speakers, especially when the interlocutor is able to understand both languages. On the other hand, fluent bilinguals switch from one language to the other and are able to separate both languages completely and without too much effort. Despite this ability, however, code-switching is also frequently observed, as bilinguals tend to introduce words from the other language into the language that they are currently using. It has been argued that both switching proficiency and code-switching are related to the degree of activation of the target and nontarget languages at a given moment (Grosjean, 1997; Paradis, 1989). In addition to a bilingual's degree of proficiency in the L2, the bilingual status of the interlocutor and the communicative setting appear to be important for the degree of code-switching. At the other end of the continuum (i.e., in the "monolingual mode"; Grosjean, 1997), a smaller number of intrusions is observed, giving the appearance of complete independence of both languages, as if the nontarget language were switched off.

It is necessary to postulate a control mechanism that regulates the activation of the different languages in bilinguals and polyglots and what the neural mechanisms might be that implement this process. Cognitive control processes enable us to adapt our actions to the ever-changing environment and to shape them in relation to our current goals. Traditionally, cognitive control has been associated mainly with two properties: the ability to filter out irrelevant information in the environment (interference suppression) and the ability to inhibit inappropriate responses or thoughts (response inhibition). Cognitive control develops gradually in infants and is thought to be related to the slow maturation of the prefrontal cortex (PFC; see review in Diamond, 2002). Bialystok (1999) hypothesized that bilingual children might develop enhanced cognitive control mechanisms compared to monolingual children because they are faced with switching and attentional control demands from early on. In fact, she demonstrated that bilingual children might become exceptionally good at ignoring distractive information and at switching between different cognitive tasks.

There is common agreement that in producing a word in a particular language, the conceptual system of a bilingual activates the lexical representation of both languages (Costa, Miozzo, & Caramazza, 1999; De Bot, 1992; Hermans, Bongaerts, de Bot, & Schreuder, 1998; Poulisse, 1999). A second question concerns whether the activation of nontarget lexical representation is transmitted onward to the phonological properties of the word.

Several pieces of evidence suggest that this is the case (Colomé, 2001; Costa, Caramazza, & Sebastian-Galles, 2000). How, then, does the system control the production of the target word if the nontarget language candidates are interfering at the levels of lexical selection and phonological representation? Roelofs (1998) has proposed a checking procedure (based on Levelt, Roelofs, & Meyer, 1999) that discards the selected lexical representation whenever the language of the selected word does not match the intended language (and the intended conceptual representation).

Whereas some bilingual language production models have postulated inhibitory processes in order to control target lexical activation, such as the inhibitory control model (IC; Green 1986, 1998), other models achieve lexical selection by increasing the level of activation of the target language (De Bot, 1992; Grosjean, 1997; Paradis, 1989; Poulisse & Bongaerts, 1994). However, in order to activate a language without activating the nontarget language at the same time, it is necessary to correctly identify the relevant items of this language. Therefore, the selection mechanisms in use for the language production system have to be clearly specified; for example, in bilingual models that rely on differential activation of the items, the most activated item would be selected. In contrast, Green (1986), following Albert and Obler (1978), proposed that words in the lexicon possess a type of "language tag" and that language, therefore, is used as a feature for selection purposes. In a similar vein, Poulisse and Bongaerts proposed that each word possesses a "language tag" linking it to a particular language, and these language tags would enable only a limited set of lexical items to be activated, mostly from the target language.

In the IC model (Green, 1986, 1998), which is inspired by the supervisory attentional system (SAS) model (Norman & Shallice, 1986), control is achieved through three mechanisms: (a) a system similar to the SAS that is dedicated to establishing and maintaining goals; (b) control mechanisms acting at the level of "language task schemas"; and (c) control mechanisms operating at the bilingual lexico-semantic system (at the lemma level).[2] In order to begin an interaction in one language, speakers have to decide and select a "specific language schema" (e.g., French) and inhibit alternative ones (e.g., English). However, selection of a word in the target language occurs at the lemma level using a type of "language tag" and is achieved by inhibiting lemmas of the other language. Thus, the selected "language task schema" is able to inhibit an activated lemma when a language "tag" does not correspond to the target language. In that case, control is executed via the inhibition of the inappropriate nontarget language items. In the Bilingual Interactive Activation (BIA) model (Dijkstra & Van Heuven, 1998), these tags are represented as language nodes, which have the role of reinforcing the lexical activation of the currently activated language and simultaneously decreasing lexical activation in the other lexicon. The competition that occurs both within and between languages is resolved via a local inhibitory mechanism (lateral inhibition). Adjacent language representations inhibit each other, and the selection of a particular response reduces the likelihood of selecting the neighboring response. Notice that the nature of the inhibitory mechanisms proposed differs between models, ranging from a general unspecific inhibitory mechanism, as in Green's model, to a local inhibitory one.

It should also be noted that in the IC model, the language task schemas are independent of the bilingual lexico-semantic system, and strong competition and conflict is predicted between them. The model considers that in order to select a language schema, speakers must first inhibit the nontarget language schema. Therefore, switching between languages will require changing the previous inhibitory status of the nontarget language, a process that will require time. In fact, a study by Meuter and Allport (1999) showed clear evidence of switching costs in naming numerals aloud. However, the switching costs were asymmetric: Speakers took longer to switch from the L2 to their L1 than the reverse switch to the nondominant language. The authors explained this finding by considering that because L1 inhibition is more demanding, it will also require more time to be reactivated. In the particular case of balanced bilinguals, in which the inhibition applied to both languages is equal, Meuter (1994, cited in Meuter & Allport, 1999) provided evidence for an identical switching cost in a category-naming task.

In a recent study, Costa and Santesteban (2004) also reported that these asymmetric switching costs are related to the degree of proficiency of the bilingual speaker in the second language, and independent of other languages learned. In this study, non-proficient Spanish-Catalan L2 speakers (1.5 years second-language learning) showed clearly asymmetric costs that replicated the previous study of Meuter and Allport (1999). In contrast, highly proficient Spanish-Catalan speakers showed nonasymmetric costs (only general switching costs). The surprising result is that this bilingual group did show nonasymmetric costs when a nonfluent third language (L3; English) was evaluated. In order to reconcile this complex pattern of results, the authors suggested that when bilinguals become highly proficient in an L2, they develop specific language-selection mechanisms that allow them to process both L1 and L2 in a very flexible way (see Costa, 2005).

These results are consistent with Bialystok's (2001) proposal that a nonspecific control mechanism is naturally tuned and developed in highly proficient and probably *early* bilinguals, which gives them an advantage in general switching-inhibitory tasks. The existence of a highly flexible neural control mechanism in bilinguals (non-language-specific) could perfectly explain the previous pattern without the need for language-specific selection and an inhibitory mechanism. In a recent study, Bialystok, Craik, Klein, and Viswanathan (2004) provided some initial evidence for the advantage of bilinguals in executive function using the Simon task.[3]

Li (1998) argued against this notion of language tags, based on a series of studies in which no cost was associated to language switching in natural speech situations (Grosjean, 1988, 1997; Grosjean & Miller, 1994; Li, 1996). According to the IC model, language switching will require time, because the process of switching to another language involves the inhibition of the previous language tags, a top-down regulated process. This idea is also contradicted by several experiments in which the two languages seem to be always activated, although their level of activation depends on the specific language situation and is modulated by several factors (e.g., proficiency level, speech mode, frequency of words; Grosjean, 1988; 1997). Li proposed an interesting alternative, in which there is no need for labels or language tags. Both lexicons become separated over time because a self-organizing network might develop localized patterns of activity in learning the different associations and mappings among phonology, orthography, morphology, and semantics. These localized patterns of activity are supposed to function as the learner's internalized representations of the two lexicons (see Li & Farkas, 2002).

Thus far, we have reviewed evidence that suggests that cognitive control is necessary to regulate bilingual language processing. However, more detailed bilingual cognitive control models are required in order to explain the previous behavioral studies and also the different clinical recovery patterns observed in bilingual aphasics, such as "alternate antagonism." In these patients (Paradis, 1995), one language is first recovered, and after a while, a switch is observed and the silent language comes back while the recovered one disappears. A specific lesion in the language control mechanisms might be able to explain this switching pattern. Additionally, several important issues are still open to controversy, such as (a) in which degree bilinguals use the same cognitive control mechanisms as monolinguals in bilingual language processing, (b) if bilinguals train specific executive functions in order to deal with language switching, inhibition, and interference or (c) if they develop a specific language control mechanism that is devoted to switching and inhibition of the different languages. In the following sections we will provide some evidence suggesting that different brain regions involved in cognitive control are active in bilinguals' language processing.

Empirical Evidence for Control in Second Language Production

The N2 noGo Component

With the aim of evaluating the interference of the nontarget language at the syntactic and phonological levels during covert naming in bilinguals, we have used the N2 noGo component.[4] The same type of task has been used in a series of studies assessing the relative timing of the retrieval of conceptual/semantic, syntactic, and phonological information during language production and single-word reading and comprehension (for a review of the different studies, see Jansma, Rodriguez-Fornells, Möller, & Münte, 2004).

A number of electrophysiological studies have assessed the effects of response inhibition using event-related brain potentials (ERPs) and fMRI. These experiments showed that the stimulus-locked ERP in noGo trials is characterized by a large negativity of about 1–4 μV in size that occurs with task-dependent onset latencies over the frontocentral scalp (Gemba & Sasaki, 1989; Kok, 1986; Pfefferbaum, Ford, Weller, & Kopell, 1985; Sasaki, Gemba, Nambu, & Matsuzaki, 1993; Simson, Vaughan, & Ritter, 1977). In Figure 1a, the N2 noGo component is depicted in two different experimental conditions. In the semantic condition, participants decided whether to press a response button depending on whether the picture presented was an animal or an object (Go/noGo decision based on conceptual information). In contrast, in the phonological condition, participants had to covertly produce the name of the picture and decide whether the first letter was a consonant or a vowel (Rodriguez-Fornells, Schmitt, et al., 2002). The right panel shows the subtraction of noGo minus Go trials. As expected from both previous studies (see Schmitt, Münte, & Kutas, 2000; Van Turennout, Hagoort, & Brown, 1997) and language production models (Levelt et al., 1999), the onset and peak latencies of the N2 noGo-Go component was earlier for the semantic decision

A. ERPs

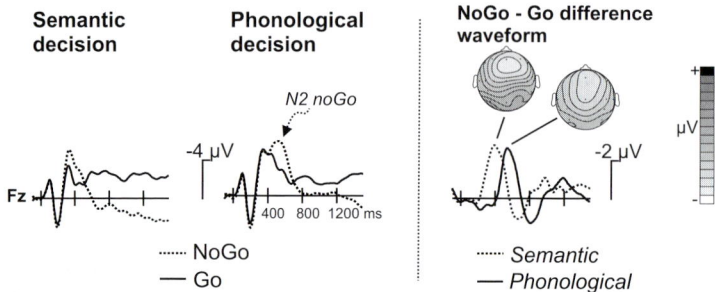

B. Event-related fMRI

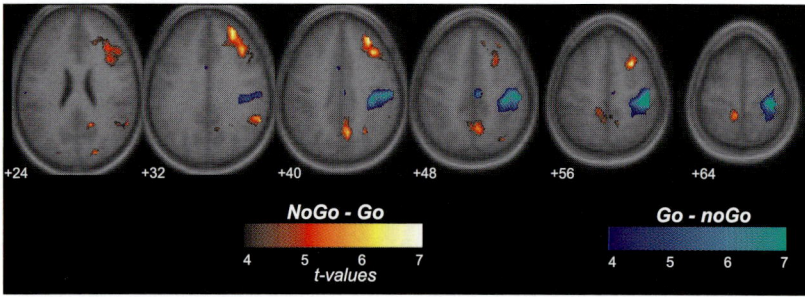

Figure 1. (a) Grand average ERPs at frontal recording locations showing the increased negativity for the noGo trials in the 250–450-ms range in the semantic condition and 450–650 ms in the phonological condition. This effect can be more clearly visualized in the noGo minus Go difference waveforms (right side). Also depicted are the isovoltage topographical maps (mean amplitude = 50 ms, centered at the latency peak; relative voltage scale) computed for both conditions. (Adapted from Rodriguez-Fornells, Schmitt, et al., 2002.) (b) Event-related fMRI activity from 22 participants (monolinguals and bilinguals responding with the left hand) for the contrast noGo–Go trials (hot-yellow colors) and Go minus noGo trials (winter-blue color). Notice the increased activation of the middle frontal gyrus (BA 9, 28/40/36) extending to the inferior frontal gyrus (BA 44, 36/16/20) for the noGo condition ($p < .0001$). (Data from Rodriguez-Fornells et al., 2005.)

than the phonological one. Notice also the standard right-central frontal distribution of this component.

Several lines of evidence link this frontal "N200" to lateral prefrontal inhibitory processes; for example, invasive studies in behaving monkeys have revealed activity related to response inhibition in the prefrontal cortex in a Go/noGo paradigm that gives rise to an N200 in humans (Sasaki, Gemba, & Tsujimoto, 1989). Moreover, Sasaki et al. were able to suppress the overt response on Go trials by electrically stimulating the prefrontal cortex at the time that an N200 would normally have developed on a noGo trial. Sasaki and Gemba (1993) presented a convincing case for an "inhibition" account of the N200 by comparing data from humans and monkeys. Moreover, destruction of the prefrontal cortex in animals has been found to lead to a profound disturbance of performance on delayed response tasks (Fuster, 1989), and to an enhancement of disinhibition and impulsive behavior (Luria, 1973). A cortical inhibitory network including the dorsolateral and inferior frontal lobe has been suggested on the basis of various neuroimaging studies in humans (Garavan, Ross, & Stein, 1999; Konishi et al., 1998, 1999; Liddle, Kiehl, & Smith, 2001; Menon, Adleman, White, Glover, & Reiss, 2001; Rubia et al., 2001; for a recent review, see Aron, Robbins, & Poldrack, 2004; Buchsbaum, Greer, Chang, & Berman, 2005). Figure 1b shows the brain network activated in the noGo condition in an event-related fMRI study (Rodriguez-Fornells et al., 2005; Go/noGo phonological decisions, all 22 subjects pooled together). In the noGo-Go comparison, clear involvement of the right middle frontal regions, including part of the inferior frontal region, is observed. In this particular paradigm, the isovoltage maps for the N2 noGo-Go effects are consistent with the fMRI activation observed (compare Figures 1a and 1b).

Thus, it is likely that the N200 elicited by noGo trials is related to inhibition processes in the prefrontal cortex (but see Donkers & Van Boxtel, 2004; Nieuwenhuis, Yeung, Van den Wildenberg, & Ridderinkhof, 2003, for a conflict-monitoring account). The presence of an N200 implies that the information, which can be used to determine whether a response is to be given, must have been analyzed (Thorpe, Fize, & Marlot, 1996). The timing of the N200 might thus provide critical information about the timing of the availability of the information that is used to determine the Go/noGo decision. As it is usual in ERPs, the peak latency of the N200 effect (i.e., the difference between Go and noGo ERPs) should be considered as an upper estimate of when the specific information must have been encoded.

The Phonological Level

The main goal of our first study was to assess the degree to which highly proficient and early German-Spanish bilinguals experienced interference from the nontarget language when naming a picture in the target language (Rodriguez-Fornells et al., 2005). The combination of ERPs and fMRI enables us to observe both the time course of interference effects during language production and the implicated brain regions. In order to avoid vocalization artifacts during EEG and fMRI acquisition, a variant of the Go/noGo picture-naming task was employed.

The phonological Go/noGo task used required subjects to access the phonological representation of a picture, as they had to decide whether the name of the depicted picture began with a consonant or a vowel (see Wheeldon & Levelt, 1995). In alternate blocks, bilinguals were required to respond when the German (Spanish) name of the picture began with a consonant and to withhold a response for words starting with a vowel. The target language was changed every block, which lasted 100 trials. Critically, the stimuli were selected such that in half the trials the names in both languages (Spanish and German) would lead to the same response (e.g., vowel coincidence *Esel-asno* "donkey," or consonant coincidence *Spritze-jeringuilla* "syringe"), whereas in the other half, responses were different for the two languages (noncoincidence condition, *Erdbeere-fresa* "strawberry"). We hypothesized that if the language-selection mechanism acts at the lemma level, no phonological activation should be observed for the nontarget language. On the other hand, if selection is operating at a later stage, the phonological form belonging to the nontarget language word should be at least partially activated, giving rise to interference effects at behavioral, electrophysiological, and brain imaging levels.

The results indeed showed clear evidence for cross-language interference at the phonological level in bilinguals (see Figure 2). Phonological interference was revealed in the form of an increased negativity for noncoincidence trials during German and Spanish conditions in bilinguals only (see Figure 2a, left panel). This increased negativity is seen in the Go trials between 300 and 600 ms, and considerably later in the ERPs to the noGo responses (Figure 2a). The difference waveforms (see Figure 3),[5] which were created subtracting the noGo waveform minus the Go waveform, surprisingly revealed that the N200 component associated with noGo trials was clearly reduced in the bilingual group compared to German monolinguals. In this group, the difference waveform showed a standard N200 component with a right-frontal distribution (see Figure 3a, left panel), a peak latency at 450 ms for the monolinguals (see Rodriguez-Fornells, Schmitt, et al., 2002; Schmitt et al., 2000), and no differences related to the coincidence and noncoincidence conditions. In the bilinguals, noGo minus Go difference waves in the noncoincidence condition were characterized by an initial positivity followed by a negative increase. An interesting pattern emerged upon subtraction of the coincidence from the noncoincidence condition (for Go and noGo trials separately; see Figure 3a, right panel). These difference waveforms show an early negativity on Go trials and a surprising early positivity on noGo trials, followed by a later negativity on noGo trials. The biphasic nature of this latter effect provides support for the notion that participants had to override their initial nontarget language lexical activation (which provided information in favor of a Go response) in the noncoincidence noGo trials and, instead, withhold a response according to the instructions in the corresponding block. This pattern of Go/noGo incongruence has also been recently shown in the study of Van der Lugt, Banfield, Osinsky, and Münte (submitted) using a different type of task.

In sum, the covert Go/noGo phonological task showed partial inhibition of the Go response in the active language by the interfering noGo response required for the nontarget language word. In contrast, in the noGo trials, the increased negativity due to

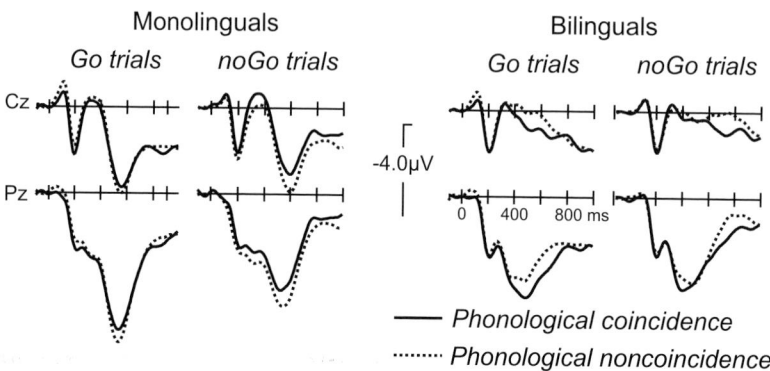

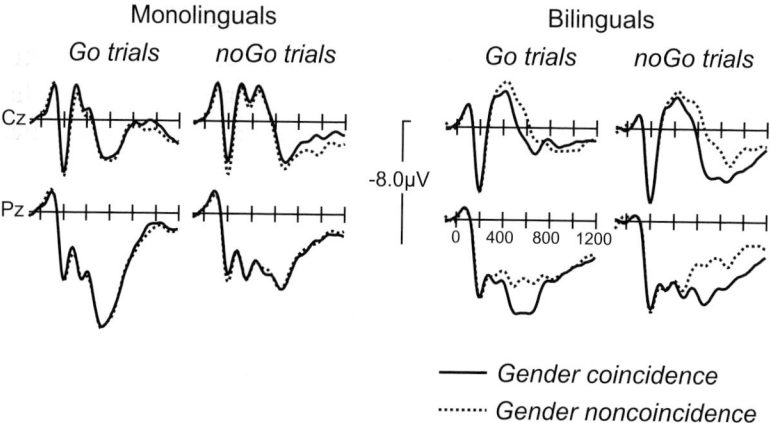

Figure 2. (a) Grand averages for monolinguals and German-Spanish bilinguals ($n = 12$) for the Go/noGo phonological interference experiment. Depicted are the waveforms for Go and noGo trials in the phonological coincidence and noncoincidence task (words that coincided or not with regard to the vowel/consonant decision across languages). The bilingual group showed an enhanced negatitivity for noncoincidence trials at frontal and central locations. This ERP effect was delayed in the noGo trials. (b) Grand averages for monolinguals and German-Spanish bilinguals ($n = 12$) for the Go/noGo gender interference experiment. In this experiment, noncoincidence trials had a different gender in both languages. As in the phonological experiment depicted above, bilingual participants showed a very similar effect in the interference condition.

noncoincidence started at about 600 ms. ERPs to noGo responses should have a reduced amplitude in the range 300–600 ms, because the nontarget language word would have required a Go response. An event-related fMRI in the same paradigm showed two regions associated with the noncoincidence effect in bilinguals when compared to monolinguals: the left dorso-lateral prefrontal cortex (DLPFC, BA 9/46) and the supplementary motor area (SMA) (Figure 4a). This result replicated previous studies in which the left middle prefrontal cortex has been found to be activated in Spanish-English bilinguals during mixed-language naming compared with blocks in which only one language was used (Hernandez et al., 2000, 2001; for a different pattern, see Price et al., 1999).

An interesting question here concerns the degree to which the previous ERP Go/noGo pattern and its interpretation could be partially supported by the fMRI study. According to our interpretation in the bilingual group, the incongruent condition has to elicit larger inhibitory activations (see Figure 1b) in Go noncoincidence trials compared to Go congruent trials. In fact, this is the case. As can be seen in Figure 4b, the comparison of noncoincidence minus coincidence in Go trials alone showed enhanced activation of the right middle frontal gyrus, anterior cingulate cortex (ACC), and SMA in the bilingual group. This pattern is clearly consistent with the ERP difference waveforms observed in Figure 3b for the Go trials in bilinguals.

The present study with alternating German and Spanish naming blocks created a rather unusual mixed-language situation, as subjects were required to name the same pictures either in Spanish or in German on alternating blocks. It is interesting to note that previous experiments (Colomé, 2001) supporting phonological interference in language production have used fluent Catalan-Spanish bilinguals who live in a strongly mixed-language context with about an equal presence of Catalan and Spanish in everyday life. In this study, Catalan-Spanish bilinguals were required to decide whether a specific phoneme (e.g., "m") was present in the Catalan name of a picture (e.g., *taula*

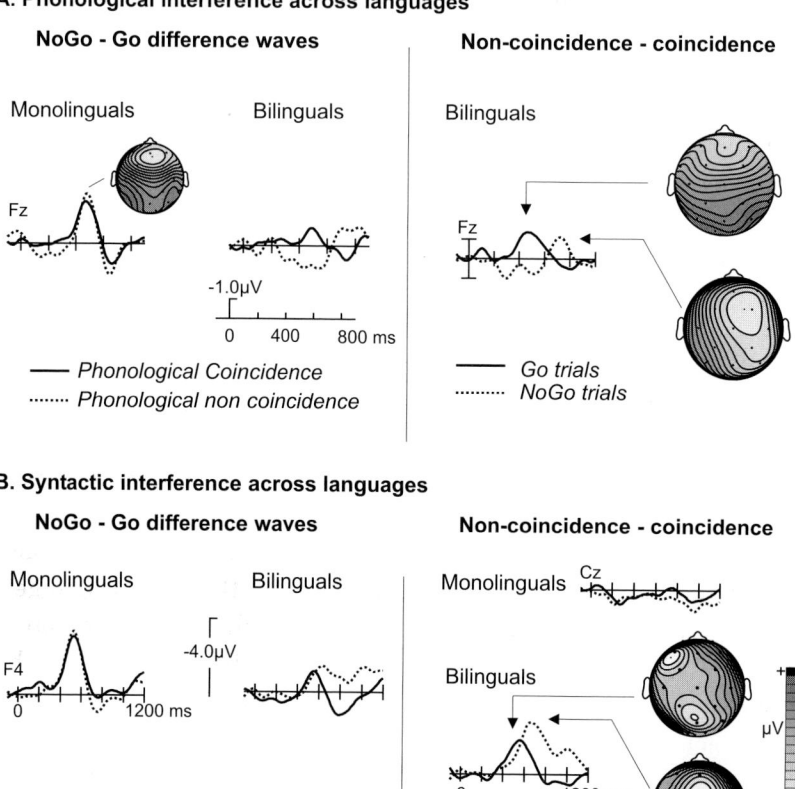

Figure 3. (a) Phonological interference task. Depicted is the difference waveform (noGo minus Go trials) for both groups in the left panel. Notice the standard N2 noGo component in the monolingual group. The isovoltage map is also depicted showing the standard right-frontal central distribution of the component. In the bilingual group, the amplitude of this N2 noGo component is reduced. In the right panel, the difference waveform (noncoincidence trials minus coincidence) is depicted. It is interesting to observe that in the bilinguals, a similar N2 component is elicited, peaking earlier in the Go trials (see also the scalp distribution of the isovoltage maps). (b) Gender interference task. The same effects as described in Figure 3a are illustrated for this new task and for a different sample of monolinguals and bilinguals.

"table"). Participants were tested only in the Catalan language mode. Interestingly, subjects were slower to answer "no" when the phoneme to be monitored was part of the Spanish (i.e., nontarget language) word (e.g., "mesa" *table*) than if it was neither part of the Catalan nor the Spanish word. This implied that the phonological representation of the nontarget language word was active. It remains to be determined to what extent the results of this study can be generalized to monolingual environments, in which bilingual speakers are only required to produce words in one of their languages.

Clearly, only models that incorporate partial activation of the nontarget language items and, more specifically, their phonological representation are consistent with the present results. To reconcile the present data with serial models with discrete processing stages (Levelt et al. 1991, 1999), it has to be assumed that lemma selection and the subsequent activation of the phonological representation occur in parallel in the two languages. Cascade or spreading activation models (Dell, 1986; Peterson & Savoy, 1998; Stemberger, 1985) can also easily accommodate the present set of results. Because semantic features of the corresponding words in the two languages of bilinguals are identical,[6] a high degree of competition would be predicted for such words by spreading activation models. The degree to which the lemma and phonological representation in the nontarget language are activated might greatly depend on individual factors of the speaker (i.e., to what degree s/he uses both languages concurrently from day to day, and on situational factors, such as mixed language or monolingual environments).

As has been shown, bilinguals seem to cope with L2 interference during language production by recruiting generic "executive function" brain areas, such as the left DLPFC and ACC. Further evidence for the role of left DLPFC in switching comes from two different lines of research. A recent case report (Fabbro, Skrap, & Aglioti, 2000) described a patient (S.J.) with a lesion encompassing the left prefrontal cortex and part of the ACC who pathologically switched between his two languages: Friulian and

Italian. In this particular case, S.J. switched to Friulian when speaking to an Italian speaker with non-Friulian knowledge. The reverse switch, from Friulian to Italian, was also observed. From neuropsychological studies, pathological switching of languages is manifested by the pervasive tendency to switch from one language to the other during verbal production, and it seems to be related to a pragmatic disorder of communication and prefrontal lesions (Fabbro, 2001). In the particular case of S.J., the involvement of the ACC and PFC could be affecting either the ability to maintain the goal of the communication (e.g., which is the language in use?) or the ability to *suppress* the interference from the nontarget language.

Although fMRI cannot provide direct evidence of the necessary participation of a brain area in a cognitive process, the use of transcranial magnetic stimulation (TMS) can be very useful to test causal implications of a particular brain region in a specific process. In a recent case report, Holtzheimer, Fawaz, Wilson, and Avery (2005) provided evidence for the participation of the left DLPFC in the process of language switching in bilinguals. Two patients, who were undergoing a therapeutic protocol for treatment-resistant major depression, experienced an unexpected language switch after high-frequency repetitive TMS (rTMS) was applied to the left DLPFC. The unexpected switch and urge to speak in the other language (not in English) appeared only in one of the 15 rTMS sessions and lasted from several minutes to 2 hr. This particular study lends support to the idea that the left DLPFC is involved in language switching. It is noteworthy that the specific TMS protocol of Holtzheimer et al. is thought to activate rather than inhibit this region via depolarization of cortical interneurons with indirect effects on pyramidal cells (Strafella & Paus, 2001).

What is the specific role of the left DLPFC/ACC-SMA network in the regulation of existing languages in multilinguals? For example, it is essential to understand how this network is related to (a) the regulation and control of language communication in switching environments (switch control and monitoring or induction of language switch), (b) the selection/activation of the target language in use ("language schema"), and (c) the inhibition of the nontarget language and, therefore, suppression of its interference. Previous imaging studies have revealed a role for these areas in different aspects of executive functioning (see Curtis & D'Esposito, 2003), such as (a) the selection of different response alternatives (D'Esposito, et al., 1995; Garavan, Ross, Li, & Stein, 2000; see also disruption of response selection with rTMS applied to the DLPFC, Hadland, Rushworth, Passingham, Jahanshahi, & Rothwell, 2001), (b) the switching between tasks (Dove, Pollmann, Schubert, Wiggins, & Von Cramon, 2000; Dreher, Koechlin, Ali, & Grafman, 2002; see also Rogers et al., 1998, who showed impairment in task switching for left-frontal patients), (c) the maintenance of a stable representation of the current task and prevention of interference (see Cohen, Botvinick, & Carter, 2000; Curtis & D'Esposito; Miller & Cohen, 2001), and (d) the inhibition of irrelevant items held in working memory (Baddeley, Emslie, Kolodny, & Duncan, 1998). In a Go/noGo lexical language decision task (Rodriguez-Fornells, Rotte, Heinze, Nösselt, & Münte, 2002), we found greater activation in the left anterior prefrontal region (BA 45/9, Figure 4) in bilingual subjects in response to Catalan words and pseudowords. This is a likely correlate of inhibition, as this region has recently been implicated in the selection of relevant information and interference resolution (Bunge, Ochsner, Desmond, Glover, & Gabrieli, 2001). The ACC has frequently been found to play a role in cognitive control and conflict tasks (Carter et al., 1998). The degree of activation of the DLPC might depend in part on the amount of conflict or interference detected, which is assumed to be signaled by the ACC (Botvinick, Nystrom, Fissell, Carter, & Cohen, 1999; Carter et al.; MacDonald, Cohen, Stenger, & Carter, 2000).

One potential and important criticism of our Spanish/German experiment concerns the extent to which the reported results would have also been obtained in a completely monolingual mode, where no language switch is present. Thus, a first cautious interpretation of the present data is that bilinguals recruit prefrontal executive brain regions in order to control the

A. Non-coincidence minus coincidence

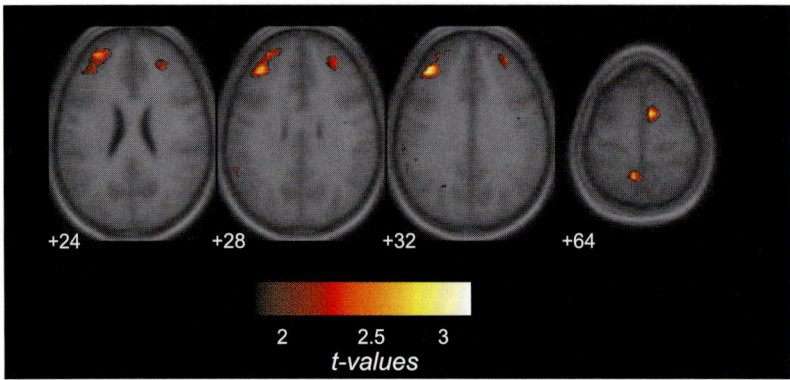

B. Go trials: non-coincidence minus coincidence

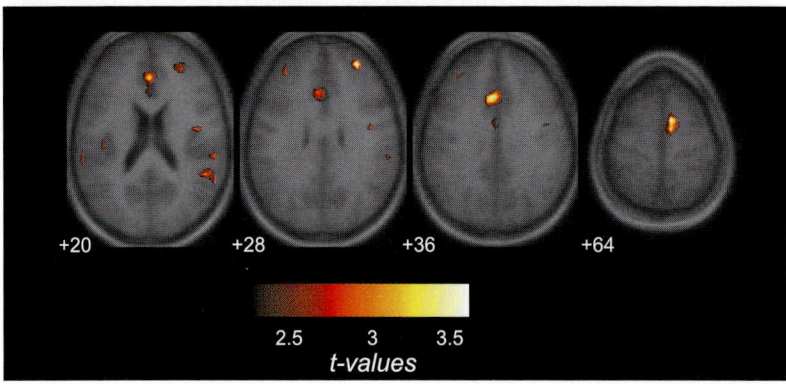

Figure 4. (a) Axial views of the group differences (German-Spanish bilinguals, $n = 11$; minus monolinguals, $n = 11$) in standard stereotactic space identified for the crucial contrast phonological noncoincidence minus coincidence condition in all trials. Values in the color scale refer to the T values of the corresponding contrast. Notice the differential recruitment of the middle frontal gyrus (left BA 46/9) and SMA (BA 6) regions in this comparison. The axial views presented were superimposed on the mean anatomical image formed by averaging, for all 22 subjects, T1 structural MRI scans mapped into normalized MNI (Montreal Neurological Institute) space. (Adapted from Rodriguez-Fornells et al., 2005.) (b) Group differences in the same contrast (noncoincidence minus coincidence) for Go trials only. Notice the increased activation observed in the ACC (BA 32, −4, 20, 30, $T = 3.5$, $p < .001$), right middle frontal gyrus (BA 9/10, 34, 48, 28, $T = 3.8$, $p < .001$) and SMA (BA 6, 12, 0, 68, $T = 4.0$, $p < .001$).

interference produced in language production in multilingual environments, for example, by monitoring internal switches and language production errors and regulating the activation and suppression levels of the target and nontarget language, respectively. We will return to this issue below in order to propose a specific activation/inhibition language system.

The Syntactic Level (Grammatical Gender)

In a follow-up study (Rodriguez-Fornells, Lutz, & Münte, submitted), we evaluated the degree of interference when accessing syntactic information in both languages in a covert naming task in another group of fluent Spanish-German bilinguals. Grammatical gender does not, in many cases, match across languages. Thus, the word "table" is masculine in Italian (*il tavolo*) and German (*der Tisch*) and feminine in French (*la table*) and Catalan (*la taula*). This example illustrates that grammatical gender of entities other than human beings, to a large extent, is arbitrary.

In this study, the names of the pictures were selected considering gender agreement between Spanish and German: 60 names had the same gender (e.g., der_{masc} *Speer*–el_{masc} *dardo* "dart") and 60 names had different genders (die_{fem} *Rakete*– el_{masc} *cohete* "rocket"). Because gender in Romance languages takes only masculine and feminine forms, neuter gender in German was not considered. Speakers were required to perform a gender decision task using a Go/noGo covert picture naming procedure. This decision conveys different processes involved in gender-marked speech production, i.e., identifying a concept, selecting and retrieving the appropriate lemma and, finally, attaching its corresponding grammatical gender (see Jescheniak & Levelt, 1994; Levelt et al., 1999; Schriefers & Teruel, 2000). Using this type of task, the onset latency of the N200 noGo component for gender decisions is approximately 400 ms after picture presentation, normally 40 ms before the onset of phonological Go/noGo decisions, and between 70 and 90 ms after conceptual Go/noGo decisions

(Schmitt, Rodriguez-Fornells, et al., 2001; Schmitt, Schittz, et al., 2001; Van Turennout et al., 1998; see Jansma et al., 2004, for a review).

Syntactic interference is expected in bilinguals, given that lemma activation is produced in all languages in bilinguals at least to some degree (Colomé, 2001; Hermans et al., 1998; Rodriguez-Fornells et al., 2005). Several bilingual studies have already proposed that lemma activation is independent of the language in use, thus predicting that interference will appear when both languages do not agree in syntactic gender.

In order to allow evaluation of the differences between the effect of language switching, on the one hand, and the interference created by noncoincidence regarding syntactic gender between languages, on the other hand, the required target language switched every 18 trials in a predictable way (the target language was signaled by a red square surrounding the pictures in one of the languages). In addition, the structure of the first three trials after the switch was always identical, comprising a coincidence/noncoincidence/coincidence triplet. This structure was repeated in between each miniblock of 18 trials for the 12th, 13th, and 14th trials. This design, therefore, allowed direct comparison of the effects of language switching with the syntactic interference effect. German monolinguals were used as a control group in this study. By necessity, the paradigm in this group did not involve switching.

The behavioral results of this study showed interference in bilinguals when naming in the noncoincidence condition compared to the coincidence condition, as evidenced by the reaction time, percentage of hits, and percentage of false alarms (see Figure 5). Noncoincidence responses were significantly delayed (by ~44 ms), with more false alarms and fewer hits when compared to the coincidence trials. Furthermore, a large language switch effect was found, especially for the first and second trial after a switch (Figure 5a). However, the interference effect of the gender noncoincidence was still visible; for example, the noncoincidence trial in the nonswitch triplet showed a delayed reaction time, fewer hits (see Figure 5a), and more false alarms. Notice

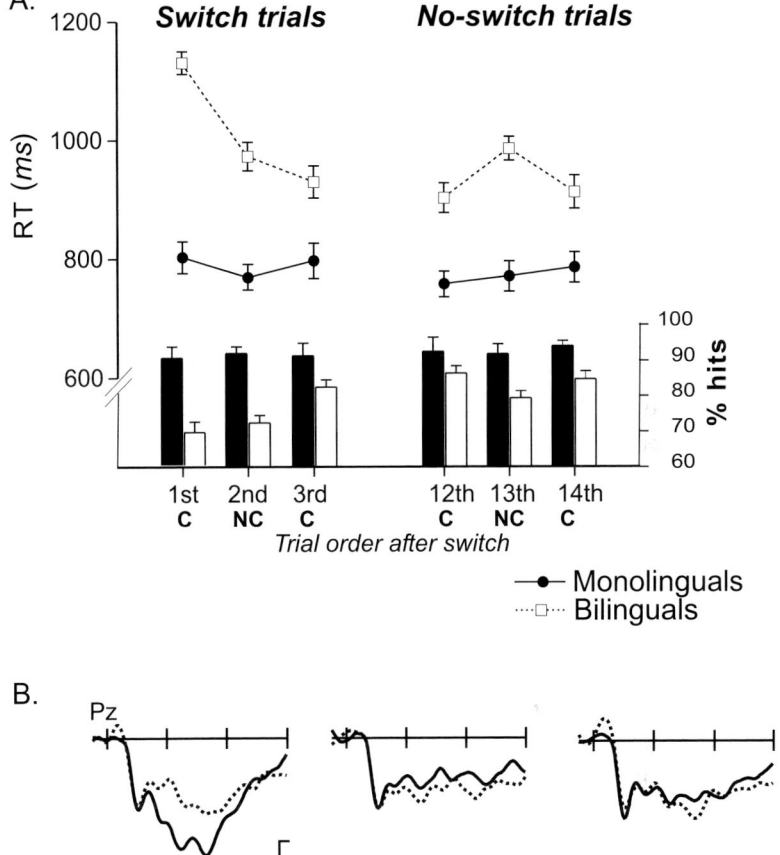

Figure 5. (a) Behavioral analysis (mean reaction time ± standard error of the mean [lines, left-side scale] and percentage of hits [bars, right-side scale]) for the three trials immediately following the language switch and the three trials in the middle of each miniblock (no-switch trials, 12th, 13th, and 14th) for monolinguals and bilinguals in the gender-interference experiment. C = Coincidence trial, NC = noncoincidence trial. (b) ERP waveforms in bilinguals comparing switch and no-switch trials. Comparisons show the averages considering the trial order of the switch and no-switch (1st trial with 12th, 2nd trial with 13th, and 3rd trials with 14th). Notice the increased late positive component (LPC) observed immediately after the language switch.

also that bilinguals showed a delayed reaction time independent of the switch or the gender-incongruence effect. Such a general delay in bilinguals had also been observed in the previous phonological interference experiment for both the ERP and fMRI parts.

The ERPs of monolinguals did not show an effect due to gender coincidence (Figures 2b and 3b, right panel), but a standard N200 noGo component was present (see Figure 3b, left panel). In contrast, and as in the previous experiment, bilinguals showed a marked incongruence effect across languages: Noncoincidence trials elicited a large negativity when compared to the coincidence condition (Figure 2b). In the noncoincidence minus coincidence difference waves (Figure 3b, right panel), a negativity can be observed for both Go and noGo trials, with a delay in the peak observed for the noGo trials. This pattern is similar to that observed in the previous phonological incongruence experiment. In addition, the N200 noGo component is attenuated for the bilinguals (see Figure 3b, left panel), who also show a different scalp distribution.

The ERP effects related to the language switch are shown in Figure 5b. An increased positivity is observed when the first trial after the switch is compared to the first nonswitch trial of the control triplet. This effect is reminiscent of similar positivities found in nonlinguistic switching tasks (Gehring, Bryck, Jonides, Albin, & Badre, 2003) and has been interpreted as an instance of the "P300." In fact, the "context updating" theory (Donchin, 1981; Donchin & Coles, 1988, 1998) would predict a larger P300 amplitude in the case of a language switch. This is because a language switch requires the participation of cognitive control for updating the contents of working memory, the participation of attention, and the decision to change the language in the present context of the performance.

Similar late positivities (in the range 400–800 ms) have also recently been found in a study of English-Spanish bilinguals, in which an unexpected language switch (code-switches) was presented while reading a sentence (Moreno, Federmeier, & Kutas, 2002; e.g., *He put a clean sheet on the cama* [Spanish for the expected "bed"]). The peak amplitude and latency of the positivity to code-switches correlated negatively with the subjects' proficiency in Spanish as measured by the naming test.

Finally, Figure 6 compares ERPs for dominant and nondominant language trials for the phonological and syntactic interference experiments. As revealed by vocabulary scores in both languages, these particular groups of bilinguals showed greater fluency in German, although their native language was Spanish. In this comparison, and regardless of the different experimental conditions, the dominant language showed a larger negativity with an onset at about 400 ms, especially at frontal and central locations. Interestingly, reaction times for dominant and nondominant language trials were not different in the two studies. Nonetheless, it is tempting to speculate that this increased negativity reflects the enhanced control required for the dominant

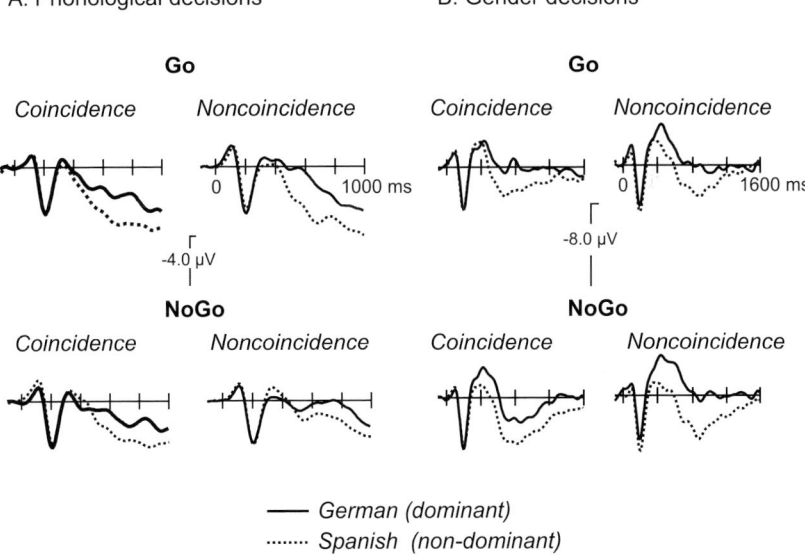

Figure 6. Grand average for the German-Spanish bilinguals in two experiments: (a) phonological interference task and (b) gender-interference task. Notice that in all conditions, the dominant language elicited an increased negative component.

language. Thus, these results would support the idea that the selection of an active item in a lexicon has to be accomplished via inhibition of the task-irrelevant language. According to the IC model (Green, 1998), access to L2 representations involves greater suppression (or inhibition) of L1; therefore, enhanced control is required to overcome the applied inhibition when naming in the L1.

Further Evidence of Enhanced Control in Bilinguals

The widespread negativity observed in bilinguals when naming in their L1 in both ERP studies has been interpreted as the existence of an enhanced control in bilinguals when engaged in L1 naming. The greater need for executive control should also show up in direct comparisons of the ERPs obtained in bilinguals with those recorded in monolingual controls. Indeed, we first observed an enhanced negativity in Catalan-Spanish bilinguals (relative to Spanish monolinguals) in a word-reading study that required them to inhibit nontarget words (see Figure 2 in Rodriguez-Fornells, Rotte, et al., 2002). We suggested that highly fluent bilinguals could strategically control the process of mapping spelling to sound in the target language when they are required to ignore words in the nontarget language. Enhanced engagement of executive control in bilinguals was also evident in the accompanying fMRI study, which showed greater activity (relative to monolinguals) in several regions, such as the left DLPFC, the anterior inferior frontal region, and the ACC.

In both of the above-discussed covert language production studies, bilinguals again show an overall enhanced negativity when compared to monolinguals (Figures 7a and 7b; see Rodriguez-Fornells et al., 2005; Rodriguez-Fornells et al., submitted). These group differences had a medial frontal maximum pointing to medial frontal generators such as the ACC (see Figure 7a). Indeed, a brain area distinguishing bilingual from monolingual subjects in the accompanying fMRI study was the ACC. Various functional neuroimaging studies have shown that

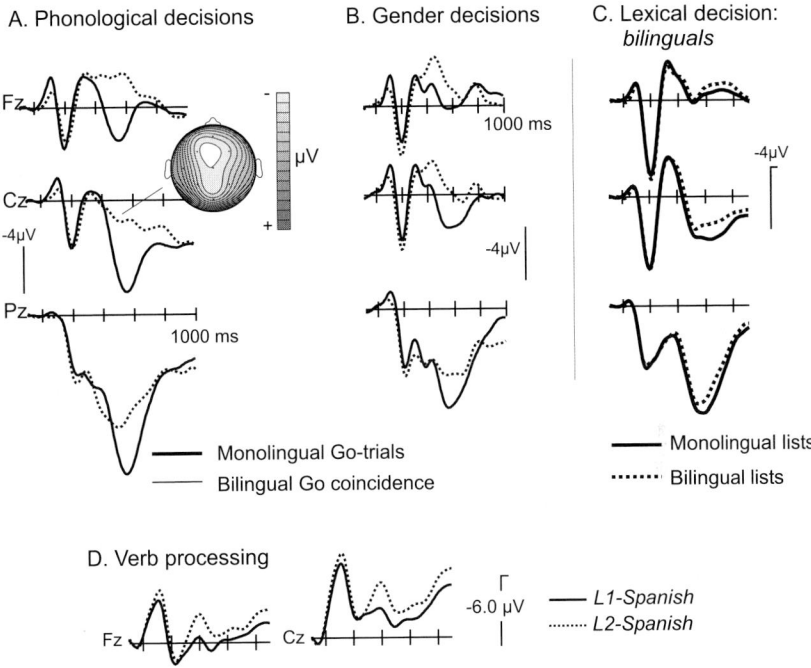

Figure 7. (a) Grand averages for monolingual ($n = 12$) and bilingual groups ($n = 12$) are plotted together for Go trials from the German naming task. For the bilinguals, only the phonological coincidence condition is presented. The topographical map for the difference wave was obtained by subtracting the monolinguals' ERPs from the bilinguals' waveforms using isovoltage mapping with spherical spline interpolation (min/max values $-6.8/-1$ μV) (from Rodriguez-Fornells et al., 2005). (b) The same grand averages showing the same contrast in the gender covert naming task. Notice the similarities for a different group of bilinguals ($n = 12$) and on a different task (Rodriguez-Fornells, Lutz, & Münte, submitted). (c) Grand average for Spanish-Catalan bilinguals ($n = 20$) for Go trials in the monolingual list condition (including only target language words and pseudowords) and the bilingual list (including target language words, nontarget language words, and pseudowords) (Rodriguez-Fornells et al., unpublished data). (d) Grand average ERPs for Spanish-Catalan bilinguals in a study on verbal morphology in Spanish (De Diego Balaguer et al., 2005). The averages show all verb conditions pooled together. Notice the increased frontocentral negativity in the L2-Spanish group (L1 = Catalan, L2 = Spanish) when compared to the L1-Spanish group (L1 = Spanish, L2 = Catalan).

this region is active (a) when a prepotent response tendency has to be overcome, such as in the Stroop task (Pardo, Pardo, Janer, & Raichle, 1990), (b) when the response is undetermined, as in verb fluency tasks (Frith, Friston, Liddle, & Frackowiak, 1991), and (c) when committing errors and or experiencing response conflict (Carter et al., 1998). A recent model considers that this region evaluates the demand or need for cognitive control by monitoring for the occurrence of conflict or interference (Botvinick et al., 1999; Carter et al., 2000). It has to be kept in mind that the ACC is only part of a larger circuit that is involved in the detection of conflict and instantiation of further processing prior to response execution (Cohen et al., 2000; MacDonald et al., 2000; Miller & Cohen, 2001). A recent proposal considers that cognitive control is supported by a brain network involving the insular, prefrontal, and ACC, which resolves interference between competing responses (Wager et al., 2005).

This suggests that the frontocentral negativity observed in bilinguals relative to monolinguals should vary as a function of the need for cognitive control. To test this hypothesis, we required Catalan-Spanish bilinguals to make Go/noGo lexical decisions in monolingual lists (low-control condition) and bilingual lists (high-control condition; Rodriguez-Fornells et al., unpublished data). In the monolingual condition, only words from the target language (either Catalan or Spanish) and pseudowords (constructed on the basis of words from the target language) were presented. Participants had to make Go decisions for target words (inhibit pseudowords). In the bilingual condition, words from the target language (e.g., Spanish) were presented as well as nontarget words (e.g., Catalan) and pseudowords (half derived from the target and half derived from the nontarget language). Participants had to press only when a word in the target language was presented, inhibiting the other nontarget words and pseudowords. An increased negativity for words was observed in the bilingual condition (Figure 7c). Finally, in a recent study with Spanish-Catalan bilinguals (De Diego Balaguer et al., 2005; see explanation of this study below), the nondominant Spanish group showed a similar enhanced negativity when doing the lexical decision task in their L2 (Figure 7d).

Another possible instance of this effect can be seen in the work of Moreno and Kutas (2005), who required participants to read sentences containing semantically congruent or anomalous final words. In alternating blocks, either their dominant or nondominant language was used. The onset and peak latency of the N400 effect (i.e., a specific ERP component associated to semantic processing) was delayed in the less proficient language when compared to the dominant one. The authors also observed (see Figure 7 in Moreno & Kutas) a larger frontal negativity during nondominant language processing. Although the authors discussed this effect in terms of working memory (viewing it as an instance of the left anterior negativity [LAN] effect[7]) or as reflecting concreteness processing (because of its similarity with ERP differences between abstract and concrete words), we suggest that it might be related to the degree of cognitive control involved in processing a language in a specific context.

A preliminary conclusion regarding these effects could be that they reflect brain mechanisms for interference resolution and conflict detection and are common to monolingual and bilingual subjects. Speakers or comprehenders might recruit them to different degrees in different tasks depending on individual differences in ability (e.g., language proficiency and the level of interference of the nontarget language). Further research is required in which individual differences in language proficiency and other relevant biographical variables could be correlated with the degree of activation of this cognitive control network.

The Role of Similarity Between Languages With Respect to the Recruitment of Control Mechanisms

One relevant issue with respect to the control mechanisms needed by the bilingual brain is the similarity between the languages spoken. Particularly in languages from the same linguistic family (e.g. Romance languages such as Spanish, Catalan, or

Italian), some commonalities at the lexical, phonological, or grammatical levels are evident. Indeed, many languages contain similar, dissimilar, and, in some cases, even identical representations. Consider, for example, so-called cognate words (e.g., *casa*, for "house" in Spanish, Italian, and Catalan) and noncognate words (e.g., *perro-cane-gos* for "dog" in Spanish, Italian, and Catalan, respectively) or, at the grammatical level, the order of arguments in the sentence (e.g., Subject-Verb-Object, Object-Subject-Verb), the number of conjugation classes, and so forth. With respect to this diversity, two possibilities are available to the system: either to consider each language independently of these similarities and differences or to take advantage of the similarities between languages. If the first strategy is adopted, then we should not see differences between those representations that are similar and dissimilar between the L1 and L2. If the latter strategy is adopted, then the control system should be flexible enough to adapt and control interference from dissimilar representations while allowing a beneficial effect of similar representations of the L2 in L1 representations.

In the following section we will comment on how the similarity at the lexical, grammatical, and phonological levels can help or interfere in the processing of the L2, and we will then discuss how the control system is able to deal with the lexical, grammatical, and phonological components when both similar and distinct representation might occur in the two languages of a bilingual.

Similarity of Lexical Items Across Both Languages (Cognate Words)

Some languages, such as Spanish and Catalan or Italian and Catalan, share many cognate words (words that have a very similar phonological structure in both languages [e.g., in Spanish-Catalan: *coche-cotxe* "car"; in Catalan-Italian: *parlar-parlare* "speak"]). Costa et al. (2000) presented pictures with cognate and noncognate words and asked Catalan-Spanish bilinguals to name them in Spanish. Bilinguals naming latencies were faster for pictures with cognate words than those with noncognate words, whereas no differences were found for monolinguals. This result is in favor of the activation of lexical items of the nontarget language and supports the idea that this activation flows to the phonological level (see also Kroll, Dijkstra, Janssen, & Schriefers, 2000).

The question addressed here is whether covert production of cognate words is facilitated, and if so, then the cognitive control required to produce these words might be reduced. Let us consider this issue with our gender-interference experiment (Rodriguez-Fornells et al., submitted). In a further condition, 20 Spanish-German cognate words were introduced (congruent gender in all cases). As expected, we observed reaction-time facilitation for the gender decisions of these cognate words (in bilinguals: noncognate words, 969 ms, and cognate words 875 ms; in monolinguals: noncognate, 785 ms, and cognate words, 816 ms). Furthermore, the ERP pattern for cognate and noncognate words was very different in the bilingual group (see Figure 8a). In a similar fMRI study performed in another group of Spanish-German bilinguals, comparison of cognate words minus noncognates revealed a clearly reduced activation in bilinguals in two critical areas: left DLPFC (BA 46/9) and ACC (BA 32). These data are thus consistent with the idea that cognitive control is reduced in the case of cognate processing in bilinguals.

It remains to be determined whether reduced cognitive control would also appear in the case of interlingual homographs such as *angel*, meaning "sting" in Dutch. Whereas cognate words share both formal and semantic information, interlingual homographs are indistinguishable in terms of form but completely different in terms of meaning. Thus, if parallel activation is produced when processing homographs, then a greater need for cognitive control will be required depending on the linguistic context (see Dijkstra & Van Heuven, 1998, 2002). In several studies, parallel activation of both languages appears to lead to facilitation effects when subjects have to decide whether the word exists in either language, whereas interference appears when the subjects have to make a lexical decision in one of the languages (Dijkstra, Van Jaarsveld, & Brinke, 1998). This parallel activation was not present in the interlingual homograph study when the material included exclusively words in one language (Dijkstra et al. 1998; see Gerard & Scarbourough, 1989), indicating that cognitive control might be even easier when semantic information is not shared, as long as no activation of the other language is forced by the material.

Learning Similar Morphological Systems

One view of the architecture of the language system distinguishes a "lexicon," storing the idiosyncratic forms of words, from a "grammatical component," which stores the information about how those lexical forms are combined to form morphologically complex forms and sentences. This distinction is crucial when it comes to adult L2 acquisition. Studies and everyday experience show that whereas lexical acquisition can be achieved at a native or near-native level, syntactic and phonological performance is generally much worse in the L2 (Birdsong, this volume; Birdsong & Molis, 2001; Flege, Yeni-Komshian, & Liu, 1999; Johnson & Newport, 1991; Weber-Fox & Neville, 1996), despite extensive practice. Thus, at the level of morphosyntactic acquisition, we should be able to observe clear similarity effects, either in the form of interference or facilitation.

The case of morphological processing in morphologically complex but similar languages poses the challenging question of how the system is able to manage lexical items that are both similar and dissimilar according to the level of processing. A direct evaluation of progressive learning on the acquisition of similar and distinct morphosyntactic rules between languages has recently been reported (Osterhout, McLaughlin, Kim, Greenwald, & Inoue, 2004). In this longitudinal study, a sample of American students who were enrolled in a formal French course was evaluated using ERPs after 1, 4, and 8 months of training. An N400 effect to semantic anomalies was already present after the first month of learning. In addition, two syntactic violations were evaluated: (a) subject-verb agreement, which is present in English as well as in French, and (b) number agreement between the article and the noun, rarely present in English. Whereas a P600 component was elicited at the fourth month of training for the subject-verb agreement violation, no P600 effect was elicited for the second type of grammatical violation, even at the last evaluation point. The P600 component, peaking over central-parietal recording sites, is elicited by syntactic violations and unexpected (complex) syntactic constructions (such as sentences with noncanonical word order, Matzke, Mai, Nager, Rüsseler, & Münte, 2002) and has been associated with syntactic reanalysis (Kaan, Harris, Gibson, & Holcomb, 2000). The results of Osterhout et al. support the hypothesis that morphosyntactic rules shared by the L1 and L2 are qualitatively different from the ones that are not common to the two languages.

A further study by Portin and Laine (2001) corroborates the idea that the limitation in L2 grammatical (morphological) learning might actually depend on the similarity between the L1 and L2 with respect to this dimension. Portin and Laine studied fluent early Swedish-Finnish bilinguals. Finnish has a very rich morphological system and native-Finnish monolinguals appear to process words in a compositional manner (in combinations of stems and affixes). Swedish, in contrast, is a Germanic language and has a much more limited morphological system. In this study, Swedish monolinguals processed monomorphemic and inflected Swedish nouns approximately equally fast. In contrast, Finnish-Swedish bilinguals processed inflected Swedish items significantly more slowly than monomorphemic words. This suggests that Finnish-Swedish bilinguals transferred the process of decomposing inflected words in their L1 (Finnish) to Swedish. A direct effect of similar compared to dissimilar structures can be found in Hahne, Mueller, and Clahsen (2006). They explored how L2 learners (L2 German, L1 Russian) process inflected words by means of two offline tasks (acceptability judgment and elicited production) and ERP recordings during the processing of morphological violations (noun plurals and past participles). Error rates for the noun plurals were higher in the L1-Russian group for the acceptability judgment task. With regard to the ERP, similar studies in monolinguals have shown a P600 following possibly a LAN peaking around 350–400 ms (Coulson, King, & Kutas, 1998; Weyerts, Penke, Dohrn, Clahsen, & Münte, 1997). In the L1-Russian group, violations of regular noun plural suffixation elicited a P600 but no LAN, whereas both components were present in a control group of native speakers of German. By contrast, a surprising result was obtained for past participle processing. In this case, the bilingual group was identical to the group of native-German speakers for the processing of regular and irregular violations in performance, as well as on the ERP measures. Interestingly, Russian and German share the same regular suffix (*-t*). Although in Russian, the selection of the *-n* and *-en* suffix participle is phonologically determined by certain stem endings, whereas in German, the *-n* suffix is determined by class membership and only applies to the subclass of strong (irregular) verbs, the similarity between the languages might have been crucial and probably facilitated the learning of this inflectional pattern in German.

In order to further explore this idea, we conducted an ERP study of Spanish morphology with two groups of highly proficient Catalan-Spanish early bilinguals with inverse profiles (de Diego Balaguer et al., 2005). One group (L1-Spanish) had learned Spanish as their primary language and Catalan as their L2. In contrast, the L2-Spanish group had learned Catalan first. Importantly, the regular suffix (*-o*) is common to both languages[8] (e.g., *menj-ar*/*menj-o* "to eat/I eat" in Catalan, *cant-ar*/*cant-o* "to sing/I sing" in Spanish), whereas the irregular alternations are totally different. Spanish and Catalan, unlike the language pairs studied by Portin and Laine (2001) or Hahne et al. (2006) have very similar morphological structures; that is, they both have the same number of conjugation classes, tenses, and persons. Therefore, learners can easily exploit the structures already available in their L1 in the process of learning the morphological system of the

L2. Regular Spanish verbs showed the same centro-parietal N400 priming effect in the second language speakers (L2) as in primary language (L1) speakers. However, ERP differences between groups were observed for irregular morphology.[9] In L1 speakers, an N400 effect appeared for both regular and irregular verbs (see Figure 8b) and had about the same size. In contrast, L2 speakers showed a reduced N400 priming effect in the irregular contrast. The size of the N400 effect in this particular study (computed by subtracting the primed from an unrelated condition) reflects the effect of morphological priming. Thus, again, similar structural properties might promote processing in L2, whereas dissimilar structures appear to cause interference.

The Phonological Level

The proverbial difficulty of native speakers of Chinese to perceive and produce the differences between /r/ and /l/ illustrates that phonological (dis)similarity between L1 and L2 is a major factor in the processing of L2. Take Spanish and Catalan for

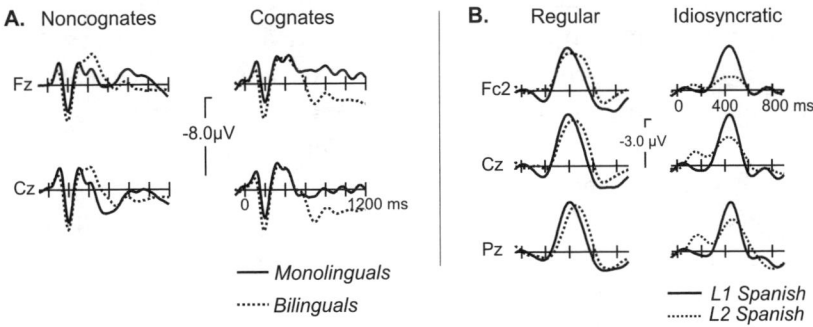

Figure 8. Similarity effects across languages. (a) ERP averages for cognate and coincidence noncognate words in monolinguals and bilinguals (from the gender-interference experiment). (b) ERP averages for regular (similar) and irregular (dissimilar) verbs in Spanish dominant (L1-Spanish) and Catalan dominant (L2-Spanish) bilinguals showing the N4 repetition priming effect (difference waveforms, unrelated minus related items). (Adapted from De Diego Balaguer et al., 2005.)

example, which differ in the number of vowels. Whereas Spanish has five vowels (/a, e, i, o, u/), Catalan has eight (/a, e, ɛ, i, o, ɔ, u, ə/). Thus, both languages have the phoneme /e/, but /ɛ/ is contrastive only in Catalan; the sound /ɛ/ is perceived as a particular instance of the category /e/ by Spanish monolinguals (Sebastian-Galles, Echevarria, & Bosch, 2005). The ambiguity of the phoneme /ɛ/ is reflected in the Spanish dominant bilingual speakers of Catalan in a particular difficulty in the perception and production of this phoneme. A previous study by Bosch, Costa, and Sebastian-Galles (2000) has shown that even highly proficient and early Spanish dominant bilinguals (before 4 years old) have trouble perceiving the /e-ɛ/ contrast. These problems have consequences also at the lexical level (Pallier, Colomé, & Sebastian-Galles, 2001). The results of a recent electrophysiological study (Sebastian-Galles, Rodriguez-Fornells, De Diego Balaguer, & Diaz, in press) show that although Spanish-dominant bilinguals have no problems in rejecting nonwords made by changing a common vowel contrast (e.g., *finestra* "window," changed to the pseudoword *finostra*), they had substantial difficulty when the change involved a Catalan-specific contrast (e.g., *finestra* changed to the pseudoword *finɛstra*). Moreover, although the Catalan group showed a clear error-related negativity (ERN) when producing an error in pseudoword decisions involving a Catalan contrast (e.g., in the case of *finɛstra*), no ERN trace was observed in the Spanish dominant bilingual group. The ERN component is obtained time-locked to erroneous decisions (Gehring, Gross, Coles, Meyer, & Donchin, 1993) and has been clearly related to the activation of the ACC and other prefrontal regions (see Carter et al., 1998; Ullsperger & Von Cramon, 2001). Although the specific relation to the ERN component and response conflict and error detection is under debate (see Botvinick, Braver, Barch, Carter, & Cohen, 2001; Holroyd & Coles, 2002; Yeung, Botvinick, & Cohen, 2004), we favored the interpretation that in this particular case, Spanish bilinguals did not experience any conflict or uncertainty when processing the incorrect Catalan pseudoword. Interestingly, the ERN might be a very useful component for evaluating the degree of response conflict in bilinguals when processing languages in different contexts (see following section).

How the Language Production System in Bilinguals Implements Inhibition

Several different bilingual models briefly discussed above posit the existence of inhibitory mechanisms that enable the deactivation or partial suppression of the language not in use. Although we have provided evidence for the involvement of frontocentral "executive" brain areas in bilingual language processing, there is, as yet, no clear or direct evidence of inhibition at the behavioral, brain region, or neuron levels involved in the regulation of both languages. How exactly does the prefrontal cortex regulate the selection and activation of the target lexicon and the corresponding grammatical rules? Inspired by findings in other domains of cognitive neuroscience we will propose how two interrelated control/inhibitory mechanisms might regulate bilingual speech production: (a) A top-down control inhibitory mechanism could be implemented by the prefrontal cortex when language schemas are activated and (b) this prefrontal selection/inhibition mechanism could interact with a more local and bottom-up inhibitory mechanism that regulates the level of activation of the nontarget language. Depending on the communicative situation (monolingual vs. mixed-language contexts), the top-down modulation could regulate the local-inhibitory system in order to control for the degree of access to nontarget representations when speaking. This control mechanism is possible due to the rich interconnectivity of the prefrontal cortex with other regions of the brain (Pandya & Barnes 1987; Pandya & Yeterian 1990).

Using single-cell recordings, functional neuroimaging, and electroencephalography, an influence of the prefrontal cortex on the selective activation and suppression of different specialized posterior regions of the brain has been observed (Barcelo, Suwazono, & Knight, 2000; Corbetta, Miezin, Dobmeyer,

Shulman, & Petersen, 1990; Hillyard, Hink, Schwent, & Picton, 1973; Moran & Desimone, 1985; Pessoa, Kastner, & Ungerleider, 2003). For example, ERP studies in patients with focal prefrontal cortex lesions have shown that primary auditory- and somatosensory-evoked responses are enhanced (Chao & Knight, 1998; Knight, Scabini, & Woods, 1989; Yamaguchi & Knight, 1990). These results suggest a disinhibition of sensory flow to primary cortical regions. In addition, damage to the lateral prefrontal cortex has been associated to marked decrements in the top-down signal accompanied by behavioral evidence of impaired detection ability (Barcelo et al., 2000). Using fMRI and ERPs, Gazzaley, Cooney, McEvoy, Knight, and D'Esposito (2005) have recently shown that the magnitude of neural activity and the speed of neural processing were clearly modulated by top-down influences. In this study, activation in the visual association cortex for faces and scenes were dependent on the instructions to attend to or ignore one or the other type of stimuli. Whereas attending to faces, for example, increased the activation of the ventral temporal lobe (fusiform gyrus, face area), decreased activation (or suppression) was simultaneously observed in the corresponding scene visual processing area (parahippocampal cortex, place area).[10] This study clearly demonstrated that top-down modulations from the prefrontal cortex can directly facilitate stimulus processing at posterior regions of the cortex and suppress activity in nonrelevant regions (for a review, see Knight & Stuss, 2002; for a similar proposal in visual attention, see Desimone & Duncan, 1995).

Invasive electrophysiological studies in patients with implanted electrodes have revealed a large face-selective N200 component, which is generated in small regions on the inferior surface of the occipito-temporal cortex (Allison, Puce, Spencer, & McCarthy, 1999). This component reflects excitatory depolarizing potentials in pyramidal neurons in these regions and has been related to face perception per se rather than later recognition stages. In other small regions of the occipito-temporal cortex, a similar N200 response has been elicited by other types of stimulus, such as letter strings (words and nonwords), numbers, complex objects, and gratings. More recently, it has been shown (Allison, Puce, & McCarthy, 2002) that the presentation of a face elicited a N200 negativity in the face-selective region, whereas it induced a positivity in word-selective regions. Letter strings led to the reverse pattern. The positivity might reflect hyperpolarizing inhibition of apical dendrites, providing strong evidence that assemblies of neurons that are activated or responsive to one stimulus category (such as faces) inhibit cell responsiveness to another category (such as letter strings), probably by a type of lateral inhibition. The authors proposed a simple model in which the mechanism of inhibition consisted in a recurrent collateral inhibition. Because the P200/N200 peaks were highly correlated, this suggests a local inhibitory process driven bottom-up by the input signal.

With regard to bilingual language processing, such a type of local inhibitory mechanisms could be the neural instantiation of the proposed inhibitory links between languages. Furthermore, top-down mechanisms might be able to influence the degree of tuning of these specific language-selective regions to the language context; for example, this top-down modulation of the prefrontal cortex in other parts of the brain will increase for weak stimulus-response mappings when they are in competition with stronger ones, specially when flexibility is required.

Future Issues in the Cognitive Neuroscience of Neural Control in Bilingualism and L2 Learning

Bilinguals seem to cope with L2 interference by recruiting typical "executive function" brain areas that have been observed in many other experimental contexts. These areas might be crucial in suppressing and inhibiting the production of the nontarget language word. Future studies should more closely examine the interplay between control and executive brain mechanisms and the selection or suppression (inhibition) of language lexicons in bilinguals. In this section we will briefly highlight several possible lines of research related to topics of discussion.

One important remaining question is how neural control is implemented in the course of learning a new language. Although several recent studies have examined the neural mechanisms involved in word learning (Cornelissen et al., 2004; Friederici, Steinhauer, & Pfeifer, 2002; Grönholm, Rinne, Vorobyev, & Laine, 2005; Laine, 2000; McLaughlin, Osterhout, & Kim, 2004; Mestres, Rodriguez-Fornells, Münte, submitted; Mueller, 2005; Sakai, 2005), the role of cognitive control in learning a L2 has been neglected so far. For example, the revised hierarchical model of bilingual memory (Kroll & Stewart, 1994) makes specific predictions about how new words (L2) are learned. This model proposes that at the early stages of word learning, the lexical link between the new (L2) word and the corresponding L1 word is stronger than the link between the concept and the new word. With the repeated use of the new word, the conceptual link is reinforced and speakers are able to use this path freely in order to comprehend and produce words without having to rely on the lexical link (new word–L1 word). Although this model provides an interesting framework to study the word-learning problem, no studies have yet addressed the brain correlates of, for example, the consolidation of the lexical and semantic links, their brain representation, and the control mechanism that regulates the differential strengths and levels of activation of the different representations during the acquisition process.

From a different perspective, Gaskell and Dumay (2003) studied behaviorally the influence of newly learned words (nonassociated to any meaning) on the processing of an existing lexical item. Using repeated exposures of new words (e.g., *cathedruke*) over the course of 5 days, these authors studied the delay in recognizing similar existing words (e.g., *cathedral*). An interesting inhibitory effect was found on the fourth and fifth day. This study demonstrated that new words gradually develop lexical representations, which subsequently lead to lexical competition during word recognition. This type of study will prove very interesting in understanding the learning of an L2 at the lexico-semantic, morphological, and syntactical levels.

A host of other interesting issues remain to be explored. An incomplete and subjective selection might include the following: (a) the impact of implicit and explicit processes in language learning and the differential involvement of memory mechanisms (Ullman, 2001), (b) assessment of the early stages of adult language acquisition using artificial languages (Friederici et al., 2002; Gomez & Gerken, 2000; Sanders, Newport, & Neville, 2002), (c) delineation of the exact contribution of the left DLPFC in cognitive control in bilinguals (selection, switching, inhibition/suppression of the nontarget language), (d) relation of activity in the DLPFC to proficiency; (e) search for the existence of two proposed selection/inhibitory mechanisms in bilinguals (e.g., using functional connectivity measures or intracortical stimulation and recording), and (f) assessment of the relation between individual differences in executive functions and mastery of an L2.

A long-term goal of such a research program would be to understand how different learning and tutoring methods, biographical circumstances, and language attitudes impact on the process of learning an L2 (Franceschini et al., 2003; Grosjean, 1998; Klein, 1996) in order to improve the acquisition of new languages.

Notes

[1] Examples are as follows: (a) access restricted to the regions of craniotomy, (b) heterogeneity of the patient samples and limited number of patients, (c) mapping mainly reflects the disruptions on the gyral surface rather than the sulcus, which is more easily mapped using fMRI (Pouratian et al., 2000), and (d) bias in the type of patients studied (i.e., problems in generalizing to functional localizations in healthy populations).

[2] In the model of language production proposed by Levelt et al. (1999), lemmas are considered to be abstract lexical entries that mediate between the conceptual and phonological/phonetic levels and are supposed to contain all syntactically relevant properties of the word to be produced. Therefore, lemmas are defined as an entry in the mental lexicon specifying a word's syntactic properties.

[3] The *Simon task* is a well-known paradigm in experimental psychology used to examine stimulus-response interference. This task is similar to other

well-known interference paradigms, like the *Stroop* and the Eriksen flanker tasks. In the Simon paradigm, the irrelevant spatial location of an object conflicts with the response required for that object (e.g., making a left-hand response to a triangle presented in the left side of a monitor).

[4]Response inhibition, which is integrated in the executive system, is frequently evaluated using the Go/noGo paradigm. This function is necessary to suppress response tendencies that could be contextually inappropriate. The Go/noGo task always involves a type of stimuli for which the volunteer has to respond as quickly as possible (Go trials) and another type that requires a withholding of the response (noGo trials). For example, in a Go/noGo visual categorization task, participants could be instructed to make a motor response if the picture presented is an animal and to inhibit their responses if it is not an animal. With the utilization of neuroimaging techniques, it is relatively easy to analyze the brain signatures and activation related to inhibition (noGo trials) and to compare it with noninhibited trials.

[5]Difference waveforms are usually computed in order to isolate a specific ERP component (in this particular case, the N2 noGo). The underlying assumption is that only one component varies between both experimental conditions and that this component could be isolated using the subtractive procedure and performing the statistical analysis on the resulting "difference waveform" (for the limitations of this approach, see Fabiani, Gratton, & Coles, 2000).

[6]The claim that semantic representations are shared across the bilingual's two languages is still a matter of debate. Support for a shared semantic system comes basically from a large number of studies showing cross-language semantic priming. However, one of the problems is that much of the research devoted to this topic has been based on materials that refer to concrete objects (as the ones used in our study). Some authors have stated that more abstract or more ambiguous types of concept might not be sharing the same meaning representation (see Kroll & Tokowicz, 2001; Van Hell & De Groot, 1998). A word in a language could be associated with a distinct pattern of semantic features or primitives and, therefore, the resulting meaning could be slightly different across languages. In a recent study, Dong, Gui, and MacWhinney (2005) proposed a new model of the semantic bilingual memory, which tries to unify different perspectives. This model proposes the coexistence of shared and separate conceptual representations in bilinguals.

[7]The LAN is an ERP component characterized by a left frontal negativity observed between 300 and 450 ms after word onset and that has been frequently encountered in morphosyntactic violations (e.g., violations of tense, number, or gender agreement; Weyerts et al., 1997; Coulson et al., 1998). However, similar LAN-like components have been associated with verbal working memory processes (King & Kutas, 1995).

[8]Although the so-called *central* variant of Catalan pronounces the -o suffix as /u/, the remaining varieties pronounce it as /o/ and the written form is also -o.

[9] Two types of irregular verb were studied (semiregular verbs with a systematic diphthong alternation, *sentir-siento*, and verbs with idiosyncratic changes, *venir-vengo*). For the sake of simplicity, we only discuss here the results for fully irregular idiosyncratic verbs.

[10] The parahippocampal cortex is a structure in the medial temporal lobe adjacent to the hippocampus. The fusiform gyrus lies lateral to the parahippocampal and lingual gyrus and is separated from these structures by the collateral sulcus.

References

Albert, M. L., & Obler, L. K. (1978). *The bilingual brain*. New York: Academic Press.

Allison, T., Puce, A., & McCarthy, G. (2002). Category-sensitive excitatory and inhibitory processes in human extrastriate cortex. *Journal of Neurophysiology, 88*, 2864–2868.

Allison, T., Puce, A., Spencer, D. D., & McCarthy, G. (1999). Electrophysiological studies of human face perception. I: Potentials generated in occipitotemporal cortex by face and non-face stimuli. *Cerebral Cortex, 9*, 415–430.

Aron, A. R., Robbins, T. W., & Poldrack, R. A. (2004). Inhibition and the right inferior frontal cortex. *Trends in Cognitive Sciences, 8*, 170–177.

Baddeley, A., Emslie, H., Kolodny, J., & Duncan, J. (1998). Random generation and the executive control of working memory. *Quarterly Journal of Experimental Psychology Section A: Human Experimental Psychology, 51*, 819–852.

Barcelo, F., Suwazono, S., & Knight, R. T. (2000). Prefrontal modulation of visual processing in humans. *Nature Neuroscience, 3*, 399–403.

Bialystok, E. (1999). Cognitive complexity and attentional control in the bilingual mind. *Child Development, 70*, 636–644.

Bialystok, E. (2001). *Bilingualism in development: Language, literacy, and cognition*. New York: Cambridge University Press.

Bialystok, E., Craik, F. I. M., Klein, R., & Viswanathan, M. (2004). Bilingualism, aging, and cognitive control: Evidence from the Simon task. *Psychology and Aging, 19*, 290–303.

Birdsong, D., & Molis, M. (2001). On the evidence for maturational constraints in second-language acquisition. *Journal of Memory and Language, 44*, 235–249.

Bosch, L., Costa, A., & Sebastian-Galles, N. (2000). First and second language vowel perception in early bilinguals. *European Journal of Cognitive Psychology, 12*, 189–222.

Botvinick, M., Braver, T. S., Barch, D. M., Carter, C. S., & Cohen, J. D. (2001). Conflict monitoring and cognitive control. *Psychological Review*, *108*, 624–652.

Botvinick, M., Nystrom, L. E., Fissell, K., Carter, C. S., & Cohen, J. D. (1999). Conflict monitoring versus selection-for-action in anterior cingulate cortex. *Nature*, *402*, 179–181.

Buchsbaum, B. R., Greer, S., Chang, W. L., & Berman, K. F. (2005). Meta-analysis of neuroimaging studies of the Wisconsin card-sorting task and component processes. *Human Brain Mapping*, *25*, 35–45.

Bunge, S. A., Ochsner, K. N., Desmond, J. E., Glover, G. H., & Gabrieli, J. D. E. (2001). Prefrontal regions involved in keeping information in and out of mind. *Brain*, *124*, 2074–2086.

Carter, C. S., Braver, T. S., Barch, D. M., Botvinick, M. M., Noll, D., & Cohen, J. D. (1998). Anterior cingulate cortex, error detection, and the online monitoring of performance. *Science*, *280*, 747–749.

Carter, C. S., Macdonald, A. M., Botvinick, M., Ross, L. L., Stenger, V. A., Noll, D., et al. (2000). Parsing executive processes: Strategic vs. evaluative functions of the anterior cingulate cortex. *Proceedings of the National Academy of Sciences of the United States of America*, *97*, 1944–1948.

Chao, L. L., & Knight, R. T. (1998). Contribution of human prefrontal cortex to delay performance. *Journal of Cognitive Neuroscience*, *10*, 167–177.

Chee, M. W. L., Caplan, D., Soon, C. S., Sriram, N., Tan, E. W. L., Thiel, T., et al. (1999). Processing of visually presented sentences in Mandarin and English studied with fMRI. *Neuron*, *23*, 127–137.

Chee, M. W. L., Hon, N., Lee, H. L., & Soon, C. S. (2001). Relative language proficiency modulates BOLD signal change when bilinguals perform semantic judgments. *NeuroImage*, *13*, 1155–1163.

Chee, M. W. L., Soon, C. S., & Lee, H. L. (2003). Common and segregated neuronal networks for different languages revealed using functional magnetic resonance adaptation. *Journal of Cognitive Neuroscience*, *15*, 85–97.

Chee, M. W. L., Tan, E. W. L., & Thiel, T. (1999). Mandarin and English single word processing studied with functional magnetic resonance imaging. *Journal of Neuroscience*, *19*, 3050–3056.

Chee, M. W. L., Weekes, B., Lee, K. M., Soon, C. S., Schreiber, A., Hoon, J. J., et al. (2000). Overlap and dissociation of semantic processing of Chinese characters, English words, and pictures: Evidence from fMRI. *NeuroImage*, *12*, 392–403.

Cohen, J. D., Botvinick, M., & Carter, C. S. (2000). Anterior cingulate and prefrontal cortex: Who's in control? *Nature Neuroscience*, *3*, 421–423.

Colomé, À. (2001). Lexical activation in bilinguals' speech production: Language-specific or language-independent? *Journal of Memory and Language, 45,* 721–736.

Corbetta, M., Miezin, F. M., Dobmeyer, S., Shulman, G. L., & Petersen, S. E. (1990). Attentional modulation of neural processing of shape, color, and velocity in humans. *Science, 248,* 1556–1559.

Cornelissen, K., Laine, M., Renvall, K., Saarinen, T., Martin, N., & Salmelin, R. (2004). Learning new names for new objects: Cortical effects as measured by magnetoencephalography. *Brain and Language, 89,* 617–622.

Costa, A. (2005). Lexical access in bilingual production. In J. F. Kroll & A. M. B. de Groot (Eds.), *Handbook of bilingualism: Psycholinguistic approaches* (pp. 308–325). New York: Oxford University Press.

Costa, A., Caramazza, A., & Sebastian-Galles, N. (2000). The cognate facilitation effect: Implications for models of lexical access. *Journal of Experimental Psychology: Learning Memory and Cognition, 26,* 1283–1296.

Costa, A., Miozzo, M., & Caramazza, A. (1999). Lexical selection in bilinguals: Do words in the bilingual's two lexicons compete for selection? *Journal of Memory and Language, 41,* 365–397.

Costa, A., & Santesteban, M. (2004). Lexical access in bilingual speech production: Evidence from language switching in highly proficient bilinguals and L2 learners. *Journal of Memory and Language, 50,* 491–511.

Coulson, S., King, J. W., & Kutas, M. (1998). Expect the unexpected: Event-related brain response to morphosyntactic violations. *Language and Cognitive Processes, 13,* 21–58.

Curtis, C. E., & D'Esposito, M. (2003). Persistent activity in the prefrontal cortex during working memory. *Trends in Cognitive Sciences, 7,* 415–423.

De Bot, K. (1992). A bilingual production model: Levelt's "speaking" model adapted. *Applied Linguistics, 13,* 1–24.

De Diego Balaguer, R., Rodriguez-Fornells, A., Rotte, M., Bahlmann, J., Heinze, H. J., & Münte, T. F. (in press). Neural circuits subserving the retrieval of stems and grammatical features in regular and irregular verbs. *Human Brain Mapping.*

De Diego Balaguer, R., Sebastian-Galles, N., Diaz, B., & Rodriguez-Fornells, A. (2005). Morphological processing in early bilinguals: An ERP study of regular and irregular verb processing. *Cognitive Brain Research, 25,* 312–327.

Dehaene, S., Dupoux, E., Mehler, J., Cohen, L., Paulesu, E., Perani, D., et al. (1997). Anatomical variability in the cortical representation of first and second language. *NeuroReport, 8,* 3809–3815.

Dell, G. S. (1986). A spreading-activation theory of retrieval in sentence production. *Psychological Review, 93,* 283–321.

Desimone, R., & Duncan, J. (1995). Neural mechanisms of selective visual attention. *Annual Review of Neuroscience, 18,* 193–222.

D'Esposito, M., Detre, J. A., Alsop, D. C., Shin, R. K., Atlas, S., & Grossman, M. (1995). The neural basis of central execution systems of working memory. *Nature, 378,* 279–281.

Diamond, A. (2002). Normal development of prefrontal cortex from birth to young adulthood: Cognitive functions, anatomy, and biochemistry. In D. T. Stuss & R. T. Knight (Eds.), *Principles of frontal lobe function* (pp. 466–503). New York: Oxford University Press.

Dijkstra, T., & Van Heuven, W. J. B. (1998). The BIA model and bilingual word recognition. In J. Grainger & A. M. Jacobs (Eds.), *Localist connectionist approaches to human cognition* (pp. 189–225). Mahwah, NJ: Erlbaum.

Dijkstra, T., & Van Heuven, W. J. B. (2002). The architecture of the bilingual word recognition system: From identification to decision. *Bilingualism: Language and Cognition, 5,* 175–197.

Dijkstra, T., van Jaarsveld, H., & Brinke, S. (1998). Interlingual homograph recognition: Effects of task demands and language intermixing. *Bilingualism: Language and Cognition, 1,* 51–66.

Donchin, E. (1981). Surprise... surprise. *Psychophysiology, 18,* 493–513.

Donchin, E., & Coles, M. G. H. (1988). Is the P300 component a manifestation of context updating. *Behavioral and Brain Sciences, 11,* 357–374.

Donchin, E., & Coles, M. G. H. (1998). Context updating and the P300. *Behavioral and Brain Sciences, 21,* 152–154.

Dong, Y., Gui, S. S., & MacWhinney, B. (2005). Shared and separate meanings in the bilingual mental lexicon. Bilingualism: *Language and Cognition, 8,* 221–238.

Donkers, F. C. L., & Van Boxtel, G. J. M. (2004). The N2 in go/no-go tasks reflects conflict monitoring not response inhibition. *Brain and Cognition, 56,* 165–176.

Dove, A., Pollmann, S., Schubert, T., Wiggins, C. J., & Von Cramon, D. Y. (2000). Prefrontal cortex activation in task switching: An event-related fMRI study. *Cognitive Brain Research, 9,* 103–109.

Dreher, J. C., Koechlin, E., Ali, S. O., & Grafman, J. (2002). The roles of timing and task order during task switching. *NeuroImage, 17,* 95–109.

Fabbro, F. (2001). The bilingual brain: Cerebral representation of languages. *Brain and Language, 79,* 211–222.

Fabbro, F., Skrap, M., & Aglioti, S. (2000). Pathological switching between languages after frontal lesions in a bilingual patient. *Journal of Neurology Neurosurgery and Psychiatry, 68,* 650–652.

Fabiani, M., Gratton, G., & Coles, M. G. H. (2000). Event-related brain potentials. Methods, theory, and applications. In J. T. Cacioppo, L. G. Tassinary,

& G. G. Berntson (Eds.), *Handbook of psychophysiology* (2nd ed., pp. 53–84). Cambridge: Cambridge University Press.

Flege, J. E., Yeni-Komshian, G. H., & Liu, S. (1999). Age constraints on second-language acquisition. *Journal of Memory and Language, 41*, 78–104.

Franceschini, R., Zappatore, D., & Nitsch, C. (2003). Lexicon in the brain: What neurobiology has to say about languages. In J. Cenoz, B. Hufeisen, & U. Jessner (Eds.), *Multilingual lexicon* (pp. 153–166). Dordrecht: Kluwer Academic Publishers.

Friederici, A. D., Steinhauer, K., & Pfeifer, E. (2002). Brain signatures of artificial language processing: Evidence challenging the critical period hypothesis. *Proceedings of the National Academy of Sciences of the United States of America, 99*, 529–534.

Frith, C. D., Friston, K. J., Liddle, P. F., & Frackowiak, R. S. (1991). A PET study of word finding. *Neuropsychologia, 29*, 1137–1148.

Fuster, J. M. (1989). *The prefrontal cortex: Anatomy, physiology and neuropsychology of the frontal lobe.* New York: Raven Press.

Garavan, H., Ross, T. J., Li, S. J., & Stein, E. A. (2000). A parametric manipulation of central executive functioning. *Cerebral Cortex, 10*, 585–592.

Garavan, H., Ross, T. J., & Stein, E. A. (1999). Right hemispheric dominance of inhibitory control: An event-related functional MRI study. *Proceedings of the National Academy of Sciences of the United States of America, 96*, 8301–8306.

Gaskell, M. G., & Dumay, N. (2003). Lexical competition and the acquisition of novel words. *Cognition, 89*, 105–132.

Gazzaley, A., Cooney, J. W., McEvoy, K., Knight, R. T., & D'Esposito, M. (2005). Top-down enhancement and suppression of the magnitude and speed of neural activity. *Journal of Cognitive Neuroscience, 17*, 507–517.

Gehring, W. J., Bryck, R. L., Jonides, J., Albin, R. L., & Badre, D. (2003). The mind's eye, looking inward? In search of executive control in internal attention shifting. *Psychophysiology, 40*, 572–585.

Gehring, W. J., Gross, B., Coles, M. G. H., Meyer, D. E., & Donchin, E. (1993). A neural system for error detection and compensation. *Psychological Science, 4*, 385–390.

Gemba, H., & Sasaki, K. (1989). Potential related to no-go reaction of go/no-go hand movement task with color discrimination in human. *Neuroscience Letters, 101*, 263–268.

Gerard, L. D., & Scarbourough, D. L. (1989). Language-specific lexical access of homographs in bilinguals. *Journal of Experimental Psychology: Learning Memory and Cognition, 15*, 305–315.

Gomez, R. L., & Gerken, L. (2000). Infant artificial language learning and language acquisition. *Trends in Cognitive Sciences, 4,* 178–186.

Green, D. W. (1986). Control, activation, and resource: A framework and a model for the control of speech in bilinguals. *Brain and Language, 27,* 210–223.

Green, D. W. (1998). Mental control of the bilingual lexico-semantic system. *Bilingualism: Language and Cognition, 1,* 67–81.

Grosjean, F. (1988). Exploring the recognition of guest words in bilingual speech. *Language and Cognitive Processes, 3,* 233–274.

Grosjean, F. (1997). Processing mixed languages: Issues, findings and models. In de A. M. B. de Groot & J. F. Kroll (Eds.), *Tutorials in bilingualism: Psycholinguistic perspectives* (pp. 225–254). Mahwah, NJ: Erlbaum.

Grosjean, F. (1998). Studying bilinguals: Methodological and conceptual issues. *Bilingualism: Language and Cognition, 1,* 131–149.

Grosjean, F., & Miller, J. L. (1994). Going in and out of languages: An example of bilingual flexibility. *Psychological Science, 5,* 201–206.

Grönholm, P., Rinne, J. O., Vorobyev, V., & Laine, M. (2005). Learning to name unfamiliar objects: A PET activation study. *Cognitive Brain Research, 25,* 359–371.

Hadland, K. A., Rushworth, M. F. S., Passingham, R. E., Jahanshahi, M., & Rothwell, J. C. (2001). Interference with performance of a response selection task that has no working memory component: An rTMS comparison of the dorsolateral prefrontal and medial frontal cortex. *Journal of Cognitive Neuroscience, 13,* 1097–1108.

Hagoort, P. (2005). On Broca, brain, and binding: A new framework. *Trends in Cognitive Sciences, 9,* 416–423.

Hahne, A., Mueller, J. L., & Clahsen, H. (2006). Morphological processing in a second language: Behavioral and event-related brain potential evidence for storage and decomposition. *Journal of Cognitive Neuroscience, 18,* 121–134.

Hasegawa, M., Carpenter, P. A., & Just, M. A. (2002). An fMRI study of bilingual sentence comprehension and workload. *NeuroImage, 15,* 647–660.

Hermans, D., Bongaerts, T., De Bot, K., & Schreuder, R. (1998). Producing words in a foreign language: Can speakers prevent interference from their first language? *Bilingualism: Language and Cognition, 1,* 213–229.

Hernandez, A. E., Dapretto, M., Mazziotta, J., & Bookheimer, S. (2001). Language switching and language representation in Spanish-English bilinguals: An fMRI study. *NeuroImage, 14,* 510–520.

Hernandez, A. E., Martinez, A., & Kohnert, K. (2000). In search of the language switch: An fMRI study of picture naming in Spanish-English bilinguals. *Brain and Language, 73,* 421–431.

Hillyard, S. A., Hink, R. F., Schwent, V. L., & Picton, T. W. (1973). Electrical signs of selective attention in human brain. *Science, 182,* 171–180.

Holroyd, C. B., & Coles, M. G. (2002). The neural basis of human error processing: Reinforcement learning, dopamine, and the error-related negativity. *Psychological Review, 109,* 679–709.

Holtzheimer, P., Fawaz, W., Wilson, C., & Avery, D. (2005). Repetitive transcranial magnetic stimulation may induce language switching in bilingual patients. *Brain and Language, 94,* 274–277.

Illes, J., Francis, W. S., Desmond, J. E., Gabrieli, J. D. E., Glover, G. H., Poldrack, et al. (1999). Convergent cortical representation of semantic processing in bilinguals. *Brain and Language, 70,* 347–363.

Jansma, B. M., Rodriguez-Fornells, A., Möller, J., & Münte, T. F. (2004). Electrophysiological studies of speech production. In T. Pechmann & C. Habel (Eds.), *Multidisciplinary approaches to language production (Trends in Linguistics)* (pp. 361–396). Berlin: Mouton de Gruyter.

Jescheniak, J. D., & Levelt, W. J. M. (1994). Word-frequency effects in speech production: Retrieval of syntactic information and of phonological form. *Journal of Experimental Psychology: Learning Memory and Cognition, 20,* 824–843.

Johnson, J. S., & Newport, E. L. (1991). Critical period effects on universal properties of language: The status of subjacency in the acquisition of a second language. *Cognition, 39,* 215–258.

Kaan, E., Harris, A., Gibson, E., & Holcomb, P. (2000). The P600 as an index of syntactic integration difficulty. *Language and Cognitive Processes, 15,* 159–201.

Kim, K. H. S., Relkin, N. R., Lee, K. M., & Hirsch, J. (1997). Distinct cortical areas associated with native and second languages. *Nature, 388,* 171–174.

King, J. W., & Kutas, M. (1995). Who did what and when: Using word-level and clause-level ERPs to monitor working-memory usage in reading. *Journal of Cognitive Neuroscience, 7,* 376–395.

Klein, D., Milner, B., Zatorre, R. J., Zhao, V., & Nikelski, J. (1999). Cerebral organization in bilinguals: A PET study of Chinese-English verb generation. *NeuroReport, 10,* 2841–2846.

Klein, W. (1996). Language acquisition at different ages. In D. Magnusson (Ed.), *The lifespan development of individuals: Behavioral, neurobiological, and psychosocial perspectives* (pp. 244–264). Cambridge: Cambridge University Press.

Knight, R. T., Scabini, D., & Woods, D. L. (1989). Prefrontal cortex gating of auditory transmission in humans. *Brain Research, 504,* 338–342.

Knight, R. T., & Stuss, D. T. (2002). Prefrontal cortex: The present and the future. In R. T. Knight & D. T. Stuss (Eds.), *Principles of frontal lobe function* (pp. 573–598). New York: Oxford University Press.

Kok, A. (1986). Effects of degradation of visual stimulation on components of the event-related potential (ERP) in go/nogo reaction tasks. *Biological Psychology, 23*, 21–38.

Konishi, S., Nakajima, K., Uchida, I., Kikyo, H., Kameyama, M., & Miyashita, Y. (1999). Common inhibitory mechanism in human inferior prefrontal cortex revealed by event-related functional MRI. *Brain, 122*, 981–991.

Konishi, S., Nakajima, K., Uchida, I., Sekihara, K., & Miyashita, Y. (1998). No-go dominant brain activity in human inferior prefrontal cortex revealed by functional magnetic resonance imaging. *European Journal of Neuroscience, 10*, 1209–1213.

Kroll, J. F., Dijkstra, T., Janssen, N., & Schriefers, H. (2000). Selecting the language in which to speak: Experiments on lexical access in bilingual production. Paper presented at the 41st Annual Meeting of the Psychonomic Society, New Orleans, LA.

Kroll, J. F., & Stewart, E. (1994). Category interference in translation and picture naming: Evidence for asymmetric connections between bilingual memory representations. *Journal of Memory and Language, 33*, 149–174.

Kroll, J. F., & Tokowicz, N. (2001). The development of conceptual representation for words in a second language. In J. L. Nicol (Ed.), *One mind, two languages: Bilingual language processing* (pp. 49–71). Malden, MA: Blackwell.

Laine, M. (2000). The learning brain. *Brain and Language, 71*, 132–134.

Levelt, W. J. M., Roelofs, A., & Meyer, A. S. (1999). A theory of lexical access in speech production. *Behavioral and Brain Sciences, 22*, 1–38.

Levelt, W. J. M., Schriefers, H., Vorberg, D., Meyer, A. S., Pechmann, T., & Havinga, J. (1991). The time course of lexical access in speech production: A study of picture naming. *Psychological Review, 98*, 122–142.

Li, P. (1996). Spoken word recognition of code-switched words by Chinese-English bilinguals. *Journal of Memory and Language, 35*, 757–774.

Li, P. (1998). Mental control, language tags, and language nodes in bilingual lexical processing. *Bilingualism: Language and Cognition, 1*, 93.

Li, P., & Farkas, I. (2002). A self-organizing connectionist model of bilingual processing. In R. Heredia & J. Altarriba (Eds.), *Bilingual sentence processing* (pp. 59–85). North-Holland: Elsevier Science Publishers.

Liddle, P. F., Kiehl, K. A., & Smith, A. M. (2001). Event-related fMRI study of response inhibition. *Human Brain Mapping, 12*, 100–109.

Lucas, T. H., McKhann, G. M., & Ojemann, G. A. (2004). Functional separation of languages in the bilingual brain: A comparison of electrical stimulation language mapping in 25 bilingual patients and 117 monolingual control patients. *Journal of Neurosurgery, 101,* 449–457.

Luria, A. R. (1973). *The working brain.* London: Penguin Press.

MacDonald, A. W., Cohen, J. D., Stenger, V. A., & Carter, C. S. (2000). Dissociating the role of the dorsolateral prefrontal and anterior cingulate cortex in cognitive control. *Science, 288,* 1835–1838.

Macnamara, J., & Kushnir, S. L. (1971). Linguistic independence of bilinguals: The input switch. *Journal of Verbal Learning and Verbal Behavior, 10,* 480–487.

Matzke, M., Mai, H., Nager, W., Rüsseler, J., & Münte, T. (2002). The costs of freedom: An ERP-study of non-canonical sentences. *Clinical Neurophysiology, 113,* 844–852.

McLaughlin, J., Osterhout, L., & Kim, A. (2004). Neural correlates of second-language word learning: Minimal instruction produces rapid change. *Nature Neuroscience, 7,* 703–704.

Menon, V., Adleman, N. E., White, C. D., Glover, G. H., & Reiss, A. L. (2001). Error-related brain activation during a Go/NoGo response inhibition task. *Human Brain Mapping, 12,* 131–143.

Mestres, A., Rodriguez-Fornells, A., & Münte, T. F. (submitted). Watching the brain during meaning acquisition.

Meuter, R. F. I., & Allport, A. (1999). Bilingual language switching in naming: Asymmetrical costs of language selection. *Journal of Memory and Language, 40,* 25–40.

Miller, E. K., & Cohen, J. D. (2001). An integrative theory of prefrontal cortex function. *Annual Review of Neuroscience, 24,* 167–202.

Moran, J., & Desimone, R. (1985). Selective attention gates visual processing in the extrastriate cortex. *Science, 229,* 782–784.

Moreno, E. M., Federmeier, K. D., & Kutas, M. (2002). Switching languages, switching Palabras (words): An electrophysiological study of code switching. *Brain and Language, 80,* 188–207.

Moreno, E. M., & Kutas, M. (2005). Processing semantic anomalies in two languages: An electrophysiological exploration in both languages of Spanish-English bilinguals. *Cognitive Brain Research, 22,* 205–220.

Mueller, J. L. (2005). *Mechanisms of auditory sentence comprehension in first and second language: An electrophysiological miniature grammar study.* Unpublished doctoral dissertation, Max Planck Institute for Human Cognitive and Brain Sciences, Leipzig.

Münte, T. F., Rodriguez-Fornells, A., & Kutas, M. (1999). One, two or many mechanisms? The brain's processing of complex words. *Behavioral and Brain Sciences, 22*, 1031–1032.

Nieuwenhuis, S., Yeung, N., Van den Wildenberg, W., & Ridderinkhof, K. R. (2003). Electrophysiological correlates of anterior cingulate function in a go/no-go task: Effects of response conflict and trial type frequency. *Cognitive, Affective, & Behavioral Neuroscience, 3*, 17–26.

Norman, D. A., & Shallice, T. (1986). Attention to action: Willed and automatic control of behavior. In R. J. Davidson, G. E. Schwartz, & D. Shapiro (Eds.), *Consciousness and self-regulation* (pp. 1–18). New York: Plenum.

Ojemann, G. A., & Whitaker, H. A. (1978). The bilingual brain. *Archives of Neurology, 35*, 409–412.

Osterhout, L., McLaughlin, J., Kim, A., Greenwald, R., & Inoue, K. (2004). Sentences in the brain: Event-related potentials as real-time reflections of sentence comprehension and language learning. In M. Carreiras & C. Clifton, *The on-line study of sentence comprehension eyetracking, ERPs and beyond* (pp. 271–308). New York: Psychology Press.

Pallier, C., Colomé, À., & Sebastian-Galles, N. (2001). The influence of native-language phonology on lexical access: Exemplar-based vs. abstract lexical entries. *Psychological Science, 12*, 445–449.

Pandya, D. N., & Barnes, C. L. (1987). Architecture and connections of the frontal lobe. In E. Perecman (Ed.), *The frontal lobes revisited* (pp. 41–72). New York: IRBN Press.

Pandya, D. N., & Yeterian, E. H. (1990). Prefrontal cortex in relation to other cortical areas in rhesus monkey-architecture and connections. *Progress in Brain Research, 85*, 63–94.

Paradis, M. (1989). Bilingual and polyglot aphasia. In F. Boller & J. Grafman (Eds.), *Handbook of neuropsychology* (pp. 117–140). Amsterdam: Elsevier.

Paradis, M. (1995). *Aspects of bilingual aphasia*. Tarrytown, NY: Pergamon.

Pardo, J. V., Pardo, P. J., Janer, K. W., & Raichle, M. E. (1990). The anterior cingulate cortex mediates processing selection in the stroop attentional conflict paradigm. *Proceedings of the National Academy of Sciences of the United States of America, 87*, 256–259.

Penfield, W., & Roberts, L. (1959). *Speech and brain mechanisms*. Princeton, NJ: Princeton University Press.

Perani, D., & Abutalebi, J. (2005). The neural basis of first and second language processing. *Current Opinion in Neurobiology, 15*, 202–206.

Perani, D., Paulesu, E., Galles, N. S., Dupoux, E., Dehaene, S., Bettinardi, V., et al. (1998). The bilingual brain: Proficiency and age of acquisition of the second language. *Brain, 121*, 1841–1852.

Pessoa, L., Kastner, S., & Ungerleider, L. G. (2003). Neuroimaging studies of attention: From modulation of sensory processing to top-down control. *Journal of Neuroscience, 23,* 3990–3998.

Peterson, R., & Savoy, P. (1998). Lexical selection and phonological encoding during language production: Evidence for cascaded processing. *Journal of Experimental Psychology: Learning Memory and Cognition, 24,* 539–557.

Pfefferbaum, A., Ford, J. M., Weller, B. J., & Kopell, B. S. (1985). ERPs to response production and inhibition. *Electroencephalography and Clinical Neurophysiology, 60,* 423–434.

Poldrack, R. A., & Wagner, A. D. (2004). What can neuroimaging tell us about the mind? Insights from prefrontal cortex. *Current Directions in Psychological Science, 13,* 177–181.

Portin, M., & Laine, M. (2001). Processing cost associated with inflectional morphology in bilingual speakers. *Bilingualism: Language and Cognition, 4,* 55–62.

Poulisse, N. (1999). *Slips of the tongue: Speech errors in first and second language production.* Amsterdam: Benjamins.

Poulisse, N., & Bongaerts, T. (1994). First language use in second language production. *Applied Linguistics, 15,* 36–57.

Pouratian, N., Bookheimer, S. Y., O'Farrell, A. M., Sicotte, N. L., Cannestra, A. F., Becker, D., et al. (2000). Optical imaging of bilingual cortical representations: Case report. *Journal of Neurosurgery, 93,* 676–681.

Price, C. J., Green, D. W., & Von Studnitz, R. (1999). A functional imaging study of translation and language switching. *Brain, 122,* 2221–2235.

Rodriguez-Fornells, A., Lutz, T., & Münte, T. F. (submitted). Syntactic interference in bilingual naming performance is modulated by language switching: An electrophysiological study.

Rodriguez-Fornells, A., Rotte, M., Heinze, H. J., Nösselt, T., & Münte, T. F. (2002). Brain potential and functional MRI evidence for how to handle two languages with one brain. *Nature, 415,* 1026–1029.

Rodriguez-Fornells, A., Schmitt, B. M., Kutas, M., & Münte, T. F. (2002). Electrophysiological estimates of the time course of semantic and phonological encoding during listening and naming. *Neuropsychologia, 40,* 778–787.

Rodriguez-Fornells, A., Van der Lugt, A., Rotte, M., Britti, B., Heinze, H. J., & Münte, T. F. (2005). Second language interferes with word production in fluent bilinguals: Brain potential and functional imaging evidence. *Journal of Cognitive Neuroscience, 17,* 422–433.

Roelofs, A. (1998). Lemma selection without inhibition of languages in bilingual speakers. *Bilingualism: Language and Cognition, 1,* 94–95.

Rogers, R. D., Sahakian, B. J., Hodges, J. R., Polkey, C. E., Kennard, C., & Robbins, T. W. (1998). Dissociating executive mechanisms of task control following frontal lobe damage and Parkinson's disease. *Brain, 121*, 815–842.

Roux, F. E., & Trémoulet, M. (2002). Organization of language areas in bilingual patients: A cortical stimulation study. *Journal of Neurosurgery, 97*, 857–864.

Rubia, K., Russell, T., Overmeyer, S., Brammer, M. J., Bullmore, E. T., Sharma, T., et al. (2001). Mapping motor inhibition: Conjunctive brain activations across different versions of go/no-go and stop tasks. *NeuroImage, 13*, 250–261.

Sakai, K. L. (2005). Language acquisition and brain development. *Science, 310*, 815–819.

Sanders, L. D., Newport, E. L., & Neville, H. J. (2002). Segmenting nonsense: An event-related potential index of perceived onsets in continuous speech. *Nature Neuroscience, 5*, 700–703.

Sasaki, K., & Gemba, H. (1993). Prefrontal cortex in the organization and control of voluntary movement. In T. Ono, L. R. Squire, M. E. Raichle, D. I. Perrett, & M. Fukuda (Eds.), *Brain mechanisms of perception and memory: From neuron to behavior* (pp. 473–496). New York: Oxford University Press.

Sasaki, K., Gemba, H., Nambu, A., & Matsuzaki, R. (1993). No-go activity in the frontal association cortex of human subjects. *Neuroscience Research, 18*, 249–252.

Sasaki, K., Gemba, H., & Tsujimoto, T. (1989). Suppression of visually initiated hand movement by stimulation of the prefrontal cortex in the monkey. *Brain Research, 495*, 100–107.

Schmitt, B. M., Münte, T. F., & Kutas, M. (2000). Electrophysiological estimates of the time course of semantic and phonological encoding during implicit picture naming. *Psychophysiology, 37*, 473–484.

Schmitt, B. M., Rodriguez-Fornells, A., Kutas, M., & Münte, T. F. (2001). Electrophysiological estimates of semantic and syntactic information access during tacit picture naming and listening to words. *Neuroscience Research, 41*, 293–298.

Schmitt, B. M., Schiltz, K., Zaake, W., Kutas, M., & Münte, T. F. (2001). An electrophysiological analysis of the time course of conceptual and syntactic encoding during tacit picture naming. *Journal of Cognitive Neuroscience, 13*, 510–522.

Schriefers, H., & Teruel, E. (2000). Grammatical gender in noun phrase production: The gender interference effect in German. *Journal of Experimental Psychology: Learning Memory and Cognition, 26*, 1368–1377.

Sebastian-Galles, N., Echevarria, S., & Bosch, L. (2005). The influence of initial exposure on lexical representation: Comparing early and simultaneous bilinguals. *Journal of Memory and Language, 52*, 240–255.

Sebastian-Galles, N., Rodriguez-Fornells, A., De Diego Balaguer, R., & Diaz, B. (in press). First and second language phonological representations in the mental lexicon. *Journal of Cognitive Neuroscience.*

Simson, R., Vaughan, H. G., & Ritter, W. (1977). Scalp topography of potentials in auditory and visual-discrimination tasks. *Electroencephalography and Clinical Neurophysiology, 42*, 528–535.

Stemberger, J. P. (1985). An interactive activation model of language production. In A. W. Ellis (Ed.), *Progress in the psychology of language* (pp. 143–186). Hove, UK: Erlbaum.

Strafella, A. P., & Paus, T. (2001). Cerebral blood-flow changes induced by paired-pulse transcranial magnetic stimulation of the primary motor cortex. *Journal of Neurophysiology, 85*, 2624–2629.

Thorpe, S., Fize, D., & Marlot, C. (1996). Speed of processing in the human visual system. *Nature, 381*, 520–522.

Ullman, M. T. (2001). The neural basis of lexicon and grammar in first and second language: The declarative/procedural model. *Bilingualism: Language and Cognition, 4*, 105–122.

Ullsperger, M., & Von Cramon, D. Y. (2001). Subprocesses of performance monitoring: A dissociation of error processing and response competition revealed by event-related fMRI and ERPs. *NeuroImage, 14*, 1387–1401.

Van der Lugt, A., Banfield, J., Osinsky, R., & Münte, T. F. (submitted). Brain potentials show extremely rapid activation of social stereotypes.

Van Hell, J. G., & De Groot, A. M. B. (1998). Conceptual representation in bilingual memory: Effects of concreteness and cognate status in word association. *Bilingualism: Language and Cognition, 3*, 193–211.

Van Turennout, M., Hagoort, P., & Brown, C. M. (1997). Electrophysiological evidence on the time course of semantic and phonological processes in speech production. *Journal of Experimental Psychology: Learning Memory and Cognition, 23*, 787–806.

Van Turennout, M., Hagoort, P., & Brown, C. M. (1998). Brain activity during speaking: From syntax to phonology in 40 milliseconds. *Science, 280*, 572–574.

Wager, T. D., Ching-Yune, C. S., Lacey, S. C., Nee, D. E., Franklin, M., & Jonides, J. (2005). Common and unique components of response inhibition revealed by fMRI. *NeuroImage, 27*, 323–340.

Walker, J. A., Quinones-Hinojosa, A., & Berger, M. S. (2004). Intraoperative speech mapping in 17 bilingual patients undergoing resection of a mass lesion. *Neurosurgery, 54*, 113–117.

Weber-Fox, C. M., & Neville, H. J. (1996). Maturational constraints on functional specialization for language processing: ERP and behavioral evidence in bilingual speakers. *Journal of Cognitive Neuroscience, 8,* 231–256.

Weyerts, H., Penke, M., Dohrn, U., Clahsen, H., & Münte, T. F. (1997). Brain potentials indicate differences between regular and irregular German plurals. *NeuroReport, 8,* 957–962.

Wheeldon, L. R., & Levelt, W. J. M. (1995). Monitoring the time course of phonological encoding. *Journal of Memory and Language, 34,* 311–334.

Yamaguchi, S., & Knight, R. T. (1990). Gating of somatosensory input by human prefrontal cortex. *Brain Research, 521,* 281–288.

Yeung, N., Botvinick, M. M., & Cohen, J. D. (2004). The neural basis of error detection: Conflict monitoring and the error-related negativity. *Psychological Review, 111,* 931–959.

On Language and the Brain—Or on (Psycho)linguists and Neuroscientists? Commentary on Rodriguez-Fornells et al.

Ton Dijkstra
Radboud University Nijmegen

Walter van Heuven
University of Nottingham

Experimental psychologists have long sought ways to incorporate effects of task demands, decision-making, and cognitive control in their models of cognitive processing (Dijkstra & De Smedt, 1996, p. 10). Nevertheless, most computational models still consist of representational networks in which executive functions play only a limited role. This situation also holds for language processing in monolinguals and bilinguals. Rodriguez-Fornells, Rotte, Heinze, Nösselt, and Münte (2002) showed that cognitive neuroscience might offer exciting insights into the relationship between executive control and the processing of multiple languages.

In our view, however, they do not make optimal use of psycholinguistic views based on behavioral studies. This is evident in several respects. The authors defined cognitive control in the context of bilingual production as the regulation of language use.

Ton Dijkstra, Nijmegen Institute for Cognition and Information, Radboud University Nijmegen, The Netherlands; Walter van Heuven, School of Psychology, University of Nottingham, UK.

Correspondence concerning this article should be addressed to Ton Dijkstra, Nijmegen Institute for Cognition and Information, Radboud University Nijmegen, P.O. Box 9104, 6500 HE Nijmegen, The Netherlands. Internet: t.dijkstra@nici.ru.nl

However, for psychologists, "executive control" is not a unitary concept. Bialystok et al. (2005) pointed out that executive control relates to various cognitive functions like the ability to focus attention, inhibit irrelevant representations, monitor ongoing activities, use working memory, and regulate switching behavior. All of these must be relevant in bilingual processing.

Furthermore, to study executive control issues, the authors argue that "brain imaging should not only be used to test predictions from psycholinguistic models but that patterns of activations might be used to derive brain-inspired hypotheses about processing differences." In support, they cite Poldrack and Wagner (2004). Although we agree that "brain storming" about activation patterns in the brain can serve as a useful source of hypotheses, we consider it a bad idea to just let the "data speak for themselves." This does not work for reaction time (RT) data and will not work for brain data either. In our view, hypothesizing is most fruitful and focused in light of a functional model or theory. Observing activity in a particular brain area only makes sense given assumptions about the causes of this activity and its relation to cognitive operations (Harley, 2004, p. 11). As Poldrack and Wagner stated: "Indeed, well-designed neuroimaging studies intended to adjucate between competing psychological hypotheses that, themselves, have been formally described are precisely the kind of studies in which the reverse-inference approach may be justified" (p. 180).

In a similar vein as in the present article, Münte and Rodriguez-Fornells (in Grosjean, Li, Münte, & Rodriguez-Fornells, 2003, pp. 161–162) argued that a collaboration between cognitive neuroscientists and psycholinguists "should not be a one-way street with neuroscientists proving theories devised by language scientists." We have two comments on this statement.

First, why would neuroscientists reinvent the cogs of cognitive science? Functional models in cognitive psychology are often based on decades of research. For instance, they have collected overwhelming evidence in favor of language nonselective lexical access (i.e., parallel activation of words from different languages during speaking, reading, and listening). This finding in particular justifies investigating bilingual executive control. If language-*specific* access held, only a simple kind of language switch would be needed.

Second, our functional models might have been built on RT studies (which is only partly true!), but they are abstract in nature and are not limited to/by such data. Rodriguez-Fornells et al. state: "We hypothesized that if the language-selection mechanism acts at the lemma level, no phonological activation should be observed for the nontarget language" (p. 146). This hypothesis is not brain based or literature independent; it was formulated and tested in earlier RT studies (Colomé, 2001; Kroll, Dijkstra, Janssen, & Schriefers, 1999; Roelofs, 2003). In many instances, behavioral studies preceded the electrophysiological or neuroimaging studies in time, but they are mentioned later in the article, if at all.

Of course, we do not deny that functional models of bilingual processing *can* profit considerably from studies measuring event-related brain potentials (ERPs; temporal data) and functional magnetic resonance imaging (fMRI; spatial data). For instance, they could clarify how later decision processes affect earlier language processing, to what extent executive and language processing functions are subserved by the same systems, and whether cognitive control affects relative language activation.

To study executive function in bilinguals, we ourselves have proposed and followed a systematic three-pronged approach. On the basis of diverse RT studies, we have built and adapted a computational model of bilingual word recognition: the Bilingual Interactive Activation + (BIA+) model (Dijkstra & Van Heuven, 1998, 2002; Van Heuven, Dijkstra, & Grainger, 1998). It consists of a word identification system and an executive control (task/decision) system. We have tested and extended this functional model to include brain data by measuring ERPs (De Bruijn, Dijkstra, Chwilla, & Schriefers, 2001; Dijkstra, Van Hell, & Brenders, in preparation; Kerkhofs, Dijkstra, Chwilla, & De Bruijn, 2006) and fMRI (Van Heuven, Schriefers, Dijkstra, &

Hagoort, submitted). Predictions for the ERP and fMRI studies were based on the functional model.

Kerkhofs et al. (2006) tested the BIA+ model equipped with a bilingual semantic priming component in RT and ERP experiments. Dutch-English bilinguals performed an English lexical decision task in which interlingual homographs like *stem* (meaning "voice" in Dutch) were preceded by primes like *root* or *fool* that were semantically related or unrelated to the English target reading. Homographs were responded to faster following semantically related primes than following unrelated primes. As predicted, responses were modulated by the relative frequencies of homograph readings: They were faster when English word frequency was high or Dutch word frequency was low. In the ERPs, N400 effects, indicative of semantic integration, were found for interlingual homographs preceded by related primes. The amplitude of the N400 effect was modulated by relative word frequency in the first and second language. The effects of Dutch and English frequency on the ERPs were in opposite directions.

Van Heuven et al. (submitted) used RT and fMRI measurements to test the BIA+ model's prediction of language conflict when bilinguals read interlingual homographs. The BIA+ model predicts stimulus-based and response-based language conflicts when Dutch-English bilinguals make lexical decisions to Dutch-English homographs in an English lexical decision task. In a language-specific lexical decision task, a Dutch-English homograph can be interpreted as an English word requiring a "Yes" response and as a Dutch word requiring a "No" response. The fMRI data revealed language conflicts, because greater activation for homographs than for control words was found in the dorsal anterior cingulate cortex (dACC) and the left inferior/middle frontal gyrus. Both of these brain regions are associated with cognitive control. Importantly, the dACC was not activated in a generalized lexical decision task, which suggests that the dACC is involved only when language conflict occurs at the response level. The results fit well with the assumption of the BIA+ model that both readings of interlingual homographs are activated in the word identification system. Furthermore, the data indicate that the executive control system links these to actions required for the task at hand, and it resolves language conflicts outside of the word identification system.

Our three-pronged approach provides coherence to what would otherwise be a bewildering set of data. Within the BIA+ framework, predictions can be formulated about the more specific functions of certain brain areas; for example, following the model, any interference caused by the nontarget language should disappear after target word identification. We miss a similar integrative framework in the article. One would also like to hear some answers, no matter how preliminary, to explicitly posed questions—for instance, about the role of the network involving the left dorsolateral prefrontal cortex (DLPFC) and the anterior cingulate cortex (ACC)-supplementary motor area (SMA); and one would welcome discussion of, for example, the empirical and theoretical basis for proposed bottom-up and top-down inhibitory mechanisms, given that "there is as yet no clear or direct evidence of inhibition at the behavioral, brain region, or neuron level involved in the regulation of both languages" (p. 171).

The three-pronged approach is by nature sensitive to the "scientific memory" formed by past studies in favor of or against certain theoretical viewpoints. For instance, the language nonselective lexical access view of the BIA+ model is supported by over 25 studies, in contrast to the selective access view, supported by only 2 studies. Remarkably, Rodriguez-Fornells et al. (2002) found evidence (that they interpreted) in favor of language-selective access. The authors should discuss *why* their conclusions are different from earlier ones. In our view, the differences in results are due to experimental differences, not to the approach followed (psycholinguistic or neuroscientific).

Finally, in spite of their "reverse inference" approach, the authors *do* link their brain data to theoretical views from the "language sciences" (Grosjean et al., 2003). However, this implies introducing controversy in a different way. Does a "language mode" (Grosjean, 1997) really exist, according to which participant expectations can change relative language activation? Are there "inhibitory links between languages" that are under top-down executive control? In our view, the authors take an implicit stand with respect to several current controversies in bilingual word production. The resolution of such issues would be helped if cognitive neuroscientists also consider available alternatives.

To summarize, in a short time, the authors and their colleagues have performed an amazing series of "brain studies" involving bilingual processing and executive control. Their report on these and other studies is interesting to read, even though we consider the contribution of behavioral research and behavioral views to be more important for the domain than the authors do. We most miss an integrative theoretical framework and a discussion of relevant issues currently under discussion. These include the difficulty of mapping subcomponent executive functions to distinct brain mechanisms; the interaction between brain regions involved in cognitive control; the difference between monitoring and conflict resolution; and dissociations among RT, electrophysiological (ERP), and neuroimaging (fMRI) data.

At the end, the authors state that "future studies should more closely examine the interplay between control and executive brain mechanisms and the selection or suppression (inhibition) of language lexicons in bilinguals" (p. 173). We agree, and we would like to see psycholinguists and neuroscientists collaborate here. In our Roundtable conference, we therefore raised our glass and proposed a toast: "To language and the brain—or to (psycho)linguists and neuroscientists!"

References

Bialystok, E., Craik, F. I. M., Grady, C., Chau, W., Ishii, R., Gunji, A., & Pantev, C. (2005). Effect of bilingualism on cognitive control in the Simon task: Evidence from MEG. *NeuroImage, 24,* 40–49.

Colomé, À. (2001). Lexical activation in bilinguals' speech production: Language-specific or language-independent? *Journal of Memory and Language, 45*, 721–736.

De Bruijn, E., Dijkstra, T., Chwilla, D. J., & Schriefers, H. J. (2001). Language context effects on interlingual homograph recognition. *Bilingualism: Language and Cognition, 4*, 155–168.

Dijkstra, T., & De Smedt, K. (Eds.). (1996). *Computational Psycholinguistics: AI and connectionist models of human language processing*. London: Taylor & Francis.

Dijkstra, T., Van Hell, J., & Brenders, P. (in preparation). Bilingual recognition of cognates and non-cognates in sentence context.

Dijkstra, T., & Van Heuven, W. J. B. (1998). The BIA-model and bilingual word recognition. In J. Grainger & A. Jacobs (Eds.), *Localist connectionist approaches to human cognition* (pp. 189–225). Hillsdale, NJ: Erlbaum.

Dijkstra, T., & Van Heuven, W. J. B. (2002). The architecture of the bilingual word recognition system: From identification to decision. *Bilingualism: Language and Cognition, 5*, 175–197.

Grosjean, F. (1997). Processing mixed language: Issues, findings and models. In A. M. B. de Groot & J. F. Kroll (Eds.), *Tutorials in bilingualism: Psycholinguistic perspectives* (pp. 225–254). Hillsdale, NJ: Erlbaum.

Grosjean, F., Li, P., Münte, T., & Rodriguez-Fornells, A. (2003). Imaging bilinguals: When the neurosciences meet the language sciences. *Bilingualism: Language and Cognition, 6*, 159–165.

Harley, T. A. (2004). Does cognitive neuropsychology have a future? *Cognitive Neuropsychology, 21*, 3–16.

Kerkhofs, R., Dijkstra, T., Chwilla, D. J., & De Bruijn, E. R. A. (2006). Testing a model for bilingual semantic priming with interlingual homographs: RT and ERP effects. *Brain Research, 1068*, 170–183.

Kroll, J. F., Dijkstra, T., Janssen, N., & Schriefers, H. J. (1999). Cross-language lexical activity during production: Evidence from cued picture naming. In A. Vandierendonck, M. Brysbaert, & K. Van der Goten (Eds.), *Proceedings of the 11th Congress of the European Society for Cognitive Psychology* (p. 92). Gent: ESCOP/Academic Press.

Poldrack, R. A., & Wagner, A. D. (2004). What can neuroimaging tell us about the mind? Insights from prefrontal cortex. *Current Directions in Psychological Science, 13*, 177–181.

Rodriguez-Fornells, A., Rotte, M., Heinze, H.-J., Nösselt, T., & Münte, T. F. (2002). Brain potential and functional MRI evidence for how to handle two languages with one brain. *Nature, 415*, 1026–1029.

Roelofs, A. (2003). Shared phonological encoding processes and representations of language in bilingual speakers. *Language and Cognitive Processes*, 18, 175–204.
Van Heuven, W. J. B., Dijkstra, T., & Grainger, J. (1998). Orthographic neighborhood effects in bilingual word recognition. *Journal of Memory and Language*, 39, 458–483.
Van Heuven, W. J. B., Schriefers, H. J., Dijkstra, T., & Hagoort, P. (submitted). Language conflict in the bilingual brain.

Novice Learners, Longitudinal Designs, and Event-Related Potentials: A Means for Exploring the Neurocognition of Second Language Processing

Lee Osterhout, Judith McLaughlin, and Ilona Pitkänen
University of Washington

Cheryl Frenck-Mestre
Centre National de la Recherche Scientifique Université de Provence

Nicola Molinaro
Università delgli Studi di Padova

Research on the neurobiology of second language (L2) learning has historically focused on localization questions and relied on cross-sectional designs. Here, we describe an alternative paradigm involving longitudinal studies of adult, novice learners who are progressing through an introductory sequence of classroom-based L2 instruction. The goal of this paradigm is to determine how much L2 exposure is needed before learners incorporate L2 knowledge into their online comprehension processes, as reflected in scalp-recorded event-related brain potentials. Our preliminary studies show that some, but not all, aspects of the

---

Lee Osterhout, Judith McLaughlin, and Ilona Pitkänen, Department of Psychology; Cheryl Frenck-Mestre, Centre National de la Recherche Scientifique Université de Provence; Nicola Molinaro, Dipartimento di Psicologia dello Sviluppo e della Socializzazione.

The authors thank Julia Herschensohn for her comments and collaboration. We received financial support from grants R01DC01947 and P30DC04661 from the National Institute on Deafness and Other Communication Disorders, National Institutes of Health.

Correspondence concerning this article should be addressed to Lee Osterhout, Department of Psychology, Box 351525, University of Washington, Seattle, WA 98195. Internet: losterho@u.washington.edu

L2 (including lexical and morphosyntactic aspects) are incorporated into the comprehension system after remarkably little L2 instruction. We discuss the benefits of this paradigm while acknowledging the limitations and potential difficulties associated with it.

One could argue that two historical facts have impeded progress in understanding how a second language (L2) is instantiated in a language learner's brain. The first historical fact is the emphasis on the "where" question (i.e., the question of which parts of the brain are involved in using the L2). Although the "where" question is a valid and interesting one, the focus on this question has led to neglect of other questions that might prove to be equally important. The second historical fact concerns how researchers have dealt with two sources of variability. The first type, variability among L2 learners, introduces potential confounding variables that are well understood but rarely adequately controlled for. The second type, variability within a single learner over time, is a potentially crucial source of evidence that has been largely ignored. Our goal in this article is to argue for an alternative paradigm for studying the neurocognition of L2 learning and usage, in which the focus is on how the processes underlying comprehension of the L2 change with increasing L2 exposure or proficiency. The paradigm minimizes the troublesome sources of between-learner variability while maximizing sensitivity to theoretically crucial within-learner variability over time.

Two primary methods have been used to identify the neural structures involved in using an L2. One method involves studying aphasic bilinguals. If a patient's facility with two languages can be damaged independently, then one reasonable interpretation is that the two languages are represented independently in the speaker's brain. A second method involves the use of neuroimaging techniques (e.g., functional magnetic resonance imaging [fMRI]) to contrast L1 and L2 localization within the brain. These tools provide a measure of the brain's metabolic activity in neurologically intact individuals and allow researchers to determine which areas of the brain are metabolically most active, given some task. Evidence that different areas of the brain become activated when a bilingual uses her first (L1) or second language could, again, be interpreted to mean that different neural areas are involved in using the two languages.

However, nearly every pattern of language loss and recovery (loss of an L1 with preservation of an L2, equal loss of both languages, etc.) has been reported in bilingual aphasics. Similarly, whereas a few neuroimaging studies seem to indicate a (partial) neuroanatomical separation of the two languages (Kim, Relkin, Lee, & Hirsch, 1997; Perani et al., 1998), others seem to indicate a complete overlap in the brain areas activated by the two languages (Chee, Tan, & Thiel, 1999; Klein, Milner, Zatorre, Zhao, & Nikelski, 1999). It is reasonable to assume that the lack of consistency across experiments reflects, in part, uncontrolled subject variability (variability in learner motivation, the type and nature of L2 exposure, the age of L2 acquisition, etc.), a problem that has vexed much of the L2 literature (Grosjean, 1998). Furthermore, the range of inferences that these methods permit is quite limited; for example, language dysfunction might result from damage to white matter tracts that course through the area, rather than from damage to the gray matter at the lesion site. A complete overlap in fMRI activation patterns for an L1 and an L2 does not necessarily imply that the two languages are processed in exactly the same way; and a partial overlap in activation patterns gives us little idea of how processing of the two languages might differ.

Rather than focus on the "where" question, the paradigm we have in mind focuses on the processes that recognize words and derive the structure and meaning of sentences. More specifically, the paradigm is designed to help us learn more about how these processes (and the neural systems that underlie them) change over time, with increasing L2 exposure or proficiency. This paradigm minimizes between-subjects variability by longitudinally studying novice L2 learners who are progressing through their first years of classroom L2 instruction. In this design, the learners are highly similar in a priori L2 proficiency (having none) and experience a highly similar learning environment. Because each learner acts as his or her own control, many of the subject variables that can potentially confound L2 studies have been eliminated within this design. By also including native speakers of the L2 in the study, we are able to compare the beginning and intermediate states of learning to the end state of native proficiency.

The longitudinal design also maximizes sensitivity to the within-subject variation that is the primary focus of our research program. Our goal is to identify changes in brain activity (and, in particular, activity that reflects online L2 processing) that accompany the earliest stages of L2 learning. We do so by recording event-related brain potentials (ERPs) from the scalp while learners read L2 words or sentences. ERPs provide an online, millisecond-by-millisecond record of the brain's electrical activity during language comprehension and, therefore (unlike fMRI and positron-emission tomography [PET]), have the necessary temporal resolution to isolate specific language comprehension processes in time. Furthermore, ERPs are multidimensional, varying in polarity, timing, morphology, and scalp distribution; this multidimensionality might in theory provide ERPs with a differential sensitivity to at least some of the processing steps underlying comprehension.

In fact, ERPs do seem to be differentially sensitive to events occurring at distinct levels of linguistic analysis (for a review, see Osterhout, McLaughlin, Kim, Greenwald, & Inoue, 2004). One ERP response, the N400 component, is sensitive to properties of words, both when they appear in isolation and when they appear in a linguistic context. A second component, the P600, is sensitive to the grammatical well-formedness of sentences. Although the precise cognitive and neural events underlying these effects are not known, their existence and particular sensitivities can be exploited to learn a great deal about language comprehension and (we hope to demonstrate here) L2 learning. We will demonstrate that although ERPs might not be tremendously informative with respect to the "where" question, they are, in some instances, extremely useful for ascertaining *what* is happening and *when* it happens. The method's sensitivity to the "what" question follows from the fact that the language-relevant ERP effects are quite specific in their sensitivities, thereby permitting fairly specific inferences about what is happening at that moment in processing. The method's sensitivity to the "when" question encompasses two timescales: the small intervals of time (tens and hundreds of milliseconds) over which a word or sentence is processed, and the large intervals of time (weeks, months, and years) over which a person progresses from no competence with a language to increasing competence. In the next section, we briefly review what has been learned about the relationship between language processing and ERPs, discussing, in particular, problems that might come up when applied to L2 research. In the subsequent sections, we provide examples of how this knowledge, when combined with the above-described research paradigm, can be used to gain some new insights about what happens during L2 learning and processing and when it happens.

ERPs as Tools for Studying First and Second Language Processing

Kutas and colleagues (Kutas & Hillyard, 1980) were the first to demonstrate that semantically anomalous words (e.g., "He spread his warm bread with *socks*") elicit an increase in the amplitude of the N400 component, a negative-going wave that peaks at about 400 ms (Figure 1a). This result is observed regardless of the position of the anomalous word within the sentence or the modality of input (visual vs. auditory; Osterhout & Holcomb, 1993). Subsequent research has shown that the N400 amplitude to words in sentences is an inverse function of the semantic congruency between the target word and the preceding context, even when the word is not semantically anomalous (Kutas & Hillyard, 1984). The N400 amplitude is also sensitive to the strength of semantic priming in a word-pair lexical decision task (Bentin, McCarthy, & Wood, 1985). These results have been interpreted as indicating

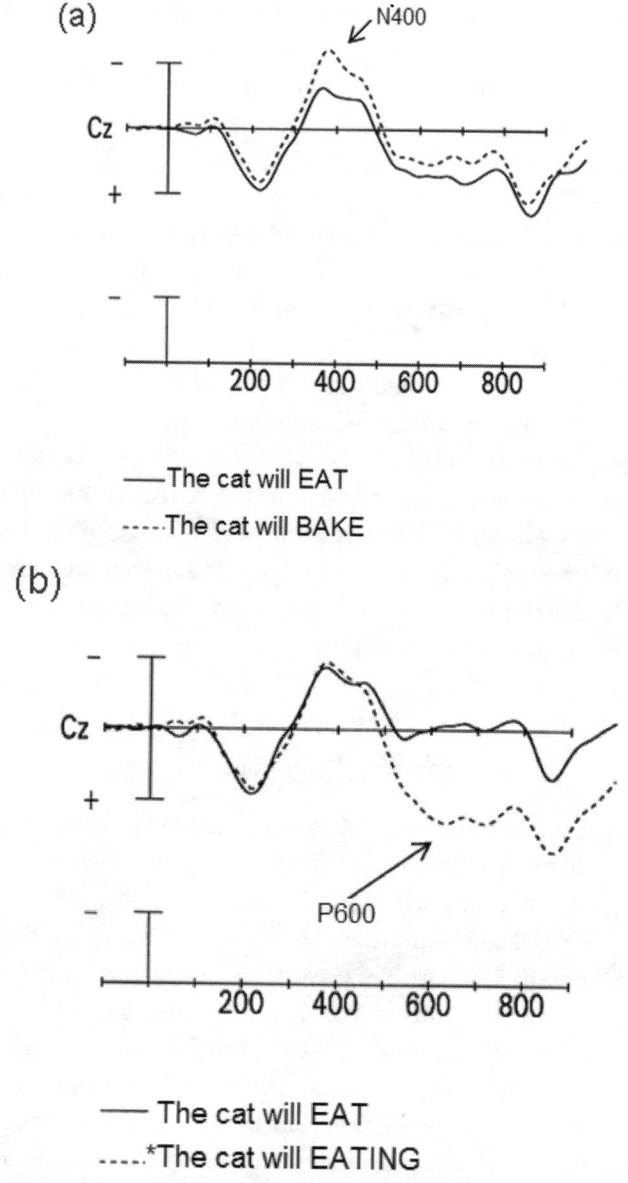

Figure 1. (a) ERPs (recorded at central midline location Cz) elicited by semantically anomalous words (dashed line) and nonanomalous control words (solid line) in sentences such as *The cat will eat/bake the food*. (b) ERPs elicited by syntactically anomalous (dashed line) words and well-formed controls (solid line) in sentences such as *The cat will eat/eating the food*. Onset of the critical word is indicated by the vertical bar. Each hashmark represents 100 ms. The vertical calibration bar is 5 μV. (Adapted from Osterhout & Nicol, 1999.)

that the N400 amplitude elicited by words is highly sensitive to the semantic relationship between the word and the preceding context. However, the N400 amplitude is not solely determined by the semantic fit between a word and its context. A variety of lexical properties affect the N400 amplitude, including lexicality (i.e., whether a letter string is a word in the language), word frequency, phonological priming, morphological content within a string (e.g., McKinnon, Allen, & Osterhout, 2003; McLaughlin, Osterhout, & Kim, 2004), and sequential probabilities concerning the likelihood of words occurring in succession (Kutas & Hillyard, 1984). A more accurate generalization is that the N400 amplitude reflects a combination of lexical and semantic/conceptual factors.

The sensitivity of the N400 component to lexico-semantic factors is strikingly contrasted with its *in*sensitivity to syntactic factors. A great deal of research conducted over the past 25 years has shown that violations of syntactic rules do not robustly affect the N400 amplitude (Allen, Badecker, & Osterhout, 2003; Osterhout & Nicol, 1999). Instead, a disparate set of syntactic anomalies elicits a large positive wave in the ERP with an onset at about 500 ms and a duration of several hundred milliseconds (labeled the P600 effect by Osterhout & Holcomb [1992; Figure 1b]). The P600 effect is highly sensitive to syntactic variables but insensitive to lexico-semantic variables (Allen et al.). Importantly, the N400/P600 dichotomy is not a function of particular combinations of stimuli, task, and language, but, instead, generalizes well across types of anomaly (including anomalies involving phrase structure, morphosyntax, and constituent movement), numerous stimulus and task conditions (e.g., rate of word presentation, subject's task), modalities (visual and auditory), and languages (including configurational languages like English and nonconfigurational languages such as Japanese). In some reports, syntactic anomalies have also elicited an anterior negativity within a window ranging from 150 to 500 ms, sometimes largest in amplitude over the left hemisphere (hence the label *left anterior negativity*, or LAN; Friederici, 1995; Neville, Nicol, Barss, Forster, & Garrett, 1991; Osterhout & Holcomb, 1992, 1993).

As is the case with any method, ERPs have their limitations. Although it is relatively easy to determine the antecedent conditions (i.e., the stimulus manipulations) that produce or modulate some effect, it is usually very difficult to identify the specific processes that are manifested by the effect; for example, even though the P600 effect is reliably elicited by violations of a syntactic rule, it does not necessary follow that the P600 effect reflects specifically syntactic (or even linguistic) processes. This ambiguity exists because ERPs (just like neuroimaging techniques such as fMRI) are correlational in nature. Consequently, the P600 effect might reflect syntactic processes directly or it might reflect some process that is highly correlated with these syntactic processes or with the processes that respond to a syntactic anomaly. Deciding which of these possibilities is correct might prove to be an intractable problem.[1] Fortunately, the correlations between manipulations of lexical and semantic variables and changes in N400 amplitude and between the presence of a syntactic anomaly and the P600 effect are very robust. One can, therefore, reasonably infer that a linguistic anomaly was perceived to be semantically anomalous if it elicited an N400 effect or syntactically anomalous if it elicited a P600 effect.

A second limitation is that ERPs usually need to be averaged twice: once for a given subject and then over all subjects. This is required to achieve the necessary signal-to-noise ratio. When the subjects are native speakers of a language, there seems to be enough consistency (within and between subjects) in the ERP responses to linguistic stimuli to produce robust and reproduceable results. However, a sample of L2 learners will almost always be more variable than a sample of L1 speakers. It seems quite likely that this variability will be expressed in the ERPs (as, e.g., greater variation in the latency or scalp distribution of particular ERP effects). If so, then the differences in variability limit the types of inference that are warranted by ERP data; for example, if an effect is present in the L1 group but absent in the L2 group, this does not necessarily mean that the process manifested by the effect is present in the L1 group but not in the

L2 group. This is because the effect might be present in both groups but obscured in the L2 group due to larger variability in the effect's timing and distribution. The increased variability would reduce the size of the effect, or eliminate it altogether, in the more variable population—even if it is present on individual trials, for individual subjects. An example of this problem is provided by a study reported by Hahne (2001). Hahne recorded ERPs to syntactic anomalies from native German speakers and from Russian L1 speakers who had learned German after the age of 10. Both groups of speakers showed a P600 effect to the anomalies, but only the native speakers also showed a LAN effect. Hahne interpreted this result as indicating that the German learners were missing the linguistic process manifested by the LAN effect. However, another possibility is that the greater variability in the L2 group simply obscured the anterior negativity, even if it was present on individual trials, in individual learners. This possibility is likeliest for effects that are small in amplitude and have a limited temporal extent (such as the LAN effect).[2]

The increased variability associated with L2 learners complicates the use of ERPs in another way. If the variability across subjects is large enough, the grand average might not reflect the response shown by any particular individual learner and might not generalize to any identifiable population; that is, in such cases, the "central tendency" represented by significant effects in the grand average might reflect only the accidental overlap of effects that were present in many learners in the sample, but it might not reflect larger and potentially more important effects that varied from learner to learner. If so, then comparisons among, for example, L2 learners who were exposed to the L2 at different ages might produce reliable differences among the groups but might not represent true effects of age on acquisition. Fortunately, the validity or invalidity of such results would become clear when the results prove difficult to replicate. Unfortunately, this problem could easily lead to the publication of many unreplicable and irreconcilable results.

Event-related brain potentials therefore offer a great deal of promise for learning more about the acquisition and use of an L2, but also carry with them certain risks. Our goal has been to reap the benefits of ERPs while minimizing our exposure to the risks. To that end, we have combined ERPs with longitudinal studies of novice language learners. Our questions have included the following: How quickly is L2 knowledge about words and sentences incorporated into the online, "real-time" processing system? What variables influence this rate? How quickly do learners' brain responses to L2 words and sentences begin to approximate their responses to L1 words and sentences? What is the relationship between brain-based manifestations of learning and a person's overt behavior? This approach might also usefully investigate some of the standard questions within the field of L2 research. The following are some examples: What influence does the learner's L1 have on acquisition of the L2? What role does a correspondence (or lack of correspondence) between phonology and morphology play in the acquisition of a second language? Does L2 learning progress in a continuous manner, or are there discrete stages of learning? Although these questions are not directly about the cognitive neuroscience of L2, we hope to show that the tools of cognitive neuroscience are particularly well suited for providing answers to them. Also, because ERPs directly reflect brain activity, the results of these studies sometimes shed light on the cognitive neuroscience of L2s even when that is not the explicit goal.

ERPs and L2 Word Learning

Research on bilingual lexical processing has primarily focused on the organization of the bilingual lexicon (Kroll & Sunderman, 2003). Most theorists assume that the representation of words is divided into two levels: a word-form (lexical) level and a word-meaning (conceptual) level. The question of interest has been the degree of independence of the two lexicons at these two levels (Kroll & Sunderman; Potter, So, Von Eckhardt, & Feldman, 1984).

However, little is known about how L2 lexical and conceptual representations develop during L2 acquisition (McLaughlin, 1998), and almost nothing is known about the neurobiological correlates of this developmental process (for one recent exception, see Raboyeau, Marie, Balduyck, Gros, Demonet, & Cardebat, 2004). In a study conducted in our laboratory, we examined word learning by measuring learning-related changes to the N400 component of the ERP (McLaughlin, Osterhout, & Kim, 2004). As noted earlier, N400 is sensitive to both lexical status (whether or not a particular wordlike form is part of the language) and word meaning (Bentin, 1987; Kutas & Hillyard, 1980). For native speakers of a given language, the N400 amplitude is largest for pronounceable, orthographically legal nonwords (pseudowords; e.g., *flirth*), intermediate for words preceded by a semantically unrelated context, and smallest for words preceded by a semantically related context. Our goal was to determine how much L2 exposure is needed before the learner's brain responses to L2 words and nonwords resemble that of a native speaker. Such results might reveal the rate at which French learners acquire information about word forms and word meanings.

Our participants were English-speaking university students progressing through their first year of classroom French instruction and members of a control group who had never received any French instruction. None of the students had significant exposure to French prior to attending class. The stimuli were prime-target pairs of letter strings. Some of the pairs contained two French words that were either semantically related (e.g., *chien-chat* "dog-cat") or unrelated (*maison-soif* "house-thirst"). Other pairs contained a target that was a pronounceable French pseudoword (*mot-naisier*). ERPs to the targets were of interest. The French learners were tested three times in a longitudinal design: once near the beginning of their French instruction (after ~14 hr on instruction), once near the middle (~60 hr), and once near the end (~140 hr). We asked the French learners to make word/nonword judgments regarding the words and pseudowords. We found that after only 14 hr of L2 instruction, the pseudowords elicited a robustly larger N400 than the words (Figure 2, top, right column). This was true even though learners were at chance levels ($d' = 0$) when deciding if the letter strings were actual words in French. Furthermore, the correlation between hours of instruction and the word/nonword N400 difference was very robust ($r = .72$), suggesting that the N400 difference was approximately linearly related to the learner's exposure to the L2. This result

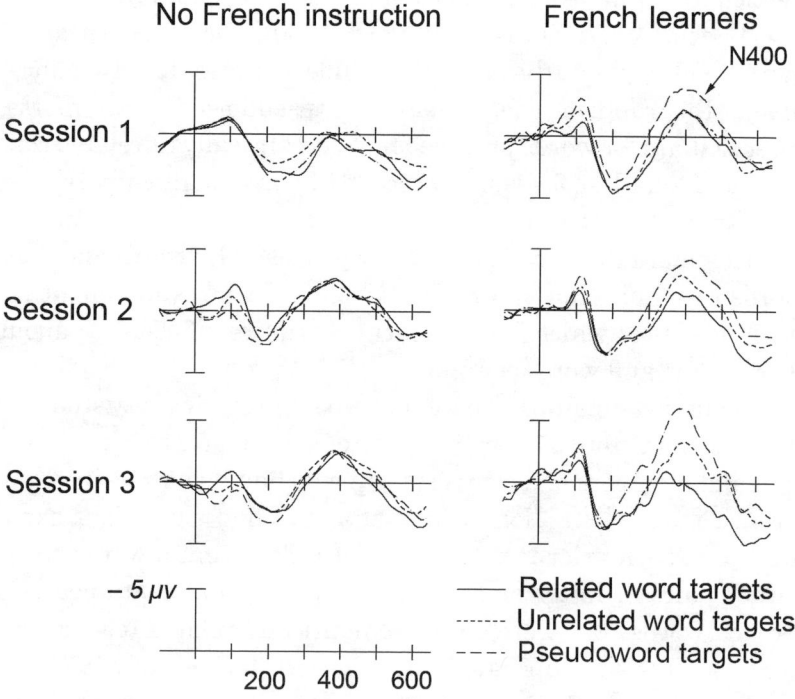

Figure 2. ERPs from the No Instruction (left panel) and French Instruction (right panel) subjects, recorded at three successive longitudinal testing sessions. ERPs are plotted for three types of target strings: target words that were semantically related to the prime word (solid line), target words that were not semantically related to the prime (small dashes), and pseudoword targets (large dashes). (Adapted from McLaughlin, Osterhout, & Kim, 2004.)

suggests that the French learners rapidly extracted enough information about French word forms so that their brains could discriminate between actual words and pseudowords, even if the learners themselves could not do so. Effects of word meaning, manifested as smaller N400s to words preceded by a related word than by an unrelated word, were observed after ~60 hr of instruction (Figure 2, middle, right column). After ~140 hr of L2 instruction (Figure 2, bottom, right column), the amplitude of the word/nonword and relatedness differences approximated that typically observed with similar stimuli in native English or French speakers, even though the learners' explicit word/nonword judgments remained very poor ($d' < 1$). No differences in the N400 amplitude across conditions were observed for a group of subjects who had received no French instruction (Figure 2, left column). These results show that the L2 learners extracted information about word form after just ~14 hr of instruction and extracted information about word meaning by at least ~60 hr. Furthermore, these changes in brain activity immediately started to approximate the responses seen in native speakers to analogous stimuli and occurred even while the learner's conscious lexicality judgments were very poor. However, what, exactly, were these learners learning about words? The very early learning observed in this study might involve elemental aspects of linguistic knowledge (e.g., knowledge about word forms) that serve as prerequisites to L2 competence. People rapidly extract co-occurrence statistics for letter and sound combinations within a language (Saffran, Johnson, Aslin, & Newport, 1999). Perhaps the word/nonword N400 effect reflected a similar type of learning. To test that idea, we computed correlations between the word/nonword N400 effect and bigram and trigram frequencies (as computed from the relevant portions of the French text). These correlations were very weak ($p > .3$). Another possibility is that learners memorized whole-word forms they were exposed to during their L2 instruction. This idea receives some support from the robust correlation between the N400 word/nonword effect and the frequency of the target words in the learners' French text ($p = .67$).

ERPs and L2 Morphosyntactic Learning

A second line of research in our laboratory is motivated by the observation, discussed earlier, that syntactic and semantic anomalies embedded within sentences elicit distinct ERP effects (the P600 and N400 effects, respectively). This finding suggests that separable syntactic and semantic processes exist. In an L2 learning context, one implication is that L2 learners must somehow segregate linguistic input into those aspects of the language that relate to sentence form and those that relate to sentence meaning; that is, learners "grammaticalize" some aspects of the L2, but not others. Such a conclusion introduces a host of interesting questions. How and when do L2 learners grammaticalize aspects of the L2? Which aspects of the syntax are grammaticalized first? What factors influence the rate and eventual success of grammaticalization? How similar is the process of grammaticalization in learning an L1 or an L2? In our work, what we mean by "grammaticalization" is specifically the instantiation of grammatical knowledge into the learner's online, real-time language processing system. Our assumption is that once a feature of the L2 has been grammaticalized, violations of that aspect of the grammar should elicit a P600 effect.

To investigate such questions, we have focused on one particular aspect of syntactic learning, namely the acquisition of grammatical features and their associated morphosyntactic rules. The centrality of grammatical features in recent formal theories has led to renewed interest in studying the acquisition of the grammatical morphemes that encode these features. Within linguistic theory (Radford, 1988), semantic properties of language are encoded in word (lexical) categories such as Noun (N), Verb (V), and Adjective (A); these categories are shared by all languages. Grammatical properties, by contrast, are encoded in the morphology of grammatical categories such as noun phrase (NP) and verb phrase (VP). Grammatical categories are associated with grammatical features such as gender, number, and verbal person. These features (and how they are involved in morphosyntax) vary across languages (Table 1); for example, with respect to grammatical features, English has a number feature in the NP (e.g., *boy* vs. *boys*), whereas French has both number and gender. With respect to morphosyntax, English and French both have sentential agreement (i.e., agreement of the verb with the subject in verbal person and number; e.g., *I like* vs. *He likes*), but only French has NP agreement (i.e., agreement between the noun and its determiner/adjective in number and gender; e.g., *le garçon* vs. *les garçons*; excluding the restricted English case of *this/those*) (Table 1). Languages also differ with respect to the explicitness of grammatical features; for example, although English and French both show agreement between subject and verb, the explicitness of the grammatical marking differs. English present tense uses only -*s* in third person singular (*he walks, they/I/etc. walk*). French present uses five orthographic forms (*je marche, tu marches* "I/you walk"), but has only three distinct pronunciations. As for NP agreement, English marks number on the noun alone (*the little boy[s]*), whereas French marks number (and gender) on the agreeing determiner/adjective (*le[s] petit[s] garçon[s]*) (Table 1). In oral French, however, plurality can only be heard on the determiner. English has no grammatical gender feature, whereas French does. What factors might inhibit or facilitate grammaticalization of these features and their morphosyntactic rules? One frequent claim is that only features that are present in the L1 can be acquired during L2 acquisition (Franceschina, 2001; Hawkins & Franceschina, 2004). Other researchers, however, argue that novel L2 features can be learned, albeit more slowly than those that are present in the L1 (White, 2003). Thus, there is no consensus about whether, or when during acquisition, L2 learners acquire L2 features that are not present in their L1. Even less is known about the acquisition of L2 morphosyntactic rules (e.g., agreement in number or gender) that are not present in the L1.

Another factor that seems likely to play a role in L2 grammatical morpheme learning is the covariation between morphology and phonology. The interaction between morphology and

Table 1

Verbal person and nominal concord in English and French
Sentential verb-subject agreement: regular verbs, present tense

	English	French
Singular		
1st	I like	Je marche
2nd	You like	Tu marche(s)
3rd	He likes	Il marche
Plural		
1st	We like	Nous marchons
2nd	You like	Vous marchez
3rd	They like	Ils marche(nt)
Noun phrase (NP) number agreement		
Singular	the boy	le petit garçon
Plural	the boys	les petit(s) garçon(s)
Noun phrase (NP) gender agreement		
Male	NA	le garçon
Female	NA	la fille

Note. Morphemes in parentheses are not expressed phonologically.

phonology can be clearly seen in written French. French has an opaque orthography due to many suffixes being phonologically silent (Table 1). Thus, the plural suffix *-s*, which marks the plural across all elements in the NP (*le-s jeune-s fille-s* "the young girls") is silent on the noun (as well as on the adjective) in almost all instances. A similar situation arises in the VP, where variations in verbal person are marked orthographically on the verb but are silent in most oral forms. Thus, the different inflections for a regular verb such as *marcher* (to walk) sound identical across three different persons/spellings (Table 1). The effect of the "missing" phonological cue is notorious on spelling. French children are drilled in the morphological variations throughout elementary school, and yet they continue to make grammatical errors, by failing to add the plural inflection in writing. Errors such as *les chien* or *Ils mange* are frequent (Negro & Chanquoy,

2000) and can be seen in adults as well. Errors are much rarer when phonology is available as a cue (Largy & Fayol, 2001); for example, in the VP, confusions are not made among persons for *nous* and *vous* ("we/you" formal or plural), as these forms are both morphologically and phonologically distinct from each other and all other persons. Errors are also far less frequent when the morphological ending is paired with a phonological difference, as is the case for certain irregular and stem-changing verbs in French. The results from French are in line with results from studies of Dutch spelling (Frisson & Sandra, 2002; Sandra, Frisson, & Daems, 1999) showing that adult Dutch writers continue to confuse the first and third person singular for verbs where these two persons have distinct morphological endings but identical oral realizations (e.g., *rijd* "drive" and *rijdt* "drives," which are both pronounced as/rɛIt/), but they do not confuse the two for verbs where they have distinct phonologies (e.g., *werkt* / wɛrkt/"works" and *werk* / wɛrk/"works"). In adult L2 acquisition, the use of oral cues has been advocated by linguists to enhance the learning of certain morphological rules, such as verb and adjective variations (Herschensohn, 1993). Arteaga, Gess, and Herschensohn (2003) found that English learners of French who were instructed to note phonological variations of masculine and feminine forms of variable adjectives in French were quicker to learn the morphological rule than were students who were not given such instruction but were taught only the spelling alterations between the two forms.

Therefore, one reasonable prediction is that L2 learners will acquire an L2 feature or morphosyntactic rule more quickly when the relevant inflectional morphology is phonologically realized. However, this possibility has received little direct attention in the recent L2 literature. It also seems likely that L1-L2 similarity and phonological-morphological covariation might have interactive effects during L2 learning; for example, L1-L2 similarity combined with phonological realization of the relevant grammatical morphemes might lead to very fast learning, whereas L1-L2 dissimilarity combined with no phonological realization might lead to very slow learning.

Finally, various theorists have suggested that L2 morphosyntactic learning involves discontinuous stages (Myles, Hooper, & Mitchell, 1998; Vainikka & Young-Scholten, 1996; Wong-Fillmore, 1976; Wray, 2002). This seems to be true of children learning their L1, who begin by memorizing particular combinations of words and only later induce general syntactic rules (Tomasello, 2000). With respect to L2 learning, one account holds that morphologically complex words are initially learned by rote and memorized as unanalyzed chunks (Myles et al., 1998; Wray, 2002). Eventually, learners unpack these chunks into roots and grammatical morphemes, induce grammatical rules governing their use, and deductively use these units to produce novel utterances. To be specific, an L2 learner might initially memorize the fact that certain subjects are followed by certain forms of the verb, without decomposing the verb into root + inflection or applying a general morphosyntactic agreement rule. In this stage of learning, the learner associates meanings with the undecomposed word and either memorizes the two words as a chunk or learns about word sequence probabilities (e.g., that *Tu* is followed by *marches*, whereas *Ils* is followed by *marchent*).[3] Unfortunately, when using conventional behavioral methods, it is very difficult to identify the transition from rote-memorization to rule use (Wray, 2002). Fortunately, ERPs might be an ideal tool for testing this theoretical claim of developmental discontinuity. As noted earlier, the N400 amplitude is highly sensitive to novel words and to word sequence probabilities, whereas the P600 effect is sensitive to grammatical rule violation. If an L2 learner initially memorizes salient word sequences, then unfamiliar word combinations (e.g., *Tu adorez*) should produce larger N400s than familiar ones (e.g., *Tu adores*). If the learner eventually decomposes the verb into root + morpheme and induces the rule, use of the wrong verbal form should elicit a P600 effect. In essence, the transition for N400 to P600 would reflect the transition from rote-memorized (i.e., *lexicalized*) knowledge to *grammaticalized* knowledge.

We investigated these predictions using a longitudinal experimental design involving 14 English-speaking novice French learners. Each learner was tested after approximately 1 month, 4 months, and 8 months of university classroom French instruction. Our stimuli were as follows:

(1) *Sept plus cinq\?livre font douze.*
semantic condition

(2) *Tu adores*adorez le français.*
verbal person agreement condition/phonologically realized

(3) *Tu manges des hamburgers*hamburger pour dîner.*
number agreement condition/phonologically unrealized

In (1), the noun *livre* is semantically anomalous. In (2), the verb *adorez* is conjugated incorrectly, given the preceding sentence fragment. In (3), the noun *hamburger* disagrees with the syntactic number of the plural article. Our stimuli were selected from the material in the textbook assigned during the first month of instruction. The anomalous items in the verbal person condition involved a grammatical rule that was present in the L1 and an orally realized contrast between inflectional morphemes. The anomalous items in the number agreement condition involved a rule that was not present in the L1 and a phonologically unrealized contrast between inflectional morphemes. Therefore, our prediction was that L2 learners would respond to the anomaly in (2) with less L2 exposure compared to the anomaly in (3).

As expected, the native-French speakers showed an N400 effect to the semantically anomalous words and large P600 effects to the two types of syntactic anomaly. The learners, as is often the case, showed striking individual differences, both in a behavioral "sentence acceptability judgment" task and in the pattern of ERPs elicited by the anomalous stimuli. We segregated the learners into upper ("fast learners") and lower ("slow learners") halves, based on their performance in the sentence-acceptability judgment task that occurred concurrently with ERP data collection, and averaged the ERPs separately for each group. Results for the "fast learner" group will be described here. At each testing session, including the initial session that occurred after just 1 month of instruction, semantically anomalous words elicited a robust N400 effect, and this effect changed minimally with increasing instruction (Figure 3, averaged over the three testing sessions). Results for the verbal person condition are shown in Figure 4. After just 1 month of instruction, the learners' brains discriminated between the syntactically well-formed and ill-formed sentences. However, rather than eliciting the P600 effect (as we saw in native-French speakers), the syntactically anomalous words elicited an N400-like effect. (This effect did not differ in distribution from the N400 effect elicited by the semantically anomalous words.) By 4 months, the N400 effect was replaced by a

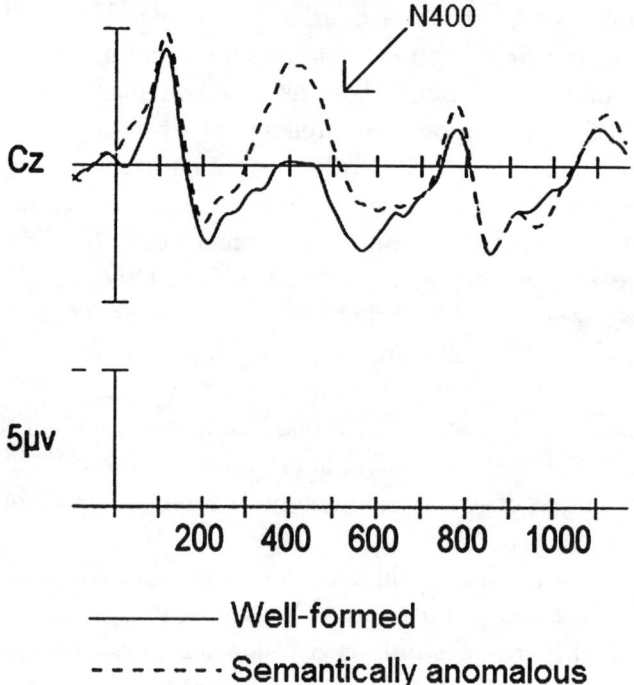

Figure 3. ERPs to critical words in the well-formed (solid line) and semantically anomalous (dashed line) conditions, collapsed over the three testing sessions.

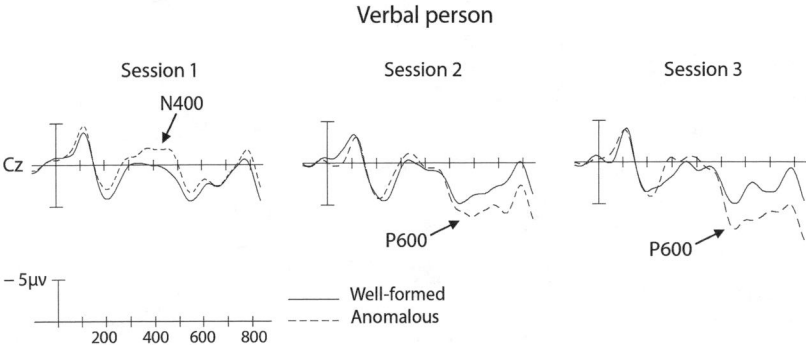

Figure 4. ERPs to critical words in the well-formed (solid line) and verbal person anomaly (dashed line) conditions, plotted separately for each of the three testing sessions.

P600-like positivity. Results for the number agreement condition can be summarized easily: Learners performed very poorly in the sentence acceptability judgment task for these materials, and there were no robust differences between the agreeing and disagreeing stimuli (Figure 5, averaged over three testing sessions).

These results are consistent with the predictions that we were testing. First, L1-L2 similarity combined with phonological realization of the relevant grammatical morphemes produced very fast L2 syntactic learning, whereas L1-L2 dissimilarity combined with no phonological realization produced very slow learning. This occurred even though our learners were drilled repeatedly on both rules from nearly the first day in class. However, the two rules we tested represent the ends of a putative continuum of morphosyntactic difficulty; without additional data, it is impossible to know whether L1-L2 similarity or phonological realization of grammatical morphemes had a larger impact on the learning rate. What can be said with more certainty is that at least some L2 syntactic rules are learned amazingly quickly. Apparently, at least some types of L2 rule are incorporated into the online sentence processing system after a very small amount of L2 instruction.

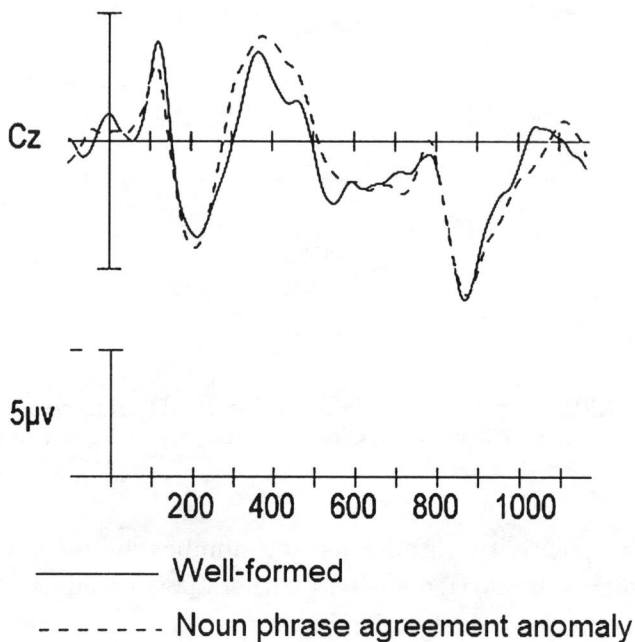

Figure 5. ERPs to critical words in the well-formed (solid line) and noun phrase agreement anomaly (dashed line) conditions, collapsing over the three testing sessions.

The second prediction was that L2 morphosyntactic anomalies would show a discontinuous pattern over time: Early in learning, such anomalies would elicit an N400 effect in learners, whereas later in learning, these same anomalies would elicit a P600 effect. This pattern is exactly what we observed for the verbal person anomalies. If our interpretation is correct, then our adult L2 learners grammaticalized this aspect of the L2 after just a few months of L2 instruction. Our results can be explained by assuming that learners (much like child L1 learners) initially memorize salient word sequences (e.g., *tu adores*). Violations of the verbal person rule (e.g., *tu adorez*) result in novel word combinations and, hence, elicit an N400 effect. After more instruction, learners induce a general verbal person rule (*tu -s, nous -ons, vous -ez,* etc.); violations of the rule elicit a P600 effect. The

P600 effect occurred at similar temporal latencies for the natives and learners. Thus, after just 80 hr of instruction, learners' and natives' ERP responses to verbal person violations were qualitatively similar.

English and French are highly similar in many respects; for example, they share many cognates and a similar (although not identical) system of morphosyntax, and they both belong to the Indo-European family. English and Finnish, by contrast, share very few cognates and come from different language families. English has a relatively impoverished morphosyntax, whereas Finnish has an extensive system: Nouns are marked for number, case, and possessiveness, and finite verbs are marked for tense or mood, number, and person. Agreement rules in Finnish exist for number, person, and case. Finnish is generally thought to have a subject-verb-object canonical structure. However, because the morphosyntactic system usually makes the syntactic and semantic functions of each constituent in a clause unambiguous, Finnish speakers have considerable flexibility in ordering words. Noncanonical structures are permitted to mark a change in focus or for other purposes. Furthermore, the Finnish morphosyntactic system is highly regular and fully realized in the phonology. Inflectional suffixes are regular, and they are added to the word stems in a fixed order. However, the addition of the endings is sometimes accompanied by sound alternations in the stem of the word.

These properties of Finnish morphosyntax provide a useful contrast with French, in the context of a native English-speaking L2 learner. If the L1-L2 similarity is the primary determinant of the rate of L2 syntactic learning, then learners should be quite slow to acquire the many novel aspects of Finnish morphosyntax. If, conversely, regularity, transparency, and phonological realization in the morphosyntax are more important, then learners might acquire the system rather quickly.

Recent work has shown that morphosyntactic and semantic anomalies elicit the P600 and N400 effects, respectively, when the subjects are native speakers of Finnish (Palolahti, Leino, Jokela,

Kopra, & Paavilainen, 2005). In our lab, we have begun to study English-speaking university students as they progress through their first year of Finnish instruction. We hope to determine how much Finnish instruction is needed before morphosyntactic violations elicit an ERP linguistic anomaly effect. We are presenting sentences that contain either a semantic anomaly and one of two kinds of morphosyntactic error, as illustrated below:

(4a) Sinä (2nd p. sing.) syö*t* (2nd p. sing.) lihaa.
"You eat meat."

(4b) *Sinä (2nd p. sing.) syö*n* (1st p. sing.) lihaa.
*"You eat meat."

(5a) He istuvat kahvila*ssa* (inessive case) illalla.
"They sit *in* a café at night."

(5b) *He istuvat kahvila*sta* (elative case) illalla.
*"They sit *from* a café at night."

In (4b), the verb is in the wrong person form. In (5b), the noun is attached to an incorrect case particle. Because English has a verbal person rule but not a system of explicit case marking, learning might be relatively fast for Finnish verbal person but very slow for case if L1-L2 similarity is the primary determinant of learning rate. Conversely, if phonological realization of the grammatical morphemes is very important, then both the similar and dissimilar aspects of the morphosyntactic system might be learned relatively quickly. In a pilot study, we tested five English speakers who had just completed their initial 9 months of Finnish instruction. The results were as follows: Semantic anomalies elicited an N400 effect; verbal person anomalies elicited a robust P600 effect; and case anomalies elicited a P600-like effect, although with a much smaller amplitude. These preliminary data suggest that even unfamiliar aspects of a complex morphosyntactic system can be grammaticalized within the first year of classroom instruction. Our ongoing research with this population might tell us more about the rate of learning and whether this type of morphosyntactic learning is continuous or discontinuous.

Conclusions and Final Comments

Our goal in this article has been to describe a research paradigm for investigating what L2 learners learn as they progress through a period of L2 instruction and when they learn it. By learning, we mean specifically the incorporation of L2 knowledge into the learner's online, real-time language processing system. We do not claim to have definitive answers to any of the questions that we hope to answer. For example, our pilot Finnish study suggests that English-speaking L2 learners might "grammaticalize" novel aspects of Finnish morphosyntax during their initial 8 months of classroom instruction. We have suggested that this might be due, in part, to the fact that the relevant grammatical morphemes are reliably (and uniquely) expressed in the language's phonology. However, of course, there are many other explanations for this effect, if it turns out to be robust. Our proposed paradigm is limited in other ways as well. Although the paradigm minimizes between-subjects variability, it cannot eliminate it. As we found in our French syntax learning study, individual differences continue to exert an influence on learning even when we longitudinally study novice learners in a homogeneous classroom environment. This is not unexpected. Our hope is that our design reduces between-learner variability to a manageable level; eventually, careful study might reveal the primary sources of the remaining variability. Finally, it is important to explicitly recognize that ERPs (like all methods of investigation) imperfectly reflect the cognitive and neural processes underlying L2 learning and processing.

Even given these caveats, our preliminary findings are promising. Our findings suggest that it is possible to ascertain *when* some aspect of the L2 has been incorporated into the set of processes that allow the learner to comprehend the language in real time. Moreover, our results suggest that developmental discontinuities exist with respect to *what* has been learned: early in the learning process, learners seem to memorize frequently occurring words or groups of morphemes. With a little more L2 exposure, learners associate the memorized lexical forms with meanings, begin to decompose the structurally complex units into roots plus grammatical morphemes, and learn the rules for agreement among these elements. This type of learning can occur with remarkably little L2 exposure, in a standard classroom setting that is often thought to provide a less-than-ideal learning environment. Some learners even seem to quickly grammaticalize novel aspects of a complex L2 morphosyntactic system that is unlike anything in their L1. Our results are preliminary, but they clearly conflict with the conventional belief that adult L2 syntactic learning is generally slow and problematic, especially for aspects of the L2 that are not present in the L1. The conventional belief about the difficulty of L2 syntactic learning is based, in part, on L2 production: Many speakers do have difficulty with grammatical aspects of an L2 when they try to speak it. However, one general rule about L1 learning is that a learner's ability to understand the language develops in advance of his or her ability to produce it. It seems likely that this maxim also applies to L2 learners. If so, then perhaps there is no contradiction between our data (indicating a rapid acquisition and implementation of some types of L2 knowledge during language comprehension) and the common impression that L2 learners have a difficult time producing certain aspects of the L2.

We should note that other researchers have come part way in implementing our proposed research paradigm, although (to the best of our knowledge) no one has implemented it completely. Several groups of researchers have used ERPs to study the incorporation of L2 knowledge into the online language processing system (e.g., Hahne, 2001; Hahne & Friederici, 2001). However, the learners in these studies were not novice learners, but had extensive (and possibly highly variable) experience with their L2. Furthermore, none of these studies provided a contrast between different levels of L2 exposure or proficiency. In other studies, researchers have recorded ERPs from subjects who were acquiring aspects of an artificial language (Friederici, Steinhauer, & Pfeifer, 2002; McCandliss, Posner, & Givon, 1997). As in our studies described earlier, the learners were novices with respect to the "language" to be learned. Although these studies are quite interesting, it is unclear to what degree the results can be generalized to the acquisition of a natural language in a typical (within the United States) L2 learning environment. These studies have also not included a contrast involving learners at different stages of acquisition.

Only a few studies have both recorded ERPs and contrasted different stages of L2 exposure or proficiency. For example, Weber-Fox and Neville (1996) examined the ERP response to linguistic anomalies in adult Chinese-English bilinguals who had been exposed to English at various ages, ranging from 1 to 16 years. They report interesting differences in the ERP responses to a variety of linguistic anomalies. However, because the study was cross-sectional in nature, there is inevitably some ambiguity about whether these different ERP patterns are in fact manifestations of age of exposure effects or, instead, reflect some other source of uncontrolled subject variability.

The approach proposed here is not a panacea and is itself open to criticism. For example, at the moment we cannot predict or explain learner differences of the type we observed in our French syntax learning study. It is also true that longitudinal designs are extremely time-consuming and are not always viable. Nonetheless, this research paradigm might produce novel results that are less subject to the ambiguity plaguing much of the existing literature. By focusing on changes in the neural manifestations of L2 processing that occur with increasing L2 instruction within a set of L2 learners, we hope to demonstrate the value of asking the what and when questions: Exactly what processes change over time with increasing exposure to or proficiency in an L2? When (in the course of understanding a word or sentence and in the course of learning a language) do those changes occur? Some progress has already been made toward answering such questions in the domain of L2 speech perception (Kuhl, 2004; Osterhout et al., 2004). Perhaps we can reasonably hope that similar advances will soon be forthcoming with respect to the online processing of L2 words and sentences.

Notes

[1] Friederici and colleagues (Hahne & Friederici, 1999; Friederici, 1995) have proposed a two-stage functional model for these ERP effects. They claim that the anterior negativity reflects a fast, automatic syntactic analyzer, and the P600 reflects syntactic reanalysis. As we have explained elsewhere (Osterhout et al., 2004, pp. 294–298), we do not believe that the available evidence strongly supports these theoretical claims.

[2] Another problematic aspect of using the LAN effect to study L2 processing is that there are a significant number of reports involving native speakers in which they are not reported (e.g., Ainsworth-Darnell, Shulman, & Boland, 1998; Allen et al., 2003; Hagoort, Brown, & Groothusen, 1993; Kim & Osterhout, 2005; Kuperberg et al., 2003; McKinnon & Osterhout, 1996; Osterhout, Bersick, & McLaughlin, 1997; Osterhout & Mobley, 1995). This makes it difficult to interpret the absence of the effect in L2 learners.

[3] A related account of L2 learning (Vainikka & Young-Scholten, 1996) proposed that L2 learners initially transfer their L1 lexical categories, but not their L1 grammatical features and categories; learners are claimed to infer these features and categories from the presence of the associated L2 grammatical morphemes. Thus, syntactic structure is projected from the acquired lexical material, and learners must go through a stage of lexical learning before they acquire the features and associated morphosyntax (see also Clahsen, Penke, & Parodi, 1994).

References

Ainsworth-Darnell, K., Shulman, R., & Boland, J. (1998). Dissociating brain responses to syntactic and semantic anomalies: Evidence from event-related brain potentials. *Journal of Memory and Language, 38*, 112–130.

Allen, M. D., Badecker, W., & Osterhout, L. (2003). Morphological analysis during sentence processing. *Language and Cognitive Processes, 18*, 405–430.

Arteaga, D., Gess, R., & Herschensohn, J. (2003). Focusing on phonology to teach morphological form in French. *Modern Language Journal, 8*, 58–70.

Bentin, S. (1987). Event-related potentials, semantic processes, and expectancy factors in word recognition. *Brain and Language, 31*, 308–327.

Bentin, S., McCarthy, G., & Wood, C. (1985). Event-related potentials, lexical decision, and semantic priming. *Electroencephalography and Clinical Neurophysiology*, *60*, 343–355.

Chee, M. W. L., Tan, E. W. L., & Thiel, T. (1999). Mandarin and English single word processing studies with functional magnetic resonance imaging. *Journal of Neuroscience*, *19*, 3050–3056.

Clahsen, H., Penke, M., & Parodi, T. (1994). Functional categories in early child German. *Language Acquisition*, *3*, 395–429.

Franceschina, F. (2001). Morphological or syntactic deficits in near-native speakers? An assessment of some current proposals. *Second Language Research*, *17*, 213–247.

Friederici, A. D. (1995). The time course of syntactic activation during language processing: A model based on neuropsychological and neurophysiological data. *Brain and Language*, *50*, 259–284.

Friederici, A. D., Steinhauer, K., & Pfeifer, E. (2002). Brain signatures of artificial languages: Evidence challenging the critical period hypothesis. *Proceedings of the National Academy of Sciences of the United States of America*, *99*, 529–534.

Frisson, S., & Sandra, D. (2002). Homophonic forms of regularly inflected verbs have their own orthographic representations: A developmental perspective on spelling errors. *Brain and Language*, *81*, 545–554.

Grosjean, F. (1998). Studying bilinguals: Methodological and conceptual issues. *Bilingualism: Language and Cognition*, *1*, 131–149.

Hagoort, P., Brown, C., & Groothusen, J. (1993). The syntactic positive shift (SPS) as an ERP measure of syntactic processing. *Language and Cognitive Processes*, *8*, 439–483.

Hahne, A. (2001). What's different in second-language processing? Evidence from event-related brain potentials. *Journal of Psycholinguistic Research*, *30*, 251–266.

Hahne, A., & Friederici, A. D. (1999). Electrophysiological evidence for two steps in syntactic analysis: Early automatic and late controlled processes. *Journal of Cognitive Neuroscience*, *11*, 193–204.

Hahne, A., & Friederici, A. D. (2001). Processing a second language: Late learners comprehension mechanisms as revealed by event-related brain potentials. *Bilingualism: Language and Cognition*, *4*, 123–141.

Hawkins, R., & Franceschina, F. (2004). Explaining the acquisition and non-acquisition of determiner-noun gender concord in French and Spanish. In P. Prevost & J. Paradis (Eds.), *The acquisition of French in different contexts* (pp. 175–206). Amsterdam: Benjamins.

Herschensohn, J. (1993). Applying linguistics to teach morphology: Verb and adjective inflection in French. *International Review of Applied Linguistics*, *30*, 97–112.

Kim, A., & Osterhout, L. (2005). The independence of combinatory semantic processing: Evidence from event-related potentials. *Journal of Memory and Language, 52*, 205–225.

Kim, K. H. S., Relkin, N. R., Lee, K.-M., & Hirsch, J. (1997). Distinct cortical areas associated with native and second languages. *Nature, 388*, 171–174.

Klein, D., Milner, B., Zatorre, R., Zhao, V., & Nikelski, J. (1999). Cerebral organization in bilinguals: Chinese-English verb generation. *NeuroReport, 10*, 2841–2846.

Kroll, J., & Sunderman, G. (2003). Cognitive processes in second language learners and bilinguals: The development of lexical and conceptual representations. In C. Doughty & M. Long (Eds.), *The handbook of second-language acquisition* (pp. 104–129). Malden, MA: Blackwell.

Kuhl, P. (2004). Early language acquisition: Cracking the speech code. *Nature Reviews Neuroscience, 5*, 831–843.

Kuperberg, G., Holcomb, P. J., Sitnikova, T., Greve, D., Dale, A. M., & Caplan, D. (2003). Distinct patterns of neural modulation during the processing of conceptual and syntactic anomalies. *Journal of Cognitive Neuroscience, 15*, 272–293.

Kutas, M., & Hillyard, S. A. (1980). Reading senseless sentences: Brain potentials reflect semantic anomaly. *Science, 207*, 203–205.

Kutas, M., & Hillyard, S. A. (1984). Brain potentials during reading reflect word expectancy and semantic association. *Nature, 307*, 161–163.

Largy, P., & Fayol, M. (2001). Oral cues improve subject-verb agreement in written French. *International Journal of Psychology, 36*, 121–131.

McCandliss, B. D., Posner, M. I., & Givon, T. (1997). Brain plasticity in learning visual words. *Cognitive Psychology, 33*, 88–110.

McKinnon, R., Allen, M., & Osterhout, L. (2003). Morphological decomposition involving nonproductive morphemes: ERP evidence. *NeuroReport, 14*, 883–886.

McKinnon, R., & Osterhout, L. (1996). Constraints on movement phenomena in sentence processing: Evidence from event-related brain potentials. *Language and Cognitive Processes, 11*, 495–523.

McLaughlin, B. (1998). Second language learning revisited: The psycholinguistic perspective. In A. F. Healy & L. E. Bourne, Jr. (Eds.), *Foreign language learning: Psycholinguistic studies on training and retention* (pp. 399–412). Mahwah, NJ: Erlbaum.

McLaughlin, J., Osterhout, L., & Kim, A. (2004). Neural correlates of second-language word learning: Minimal instruction produces rapid change. *Nature Neuroscience, 7*, 703–704.

Myles, F., Hooper, J., & Mitchell, R. (1998). Rote or rule? Exploring the role of formulaic language in classroom foreign language learning. *Language Learning, 48*, 323–363.

Negro, I., & Chanquoy, L. (2000). Subject-verb agreement with present and imperfect tenses: A developmental study from 2nd to 7th grade. *European Journal of Psychology of Education, 15*, 113–134.

Neville, H. J., Nicol, J. L., Barss, A., Forster, K. I., & Garrett, M. (1991). Syntactically based sentence processing classes: Evidence from event-related brain potentials. *Journal of Cognitive Neuroscience, 3*, 151–165.

Osterhout, L., Bersick, M., & McLaughlin, J. (1997). Brain potentials reflect violations of gender stereotypes. *Memory and Cognition, 25*, 273–285.

Osterhout, L., & Holcomb, P. J. (1992). Event-related brain potentials elicited by syntactic anomaly. *Journal of Memory and Language, 31*, 785–806.

Osterhout, L., & Holcomb, P. J. (1993). Event-related potentials and syntactic anomaly: Evidence of anomaly detection during the perception of continuous speech. *Language and Cognitive Processes, 8*, 413–438

Osterhout, L., McLaughlin, J., Kim, A., Greenwald, R., & Inoue, K. (2004). Sentences in the brain: Event-related potentials as real-time reflections of sentence comprehension and language learning. In M. Carreiras & C. Clifton, Jr. (Eds.), *The on-line study of sentence comprehension* (pp. 271–308). New York: Psychology Press.

Osterhout, L., & Mobley, L. A. (1995). Event-related brain potentials elicited by failure to agree. *Journal of Memory and Language, 34*, 739–773.

Osterhout, L., & Nicol, J. (1999). On the distinctiveness, independence, and time course of the brain responses to syntactic and semantic anomalies. *Language and Cognitive Processes, 14*, 283–317.

Palolahti, M., Leino, S., Jokela, M., Kopra, M., & Paavilainen, P. (2005). Event-related potentials suggest early interaction between syntax and semantics during on-line sentence comprehension. *Neuroscience Letters, 384*, 222–227.

Perani, D., Paulesu, E., Galles, N. S., Dupoux, E., Dehaene, S., Bettinardi, V., et al. (1998). The bilingual brain: Proficiency and age of acquisition of the second language. *Brain, 121*, 1841–1852.

Potter, M. C., So, K.-F., Von Eckardt, B., & Feldman, L. B. (1984). Lexical and conceptual representation in beginning and more proficient bilinguals. *Journal of Verbal Learning and Verbal Behavior, 23*, 23–38.

Raboyeau, G., Marie, N., Balduyck, S., Gros, H., Demonet, J., & Cardebat, D. (2004). Lexical learning of the English language: A PET study in healthy French subjects. *NeuroImage, 22*, 1808–1818.

Radford, A. (1988). *Transformational grammar*. Cambridge: Cambridge University Press.

Saffran, J. R., Johnson, E. K., Aslin, R. N., & Newport, E. L. (1999). Statistical learning of tonal sequences by human infants and adults. *Cognition, 70,* 27–52.

Sandra, D., Frisson, S., & Daems, F. (1999). Why simple verb forms can be so difficult to spell: The influence of homophone frequency and distance in Dutch. *Brain and Language, 68,* 277–283.

Tomasello, M. (2000). The item-based nature of children's early syntactic development. *Trends in Cognitive Sciences, 4,* 156–163.

Vainikka, A., & Young-Scholten, M. (1996). Gradual development of L2 phrase structure. *Second Language Research, 12,* 7–39.

Weber-Fox, C., & Neville, H. J. (1996). Maturational constraints on functional specializations for language processing: ERP and behavioral evidence in bilingual speakers. *Journal of Cognitive Neuroscience, 8,* 231–257.

White, L. (2003). *Second language acquisition and Universal Grammar.* Cambridge: Cambridge University Press.

Wong-Fillmore, L. (1976). *The second time around: Cognitive and social strategies in second language acquisition.* Unpublished doctoral dissertation, Stanford University, Stanford, CA.

Wray, A. (2002). *Formulaic language and the lexicon.* Cambridge: Cambridge University Press.

Strategies for Longitudinal Neurophysiology. Commentary on Osterhout et al.

Doug Davidson
F. C. Donders Centre for Cognitive Neuroimaging
and
Max Planck Institute for Psycholinguistics

Osterhout et al. outline a research strategy to better understand the neurophysiological correlates of second language (L2) comprehension. Previous work has often employed cross-sectional experimental designs and group comparisons (e.g., the first language [L1] of native speakers vs. the L2 of bilinguals), and this work has shown variability in the response patterns in bilinguals. To reduce this variability, Osterhout et al. suggest an alternative research strategy in which novice learners are followed over time as they learn an L2. They argue that the common starting point of the novice learners will reduce the extra variability between individuals that would otherwise exist due to their prior L2 knowledge and skill and, furthermore, that by tracking individuals' rate of learning over time, subjects can serve as their own controls. This should then allow for greater experimental sensitivity to L1-L2 differences as L2 proficiency develops. They provide evidence from two series of studies: one concerning lexico-semantic acquisition and another concerning semantic and grammatical sentence processing using this type of design.

The second part of Osterhout et al.'s recommendation is to employ event-related potential (ERP) responses to study L2

Correspondence concerning this article should be addressed to Doug Davidson, F. C. Donders Centre for Cognitive Neuroimaging, P. O. Box 9101, 6500 HB Nijmegen, The Netherlands. Internet: Doug.Davidson@fcdonders.ru.nl

acquisition and comprehension. They suggest that ERP responses can reveal real-time sensitivity to semantic contrasts, as demonstrated by the N400 effect, and grammatical contrasts, via the P600 effect. It is important to emphasize the utility of this technique for understanding the neurophysiology of sentence processing. Much linguistic information, whether L1 or L2, appears in the form of sentences or clauses, and because words appear sequentially over time, a response measure is required that will show a response to each individual word as it is encountered. EEG or MEG measurements have the required temporal resolution for this measurement. As Osterhout et al. point out, other investigations of L2 processing using techniques such as functional magnetic resonance imaging (fMRI) have coarser temporal resolution and are thus less well suited for addressing questions about the time course of responses. Nevertheless, there are important limitations with electrophysiological techniques as well; for example, active areas that are located in the deep cortex or that produce closed fields, such as nuclei, might be difficult to observe using techniques such as EEG or MEG. Some evidence from language learning tasks using fMRI has shown involvement of thalamic nuclei (Opitz & Friederici, 2003). Especially for experimental designs in which changes are expected to occur over an extended period of time, it might be advantageous to have participants perform the same task in both fMRI and EEG to understand how changes in activity that is registered in fMRI are linked to changes observed with electrophysiological techniques (Dale & Halgren, 2001).

Osterhout et al. observed striking changes in the ERP response patterns in both the lexical and sentence processing tasks. In the first sentence processing study reported by Osterhout et al., however, they noted that variability between subjects remained as a dominant factor (a point they raise in the Discussion). Despite employing novice learners from the same university class and using textbook materials obtained from the classroom, the initial change in the response to grammatical violations was observed in only a subset of subjects. However, it should be emphasized that a group of subjects such as a university class is already quite homogenous. Students in a classroom setting are likely to have similar educational backgrounds with relatively high achievement in order to enroll in a university. The students would have the same instructor, the same homework assignments, and the same textbook. Thus, these subjects might be more homogenous, in terms of their educational background, prior language history, and classroom experience, than other groups of learners. Despite this, between-subject variability remained an important factor. Thus, it appears that other methods for assessing the impact of individual differences would be helpful, in conjunction with longitudinal designs. One possible approach would be to collect data from a group of tasks that measure a range of cognitive abilities and to use these as covariate measures for the changes in response patterns that are observed. Although this method does not eliminate individual differences, it could allow their influence to be modeled.

Finally, longitudinal experimental designs using students from a classroom setting could be used to examine learning on several different timescales. From an experimental point of view, there are many potential advantages to examining how responses change in students in a classroom setting. For example, the learning materials and vocabulary of the students can be used to construct experimental materials. Detailed information about teaching methods might be available, as well as information from a syllabus about when topics occur and when exercises and drills are performed. In addition, many behavioral and ERP experimental tasks share some similarity to the exercises in a classroom setting. In certain respects, an ERP experiment could be designed to be similar to a grammar learning exercise, as has already been done in some artificial grammar learning experiments (Opitz & Friederici, 2003). Thus, with this type of design, it might be possible to link short-term changes that occur within the time period of a single experiment (roughly corresponding to a classroom exercise) to the lasting longer-term changes in responses that occur over the duration of an entire course; for example, research on word repetition has shown that delayed repetition can lead to later responses than observed during immediate repetition (Rugg & Nagy, 1989) and that successful delayed recognition can lead to earlier responses than unsuccessful delayed recognition (Dhond, Witzel, Dale, & Halgren, 2005).

References

Dale, A. M., & Halgren, E. (2001). Spatiotemporal mapping of brain activity by integration of multiple imaging modalities. *Current Opinion in Neurobiology, 11*, 202–208.

Dhond, R. P., Witzel, T., Dale, A. M., & Halgren, E. (2005). Spatiotemporal brain maps of delayed word repetition and recognition. *NeuroImage, 28*, 293–304.

Opitz, B., & Friederici, A. (2003). Interactions of the hippocampal system and the prefrontal cortex in learning language-like rules. *NeuroImage, 19*, 1730–1737.

Rugg, M. D., & Nagy, M. E. (1989). Event-related potentials and recognition memory for words. *Electroencephalographic and Clinical Neurophysiology, 72*, 395–406.

L2 in a Nutshell: The Investigation of Second Language Processing in the Miniature Language Model

Jutta L. Mueller
Max Planck Institute for Human Cognitive and Brain Sciences

The present chapter bridges two lines of neurocognitive research, which are, despite being related, usually discussed separately from each other. The two fields, second language (L2) sentence comprehension and artificial grammar processing, both depend on the successful learning of complex sequential structures. The comparison of the two research directions will be taken as the starting point for the attempt to study L2 sentence comprehension using a miniature language model. The experiments that will be presented made use of the event-related potential (ERP) method, which provides a sensitive tool for the fast and multidimensional processes that characterize online processing of sequential structures. The first part of the article provides an overview of ERP studies on first and second language sentence comprehension and compares these to findings from the field of artificial grammar processing. In the second part, two studies in which native and nonnative sentence processing in the model of a miniature version of Japanese was investigated is presented. The findings of these studies suggest that miniature languages can serve as useful tools for capturing the

---

The research reported here was supported by the Max Planck Institute for Human Cognitive and Brain Sciences, Leipzig. Specifically I would like to thank Angela D. Friederici, Yugo Fujii, Anja Hahne, and Shirley-Ann Rueschemeyer for their advice in the preparation of the experiments and the manuscript.

Correspondence concerning this article should be addressed to Jutta L. Mueller, Max Planck Institute for Human Cognitive and Brain Sciences, Stephanstrasse 1a, 04103 Leipzig, Germany. Internet: muellerj@cbs.mpg.de

impressive degree of plasticity as well as certain limitations in language comprehension mechanisms applied by learners.

Compared to children's ultimate success in language learning, the achievements of late learners of a second language (L2) seem often very poor. The prototypical late L2 learners' speech is riddled with grammatical errors and usually identifiable by its foreign-sounding accent. For this reason, many researchers emphasize the determining role of age of acquisition (AoA) specifically for the acquisition of syntax and phonology and assume that a critical period constrains the ultimate level of L2 proficiency (e.g., Johnson & Newport, 1989; Pulvermüller & Schumann, 1994). Contradictory to this possible restriction, some learners seem to become so skillful in their L2 that they are no longer distinguishable from native speakers on the basis of their linguistic performance (e.g., Birdsong, 1992; White & Genesee, 1996).

With the advent of neurophysiological methods that allow for the recording of neural activity during language processing, researchers have begun exploring differences between L1 (first language) and L2 processing at the level of brain functions. Many studies using electrophysiological (e.g., electroencephalography [EEG], magnetoencephalography [MEG]) or hemodynamic measures (e.g., functional magnetic resonance imaging [fMRI], positron-emission tomography [PET]) have found that brain activity during L1 processing is different in some aspects to brain activity during L2 processing. In the first part of this article it will become evident that absolutely nativelike patterns in L2 learners were rarely found in neurophysiological studies, specifically in the domain of syntactic processing. However, there is a lack of neurophysiological studies testing L2 processing at the highest possible proficiency levels.

The miniature language studies reported later in this chapter are aimed to contribute to the question of whether neurophysiological correlates of language comprehension in L2 learners at very high proficiency levels converge with those in natives by using event-related potentials (ERPs). With their time resolution in the range of milliseconds and their sensitivity to specific subprocesses involved in language comprehension, such as phonology, prosody, semantics, and syntax, ERPs are a tool well suited for the study of language processing in L1 as well as in L2 speakers. By conjoining aspects from two separate lines of research, namely natural language and artificial grammar[1] studies, a paradigm will be derived that allows for the examination of specific language comprehension mechanisms in the model of a miniature version of a real language. It will be argued that those artificial languages that follow universal rules of language and miniature versions of real languages in particular can serve as valid models for full natural languages with respect to the neurophysiological processes that they entail. As a possible advantage to natural languages, those restricted systems are learnable within a very short period of time. In the approach that is taken here, a miniature language that is a subset of a real language was applied for the study of native versus nonnative processes of sentence comprehension. Listeners engage in multiple processes of structure building and interpretation before the end of a sentence is reached (Marslen-Wilson, 1973; Mitchell, 1994). These processes can comprise syntactic structure building as well as the establishment of thematic relationships in order to determine "who is doing what to whom." Violations of both information types were shown to elicit specific electrophysiological responses during sentence processing (e.g., Friederici, Pfeifer, & Hahne, 1993; Frisch & Schlesewsky, 2001). It is not resolved whether nonnatives can become capable and as efficient as natives in the application of the mental operations involved in syntactic and thematic analyses and develop similar patterns of electrophysiological brain activity. Before turning to the miniature language studies, however, some theoretical and empirical background will be provided in the next sections (cf. Birdsong, this volume, for a much more detailed discussion of relevant theoretic concepts in the investigation of L2 processing at the upper level of proficiency).

Theoretical Approaches to Differences Between L1 and L2 Speakers

Since Lenneberg (1967) suggested the critical period hypothesis for L2 acquisition, conceptions of how late learned L2s differ from early learned L1s have been elaborated within the different branches of biological and cognitive sciences. Some approaches stress the role of maturational parameters such as changes in brain plasticity (Lenneberg; Pulvermüller & Schumann, 1994) or even hormonal changes (Ullman, 2004) in explaining difficulties of adult L2 learners. Other theorists focus on general cognitive (Elman, 1993, Newport, 1988, 1990) or linguistic factors (Bley-Vroman, 1989; Clahsen & Felser, 2006; Eubank & Gregg, 1999) with a disadvantageous influence on the ability to learn an L2. Whereas some theorists explicitly assume a critical period restricting the age up to which an L2 can be possibly learned to full proficiency (Bley-Vroman; Pulvermüller & Schumann), others put the concept of critical period into question as a whole and favor the notion of a more steady age-related decline of possible L2 attainment (e.g., Bialystok & Hakuta, 1999; Birdsong & Molis, 2001). It is uncontroversial, though, that the ability to learn an L2 declines with age, although the degree of attainable "nativelikeness" remains a matter of debate. Nearly all models assume that the area of syntax is more adversely affected by delays in language learning than the area of semantics. This implies that neural correlates of syntactic processing should be expected to deviate in L2 speakers more from a native norm than neural correlates of semantic processing (e.g., Clahsen & Felser; Ullman, 2001). How the different factors that impact proficiency in L2 learning influence neural correlates of L2 processing and how these develop during the course of learning is still widely unexplored. In order to test and further specify models of L2 processing at the neurophysiological level, data from L2 speakers at different stages of proficiency are needed—from beginner state up to the near-native level. As near-natives are a rather small population (cf. Birdsong, 1999) and longitudinal studies are very resource-demanding, most of the currently available neurophysiological experiments have tested participants at an advanced, but not nativelike proficiency stage. Just recently, some studies have tested L2 learning in longitudinal settings (cf. Indefrey, this volume; Osterhout et al., this volume) or in paradigms or groups, which allow the testing of processing mechanisms at upper levels of proficiency (Barber-Friend, Gillon-Dowens, & Carreiras, 2004; Friederici, Steinhauer, & Pfeifer, 2002). These lines of research will certainly help specify what late L2 learners are capable of during the process and at ultimate stages of learning.

ERPs as a Measure of Human Sentence Comprehension Mechanisms

Event-related potentials are small scalp-recorded voltage changes in the continuously measured EEG. They are precisely time-locked to an external event and averaged over many trials in order to enhance the signal-to-noise ratio. The positive and negative deflections in the ERP, which are typically observed in relation to specific experimental conditions, are referred to as ERP components. ERP components can vary in amplitude, topography, latency, and polarity. Because ERP components have proven to be sensitive to different types of linguistic information, they provide information that can help disentangle different subprocesses involved in language comprehension, such as syntactic and semantic processing. A paradigm that is frequently, although not exclusively, used for this purpose is the violation paradigm, in which a specific property or information type is presented in a correct as well as an incorrect version. Differences in the ERP patterns between correct and incorrect stimuli are then related to processing characteristics of the violated information type.

The most intensively studied language-related component, the N400, was first reported by Kutas and Hillyard (1980), who visually presented normal sentences (e.g., *He spread the warm bread with butter.*), sentences containing a semantic violation (e.g., *He spread the warm bread with socks.*), and sentences ending with a semantically congruent but physically deviating word, written in capitalized letters (e.g., *He spread the warm bread with BUTTER.*). They found an enhanced negativity peaking at around 400 ms poststimulus onset time-locked to the semantically incongruent word. The negativity was not present for the physical deviation. Here, a positivity (P300) was observed, a component that has been reported in many other studies in the case of infrequent and task-relevant attended stimuli (cf. Donchin, 1981). A rich body of subsequent research has shown that the N400 occurs whenever a word is semantically unexpected in its given context, be it within a word list, a sentence, or a text passage, and thus, it is an established marker of semantic/conceptual integration processes (for an overview, see Kutas & Federmeier, 2000). The N400 has been elicited not only by visually presented stimuli but also by auditory sentence presentation, although often with a more frontal distribution (Conolly & Phillips, 1994; Conolly, Stewart, & Phillips, 1990; Holcomb & Neville, 1990). It should be noted that the N400 in auditory experiments is sometimes accompanied by modulations of the earlier N200 component. Conolly and colleagues could demonstrate that this earlier effect, which they termed PMN (phonological mismatch negativity), was related to the phonological expectancy of a word independently of its meaning (Conolly & Phillips). Some studies have shown that the N400 is further modulated by a mismatch of case information, which specifies the thematic roles within a sentence (Friederici & Frisch, 2000; Frisch & Schlesewsky, 2001, 2005). Thematic role information relates to nonsyntactic, semantically very general hierarchical structure building processes. Thus, Frisch and colleagues' findings possibly extend the range of processes known to influence the N400. A number of language-related studies focusing on syntactic processing reported negativities that were more pronounced on anterior or left anterior electrodes occurring between 300 and 500 ms. This component, known as the LAN (left anterior negativity), has been shown for different kinds of morphosyntactic violation conditions, such as tense, number, and gender agreement errors (Coulson, King, & Kutas, 1998; Gunter, Friederici, & Schriefers, 2000). Furthermore, LAN effects were found in response to function words as compared to content words in the absence of any violations (Brown, Hagoort, & Ter Keurs, 1999; Neville, Mills, & Lawson, 1992). For a specific class of syntactic manipulations, namely word category violations, an early left anterior negativity (ELAN) between 100 and 300 ms after stimulus onset has been reported (Friederici et al., 1993; Neville, Nicol, Barss, Forster, & Garrett, 1991), which has been interpreted as an index of first-pass parsing processes, which enable initial phrase structure building based solely on word category information (Friederici, 2002).

Another ERP component that has frequently been reported in the syntactic domain is a centro-parietally distributed positive wave, the P600 or SPS (syntactic positive shift). The P600 has been reported for syntactic violation conditions (Friederici et al., 1993; Neville et al., 1991; Osterhout & Holcomb, 1992), for noncanonical, less preferred structures (Osterhout, Holcomb, & Swinney, 1994), and for syntactically complex structures (Kaan, Harris, Gibson, & Holcomb, 2000). Thus, the P600 has often been interpreted as reflecting processes of reanalysis and syntactic repair (Friederici, 2002; Osterhout & Holcomb; Osterhout et al.) or as a more general index of the complexity of syntactic integration (Kaan et al.).

Rather recently, another ERP component was discovered, which reflects sentence-level phrasing on a prosodic rather than a syntactic level. Steinhauer, Alter, and Friederici (1999) have shown that the processing of intonational phrase boundaries[2] is accompanied by a centro-parietally distributed positive component, which has been termed "closure positive shift" (CPS). The CPS has been replicated for the processing of spoken, written, delexicalized, and hummed language (Pannekamp, Toepel, Alter, Hahne, & Friederici, 2005; Steinhauer & Friederici, 2001).

ERP Studies on L2 Sentence Processing

With respect to L2 comprehension, ERP evidence suggests that sentential lexico-semantic information processing in proficient L2 speakers and native speakers of a language are not qualitatively different. Ardal, Donald, Meuter, Muldrew, and Luce (1990) found an N400 in response to semantic violations in sentences that peaked later in the L2 compared to the L1. Subsequent studies using comparable paradigms obtained similar results, namely N400s similar to those for native speakers that were delayed in some cases (Hahne, 2001; Hahne & Friederici, 2001; Moreno & Kutas, 2005; Weber-Fox & Neville, 1996). Together, the studies suggest a possible slowdown of lexico-semantic integration processes in late L2 learners. The study of Moreno and Kutas suggests that both age of exposure and attained vocabulary proficiency might play an important role in determining the observed N400 latency differences. In strong contrast to this, ERP components related to syntactic processing for L2 differ significantly from those for native speakers. Weber-Fox and Neville found differential ERP patterns in response to syntactic (word category) violations in L1 and L2 speakers of English, who had started L2 learning at different ages (1–3, 4–6, 7–10, 11–13, >16 years of age). Native speakers displayed an early left-lateralized negativity (N125) that was followed by a later left-lateralized negativity (N300-500) and a P600. When the L2 was learned after the age of 3 years, the early left negativity (N125) was not observed, except for the group with AoA between 11–13 years who showed a reversed left-right topographical distribution. The later negativity (N300-500) was present in all L2 groups, but it was bilaterally distributed when the AoA was later than 11 years of age. The P600 effect was similar to that of native speakers in the groups up to the AoA of 10 years. The "11–13" years group displayed a very delayed positivity starting at about 700 ms, and in the ">16" years group, no positivity could be found in response to word category violations. The authors argue in favor of a biologically determined sensitive period specific to syntax acquisition indicated by the differential sensitivity of the syntactic processing system to AoA effects, which was not seen for semantic processing. In two similar studies Hahne (2001) and Hahne and Friederici (2001) found native speakers of German to show an ELAN followed by a P600 for word category violations. When native Japanese late L2 learners of German were tested with the same material, they displayed neither of the two ERP effects in response to word category violations (Hahne & Friederici). Native Russian L2 speakers of German showed a P600 for syntactically incorrect sentences with a slightly delayed peak latency, but no ELAN (Hahne). The difference between the Russian and the Japanese L2 speakers of German might be due to proficiency, as the Russian group was more proficient than the Japanese group, which was evident in behavioral and questionnaire data (Hahne).

Neville and colleagues (Neville et al., 1992; Weber-Fox & Neville, 2001) also compared the processing of open- and closed-class words in L1 and L2 speakers. In native participants, open-class words (as compared to closed-class words) elicited a more negative-going waveform at 350 ms, whereas closed-class words (as compared to open-class words) led to a more negative ERP at 280 ms. Thus, the two word types go along with temporally different ERPs. Wheras the negativity for open-class words was broadly distributed and peaked at 350 ms, the negativity for closed-class words was stronger on left anterior electrode sites and peaked earlier (280 ms). Neville et al., who tested native speakers of American Sign Language (ASL) in their L2 English, did not observe the N280 for closed-class words, whereas the N350 for open-class words was similar to that of English native speakers. Weber-Fox and Neville tested native Chinese speakers (L2 English) with the same material. The negativities for both word classes were observed in the L2 speakers, but the peak latency of the N280 for closed-class words was delayed if the L2 was learned after the age of 7 years. This was not the case for the N350 for open-class words. Weber-Fox and Neville noted that possible differences in the proficiency levels between the ASL and the

Chinese group might account for the observed differences in the N280.

In a recent study Hahne, Mueller, and Clahsen (2006) investigated the processing of regular and irregular participles and plurals in late L2 learners of German. In native speakers of German, it has been found that incorrect applications of regular and irregular suffixes lead to differential ERP effects, namely an LAN (and a P600) for the incorrect use of regular suffixes, and an N400 for incorrect use of irregular suffixes (Penke et al., 1997; Weyerts, Penke, Dohrn, Clahsen, & Münte, 1997). Because the LAN can be interpreted as indicating morphosyntactic processes and the N400 as indicating lexico-semantic processes, these findings suggest a dual-route model that assumes rule-based decomposition for regular words and lexical storage for irregular words (Clahsen, 1999; Pinker, 1991). Hahne et al. found that participles and plurals with incorrect irregular suffixes both elicited an N400 in L2 speakers. Whereas incorrect regular suffixes led to a LAN and a P600 effect in the participle system, only a P600 was found for plurals with incorrect regular suffixes. The authors argued that, as for native speakers, two processing routes are, in principle, available to the L2 speakers. However, the absence of the LAN for incorrectly applied regular plural suffixes points to increased difficulties of the L2 speakers while processing the relatively complex plural system compared to the morphologically simpler participle system (Hahne et al.). Therefore, the development of relatively automatic morphosyntactic processes seems possible depending on the complexity of the morphosyntactic domain.

The overall picture of language-related ERP components in L2 learners can be looked at from different perspectives. The above-reviewed studies show that ERP patterns in L2 processing often resemble those observed in L1 processing, as the N400 and also the P600 are typically found in late L2 learners. This suggests that, at an advanced proficiency stage, the processing mechanisms in these domains are functionally equivalent to native speakers'. However, other processes, specifically those relatively automatic syntactic processes reflected in the ELAN component, seem to be very difficult to acquire in late L2 learning. What is underlying this difficulty and if and how it might be overcome are still subject to further research.

ERP Studies on Sequence Learning and Artificial Language Processing

There are several ERP studies in which cognitive processes involved in the processing of artificially constructed sequences or languages have been tested. These have interesting implications for research on language processing, although this was not necessarily the main focus of some of the studies. ERP studies in which the processing of nonlinguistic sequentially ordered events was tested were, for example, reported by Eimer, Goschke, Schlaghecken, and Stürmer (1996), Baldwin and Kutas (1997), Schlaghecken, Stürmer, and Eimer (2000), and Rüsseler, Henninghausen, Münte, and Rösler (2003). Baldwin and Kutas, for example, used the implicit structured sequence learning task in which participants respond to rule-governed motions of a visual object on the computer screen. The sequences in the experiment consisted of movements of a small green square between different locations within a small grid displayed on the screen. Those movements could either be grammatical or ungrammatical with respect to a finite-state grammar.[3] The participants were given the task of responding to specific target movements. Ungrammatical events led to a delayed P3b component in the case of target movements and to an enhanced positivity in the case of nontarget movements. Other studies on the processing of nonlinguistic sequences similarly reported modulations of the P3b component in response to violated items and, depending on conscious knowledge of the sequential input structure, additional N2b effects (Eimer et al.; Schlaghecken et al.; Rüsseler et al.). Importantly, LANs, as known from syntactic violations in language, were not reported in these studies. One reason for this might be, apart from differences in the complexity of the sequential structure, that syntactic dependencies in language often involve different word categories rather than specific items of the same category.

There are only a few ERP studies in which the investigated sequential structures comprise elements that correspond to syntactic categories of language. There is one study conducted by Hoen and Dominey (2000) in which participants were visually presented with sequences of letters that corresponded or did not correspond to previously learned rules. The rule was either a transformation rule, whereby a specific letter predicted a structural transformation of the subsequent letters (element X in position 4 indicates transformation of the type 123-312: ABCXCABX), or a repetition rule, whereby a letter in a specific position had to be repeated in the final position of the sequence (element Z in position 4 has to be repeated in position 8: ABCZDEFZ). The participants had to give a grammaticality judgment. The ERP in response to letters indicating the transformation rule versus the repetition rule strongly resembled the LAN response that is elicited by function words versus content words in language.

Friederici et al. (2002) studied miniature artificial grammar processing in an auditory paradigm. They examined the processing of an artificially constructed miniature language (Brocanto) consisting of only 14 words and a phrase structure grammar[4] generating legal sentence strings. The words belonged to one of six different grammatical categories. Participants learned the rules and the meaning of Brocanto sentences by playing a computer game that included a comprehension and a production task until they reached a high-proficiency criterion. In the ERP experiment, correct sentences of Brocanto and sentences containing word category violations were presented. Word category violations elicited an early anterior negativity and a P600 in trained listeners, whereas listeners without knowledge of the Brocanto rules did not show any ERP effect. The authors argued that the early negativity represents, similar to the ELAN component, automatic processing of word category information, and the P600 reflects syntactic repair processes.

In sum, the sequence learning and artificial grammar learning studies have two important implications. First, ERP patterns observable in sequence processing and artificial grammar studies can be highly similar to those observed for syntactic processing in natural language provided that the investigated structures are formally equivalent to syntactic structures in language (Friederici et al., 2002; Hoen & Dominey, 2000). Taking this restriction into consideration, artificial sequences seem to be interesting models for "full" natural languages. Second, the finding of ERP components that indicate relatively automatic syntactic processes, namely anterior negativities (Friederici et al.; Hoen & Dominey), provides evidence against the suggestion that the underlying processes cannot be acquired during the acquisition of an L2 after puberty, as one might conclude from previous ERP studies using natural language material (Hahne, 2001; Hahne & Friederici, 2001; Weber-Fox & Neville, 1996). As the number of empirical studies is still very small and as the stimuli that were used in the above-described experiments are far less complex than in language, these conclusions should be considered as preliminary and need further empirical testing. In the following sections, two studies will be reported in which we tried to bridge artificial and natural language processing research by using a miniature version of real Japanese, which we termed Mini-Nihongo (*Nihongo* is the Japanese expression for "Japanese").

Mini-Nihongo: Testing L2 Processing in the Model of Miniature Japanese

Based on the above-mentioned conclusions about the comparability of artificial and natural language processing, it is assumed here that a miniature version of Japanese serves as a valid model of natural Japanese. Thus, native Japanese speakers should process such a miniature version of their native language in a similar way as natural Japanese, and nonnative learners of Mini-Nihongo should apply processes that are relevant when processing a natural foreign language. We chose Japanese as the target language for our miniature language learning experiments because of its typological distance to Indo-European languages and its rather restricted phonology. Japanese makes use of only 24 different phonemes (Katsuki-Pestemer, 1991), which all belong to different phoneme categories in German. Therefore, we assumed that phonology would not be a great obstacle to the learning of syntactic rules in native German speakers. Furthermore, there is already a small number of studies on Japanese sentence processing that have shown that semantic and syntactic manipulations lead to characteristic ERP effects similar to the more extensively studied languages: English, Dutch, and German. The N400, the P600, and the LAN components have all been reported for processing of Japanese as an L1 (Miyamoto, Katayama, & Koyama, 1998; Nakagome et al., 2001; Takazawa et al., 2002; Ueno & Kluender, 2003). For several reasons, it was planned to present the sentences of Mini-Nihongo only in the auditory modality. First, the auditory channel is used most frequently for communication in daily life, which makes the study of its functioning interesting for reasons of ecological validity. Second, the different script types used in Japanese and German[5] precluded the unconfounded comparison of processing mechanisms in native Japanese and trained German participants in the visual domain. Third, the presence of a natural prosodic structure ensures a high degree of comparability to natural language acquisition and provides an additional cue to phrase structure. In particular, previous studies, mostly concerned with L1 acquisition, have shown that prosodic information might assist in the learning of syntactic dependencies (Kemler Nelson, Hirsh-Pasek, Jusczyk, & Cassidy, 1989; Morgan, Meier, & Newport, 1987; Soderstrom, Seidl, Kemler Nelson, & Jusczyk, 2003). In the following, the linguistic structure of Mini-Nihongo and the range of processes that can be tested within its framework are outlined.

Lexicon and Rules of Mini-Nihongo

Mini-Nihongo contains seven syntactic categories consisting of very few words that can, however, be combined to form a total of 2,048 different grammatical sentences. Figure 1 illustrates the structure and elements of Mini-Nihongo. The elements of Mini-Nihongo are four nouns, four verbs, three postpositions, two numeral classifiers, two numerals, one adjective, and one temporal adverb. Within the structural constraints implemented in Mini-Nihongo, both canonical sentences as in (1) and scrambled sentences as in (2) are grammatical examples.

(1) Ni hiki no neko ga ichi wa no hato o tsukitobasu tokoro desu.
2 [small-animal][gen.] cat [nom.] 1 [bird] [gen.] pigeon [acc.] push away take place.
"Two cats are pushing away one pigeon."

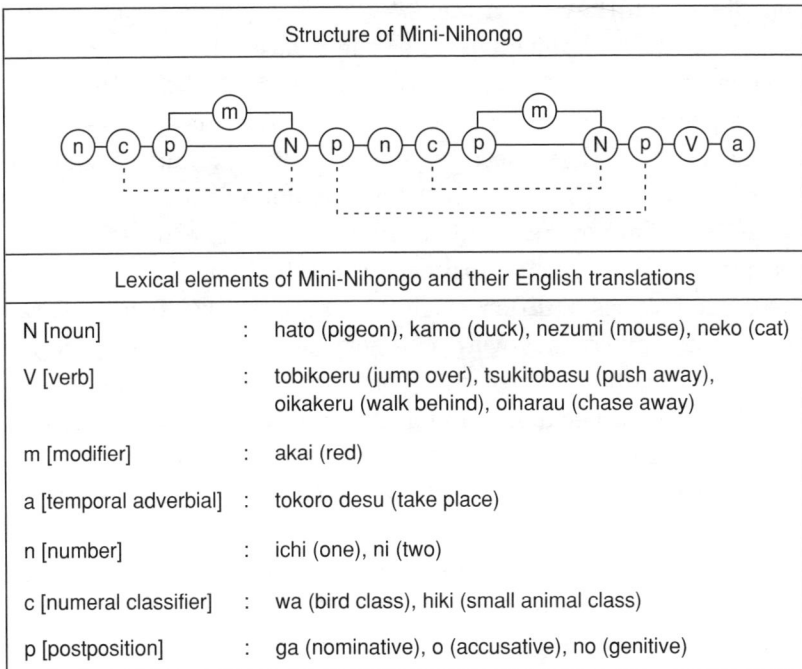

Figure 1. Schematic representation of structure and lexicon of Mini-Nihongo. The nodes in the upper box denote word classes, solid lines mark legal transitions between elements of the classes, and dotted lines indicate dependencies between specific class members. Sentences are generated from left to right.

(2) Ichi wa no akai kamo o ni hiki no nezumi ga oikakeru tokoro desu.
1 [bird][gen.] red duck [acc.] 2 [small-animal] [gen.] mouse [nom.] follow take place.
"Two mice are following one red duck."

A nominal phrase (NP) of Mini-Nihongo consists of a case-marked noun that is modified by a numeral, a classifier, and optionally by a color adjective. Number information is not encoded in Japanese noun morphology. Thus, the same noun form can refer either to the singular or to the plural meaning. In order to count Japanese nouns, they have to be combined with numeral classifiers, which serve as individualizers (Greenberg, 1972). The numerals and the classifier always precede their head noun to which they are attached with the genitive postposition "no." The optional adjective takes the position between "no" and the head of the NP. The accusative postposition "o" and the nominative postposition "ga" mark the end of nominal phrases. The verbal phrase (VP) is located sentence finally and consists of a verb and a temporal adverb, which denotes the continuous mode of the action referred to by the verb. The present studies investigate language processing mechanisms by measuring the consequences of different violation types. Specifically, it was planned to investigate word category processing, case processing, and classifier processing. Word category processing was of specific interest because it involves relatively automatic computational operations in native speakers and it is unclear if these processes can become nativelike in the course of L2 learning. It has been suggested that the development of (automatic) syntactic processes is subject to maturational effects and, thus, might not be possible in adulthood (Neville et al., 1992; Ullman, 2001; Weber-Fox & Neville, 1996, 2001). The observation of indications for automatic processing of word category information in the above-mentioned artificial language study by Friederici et al. (2002) motivated the study of these processes within the model of Mini-Nihongo.

Case processing was chosen as a target of investigation because it allows testing of L2 processing in the thematic domain, an area that has been neglected in previous ERP studies on L2 processing. Overt case marking, present in Japanese as well as in Mini-Nihongo, might trigger the thematic interpretation of the constituents before encountering the verb, which is always located sentence finally. Previous psycholinguistic studies provided evidence that Japanese case markers are used incrementally during sentence processing (Miyamoto, 2002; Yamashita, 1997). In a recently proposed neurocognitive model of sentence comprehension (Schlesewsky & Bornkessel, 2004), thematic processing is seen as initially independent of syntactic and general semantic processes, although some specific semantic features (e.g., animacy) might be considered for the building of thematic dependencies. It is known that semantic processing is relatively unimpaired in proficient L2 speakers, but it is not clear if this extends to the processing of thematic information. A special area of difficulty in L2 acquisition might be the development of nativelike processing mechanisms for the computation of grammatical features that are not present in the learners' L1. As a test case for such a process, we included Japanese classifiers in our miniature language that are not a grammatical feature inherent to German. Although the functional characteristics of classifier processing are yet to be specified, it is interesting to compare native and nonnative processing of this grammatical feature, which is not transferable from the nonnatives' L1.

Mini-Nihongo Study 1

In a first study using Mini-Nihongo by Mueller, Hahne, Fujii, and Friederici (2005), the processing of word category, case, and classifier information was assessed in Japanese native speakers and German participants (referred to as "nonnative participants" in the following) before and after intensive training. The training, which was provided to nonnative participants after a first pretraining ERP measurement, was a computerized game, which included auditory comprehension and oral production tasks. Participants had to reach a high-proficiency criterion in order to be admitted to the second, posttraining ERP measurement. The average training time was 7.2 hr (range: 4–10 hr).

In the ERP experiment, participants listened to correct and incorrect sentences of Mini-Nihongo (50% correct) and gave a grammaticality judgment by button-press after every sentence. The different experimental conditions are exemplified in Table 1. During the ERP experiment, the accuracy rates in the grammaticality judgment task were equally high for native and trained nonnative participants for the correct condition, the classifier violation, and the word category violation condition (classifier violation: 89% correct; word category violation: 98–99% correct; correct condition: 97% correct). Only in the case violation condition did the nonnative trained participants perform worse

Table 1

Examples of the stimuli used in the Mini-Nihongo study 1

Correct condition		
Ichi wa no kamo ga	ni hiki no neko o	tobikoeru tokoro desu.
1 [bird][gen.] duck [nom.]	2 [small-animal][gen.] [acc.]	jump over take place.
Word category violation		
Ichi wa no kamo ga	ni hiki no	<u>tobikoeru</u> tokoro desu.
1 [bird][gen.] duck [nom.]	2 [small-animal][gen.]	jump over take place.
Case violation		
Ichi wa no kamo ga	ni hiki no <u>neko ga</u>	tobikoeru tokoro desu.
1 [bird][gen.] duck [nom.]	2 [small-animal][gen.] cat [nom.]	jump over take place.
Classifier violation		
Ichi wa no kamo ga	ni wa no <u>neko</u> o	tobikoeru tokoro desu.
1 [bird][gen.] duck [nom.]	2 [bird][gen.] cat [acc.]	jump over take place.

Note. gen. = genitive, nom. = nominative, acc. = accusative; incorrect elements are underlined.
Source. Mueller et al., 2005.

than natives (native participants: 98% correct; trained nonnative participants: 84% correct). Thus, the behavioral data confirm a learning effect for the trained nonnative participants who were not able to discriminate correct from incorrect sentences before training. The three violation conditions elicited three different ERP patterns in Japanese native speakers. The word category violation, in which a verb occurred in the structural position of the noun, elicited an anteriorly focused, broadly distributed negativity (100–300 ms) followed by a P600 (500–800 ms). This was taken to suggest that Japanese natives apply relatively automatic syntactic mechanisms and more controlled processes of repair sequentially during the analysis of the syntactic phrase structure. The unusually broad distribution of the early negativity was interpreted as an overlap between an anteriorly focused "syntactic" portion of the negativity and a centro-parietally focused "prosodic" portion. The interpretation of the centro-parietal effect as prosodic was derived from the observation that correct sentences seemed to elicit a CPS (cf. Pannekamp et al., 2005; Steinhauer et al., 1999) directly before the onset of the verb. This probably reflects the processing of the prosodic boundary present in correct sentences at this position. In the word category violation, however, the verb is placed in the middle of a prosodic phrase, where no such boundary is expected. Therefore, there was no CPS observable at verb onset in the incorrect sentences. The early, more negative-going ERP for incorrect sentences can thus be interpreted as the lack of the CPS at verb onset. Interestingly, a similar early negativity was observed in the nonnative participants already before learning. However, the topographical distribution of the effect was different, as it was more centrally distributed than the early negativity in Japanese native speakers. Because a CPS before verb onset was also observable in the nonnative participants before training, the early negativity was seen as being related to prosodic phrasing processes in a similar way as in the native participants. This points to a potentially important role of prosodic information not only during L1 acquisition but also during L2 learning. Prosody might be particularly useful in beginning stages of learning, where it could be used to chunk language input before it is fed into higher order analyses (cf. Morgan et al., 1987). After training, however, the nonnative participants displayed both an early negativity and a P600 effect for word category violations. Although the late processes of syntactic repair as reflected in the P600 seemed to be applied by learners in a way similar to native speakers, topographical differences in the time window of the early negativity indicated differences between learners and natives. The early negativity in learners was more posteriorly focused, which was interpreted as an indicator of enhanced reliance on prosodic processes as compared to syntactic processes. The case violation condition, which consisted of an incorrect double nominative case marking, elicited a biphasic pattern consisting of an N400-like negativity (300–500 ms) and a P600 (500–800 ms) in Japanese native speakers. Corresponding to similar results from the German language (Frisch & Schlesewsky, 2001, 2005), the negativity was interpreted as an instance of an N400 and seen as a reflection of difficulties in thematic hierarchizing of the case-marked arguments. The P600 was interpreted as indicating syntactic processes of repair. Additionally, case violations elicited an earlier negativity in Japanese native speakers (100–300 ms), which was interpreted as related to processes of phonological expectation similar to the PMN reported by Conolly and colleagues (Conolly & Phillips, 1994; Conolly, Phillips, Stewart, & Brake, 1992). Such processes could have occurred, as case violations always occurred at the same position in the sentence and were always marked by one specific unexpected phoneme (the consonant "g" instead of the vowel "o"). In the nonnative group, there was no ERP effect observed before training. After training, only the P600 effect was observed in nonnative participants, demonstrating, once more, that the processes reflected in the P600 can be acquired relatively easily when a sufficient amount of training is provided. However, the thematic processes indicated by the N400 component do not seem to be established at this stage of learning, at which trained participants were at a high, but not nativelike, behavioral performance level. In a similar way, earlier phonological processing might not be as efficient as in native speakers, as indicated by the absence of the earlier negativity in the trained nonnative participants. The classifier violation, consisting of an agreement error between the classifier and its head, solely led to a relatively late left anterior negativity in Japanese native speakers (500–800 ms), whereas it did not elicit any ERP effect at any measuring time in the nonnative participants. Because of its temporal characteristics, the left anterior negativity in natives was seen as related to working memory costs, potentially brought about by some post hoc agreement checking process, rather than by purely morphosyntactic processes (Mueller et al., 2005). However, it remains unclear why the effect was lacking in the nonnative group, as there was no behavioral difference between native and nonnative participants after training. Typological differences between Japanese and German, which does not make use of classifiers, might have contributed to the application of different processing strategies in learners as compared to native speakers. This first ERP study, in which L2 processing was investigated in the model of a miniature version of a real language, compares to previous ERP studies using natural and artificial languages in interesting ways. In Table 2, the various ERP patterns found for native and nonnative participants are summarized.

On the one hand, some striking similarities to native speakers' ERP patterns with respect to relatively late, controlled syntactic processes as reflected in the P600 component emerged in nonnative learners after only a few hours of training. This underlines the important role of proficiency as the determining variable for this ERP component. On the other hand, nonnative participants seemed to underuse automatic syntactic processes, indicated by the deviant topographical distribution of the early negativity, as well as thematic processes, as reflected in the absence of the effect. Thus, the findings of the artificial language study reported by Friederici et al. (2002) cannot be simply generalized to more natural language input. There are several differences between the two studies that potentially account for the less

Table 2

Summary of main ERP results for native Japanese and trained and untrained nonnative participants in the Mini-Nihongo study 1

	Japanese native	Trained nonnative	Untrained nonnative
Word category violation	N100-300, frontal P600	N100-300, central P600	N100-800, central
Case violation	N100-300, central N400, P600	P600	no effect
Classifier violation	N500-800, left frontal	no effect	no effect

nativelike pattern in the present study. First, in contrast to the Friederici et al. study, we used connected speech, which probably makes higher demands on speech segmentation and speed of processing. Second, the Mini-Nihongo phonology was Japanese, not German as in the Friederici et al. study. Furthermore, it is difficult to determine how nativelike the ERP patterns in the Friederici et al. study really were, as a direct comparison between native and nonnative speakers of Brocanto is simply not possible. The study that will be reported next was aimed to provide more favorable conditions for nonnative participants to apply nativelike processing mechanisms in order to test if the differences to native language processing can be "overcome."

Mini-Nihongo Study 2

The detection of case violations was surprisingly difficult for the learners in the Mini-Nihongo study 1, and the N400 indexing the immediate use of thematic information in native speakers was not observed in the learners' ERPs. Furthermore, topographical differences in the early time window in the word category violation condition indicated that the learners processed syntactic phrase structure in a less "syntactic" but more "prosodic" manner as compared to native speakers despite comparable behavioral performance (Mueller et al., 2005). The Mini-Nihongo study 2 (Mueller, 2005) set out to further assess case and word category processing in native speakers and the same nonnative participants as in the first Mini-Nihongo study after some additional training. The use of canonical and scrambled sentences aimed to enforce hierarchical processing procedures, which could possibly have been avoided by the learners in the first study and replaced by a simpler linear processing strategy. Thus, case and word category violations were presented both in sentences with canonical and scrambled word order. Example sentences for each experimental condition are listed in Table 3.

In order to refresh and improve behavioral skills, the same native German speakers as in experiment 1 received additional training until they again reached the proficiency criterion. The training time ranged between 1 and 3.5 hr (average: 1.7 hr). This time, there were no differences in any of the conditions between native Japanese and trained nonnative participants in the grammaticality judgment task during the ERP experiment. Accuracy rates were between 95% and 99%. The ERP results for Japanese native speakers were similar to those of the Mini-Nihongo study 1.

Figure 2 illustrates that word category violations elicited an early anteriorly enhanced negativity followed by a P600 for both canonical and scrambled sentences. Figure 3 depicts the ERPs found for case violations in scrambled (double accusative) and canonical (double nominative) sentences, which elicited a pattern of an N400 and a subsequent P600 in both cases. Importantly, in contrast to the findings for Japanese native speakers in the first study, there was no evidence for an additional negativity in an earlier time window for the case violation condition. In the Mini-Nihongo study 1, the earlier negativity, which was marked by an additional peak, was hypothesized to reflect the violation of phonological expectations related to a specific position in the sentence. The fact that this negativity was no longer present in the Mini-Nihongo study 2 supports this hypothesis, as such specific

Table 3

Examples of the stimuli used in the Mini-Nihongo study 2

Correct condition - canonical	
I chi wa no kamo ga	tobikoeru tokoro desu.
1 [bird][gen.] duck [nom.]	jump over take place.
Correct condition - scrambled	
I chi wa no hato o	tobikoeru tokoro desu.
1 [bird][gen.] pigeon [acc.]	jump over take place.
Word category violation - canonical	
I chi wa no kamo ga	tobikoeru tokoro desu.
1 [bird][gen.] duck [nom.]	jump over take place.
Word category violation - scrambled	
I chi wa no hato o	tobikoeru tokoro desu.
1 [bird][gen.] pigeon [acc.]	jump over take place.
Case violation - canonical	
I chi wa no kamo ga	tobikoeru tokoro desu.
1 [bird][gen.] duck [nom.]	jump over take place.
Case violation - scrambled	
I chi wa no hato o	tobikoeru tokoro desu.
1 [bird][gen.] pigeon [acc.]	jump over take

ni hiki no nezumi o	
2 [small-animal][gen.] mouse [acc.]	
ni hiki no neko ga	
2 [small-animal][gen.] cat [nom.]	
ni hiki no	
2 [small-animal][gen.]	
ni hiki no	
2 [small-animal][gen.]	
ni hiki no nezumi ga	
2 [small-animal][gen.] mouse [nom.]	
ni hiki no neko o	
2 [small-animal][gen.] cat [acc.]	

Note. gen. = genitive, nom. = nominative, acc. = accusative; incorrect elements are underlined.

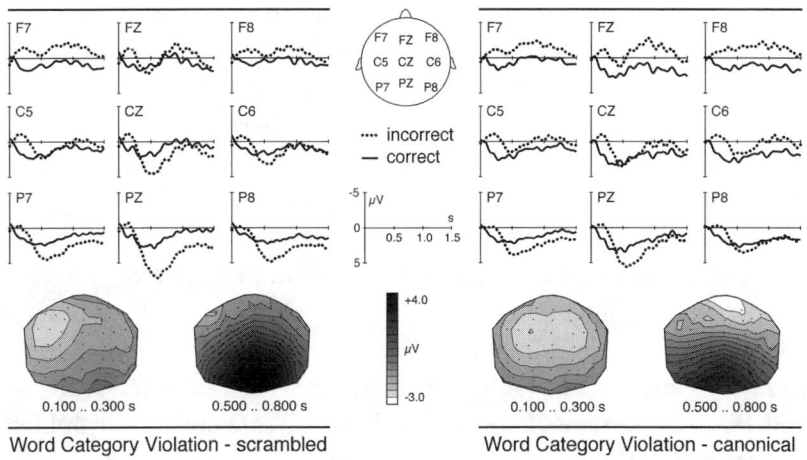

Figure 2. ERPs and isovoltage difference maps of the word category violation conditions in scrambled and canonical sentences for Japanese native speakers.

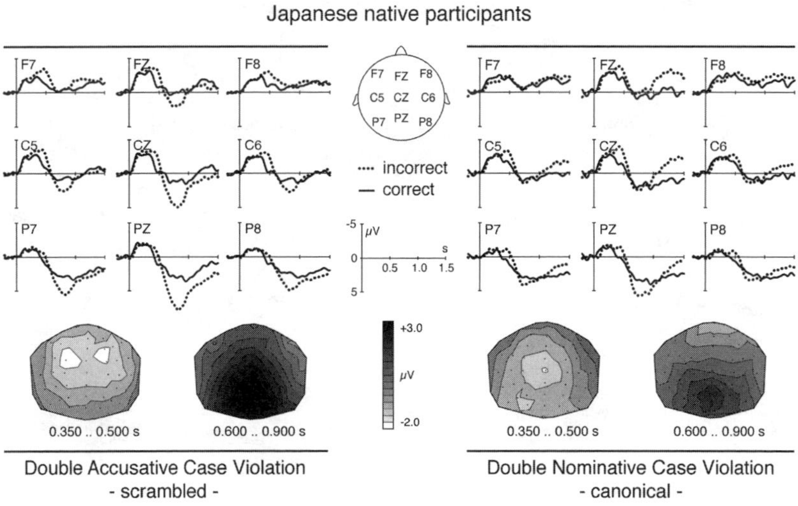

Figure 3. ERPs and isovoltage difference maps of the case violations in scrambled (double accusative violation) and canonical (double nominative violation) sentences in native Japanese speakers.

positional expectations were no longer possible due to the presence of word-order variations. In both the canonical and the scrambled word category violation conditions, trained nonnative participants displayed the same ERP pattern as in Mini-Nihongo study 1 (cf. Figure 4), namely a broadly distributed, early negativity. Compared to native speakers' early negativity, the effect was more posteriorly distributed in nonnative participants. Additionally, a P600 was observed. This indicates that the additional training did not suffice to induce more similarity between native and nonnative participants during the phase of early syntactic phrase structure building. In trained nonnative participants, case violations elicited an N400-like negativity (with an anterior focus) and P600 in canonical sentences, but only a P600 in scrambled sentences (cf. Figure 5).

The emergence of the N400-like negativity in learners illustrates that case information was used in the same time window as in native speakers for canonical sentences. The anterior topographical distribution of the negativity might have resulted from

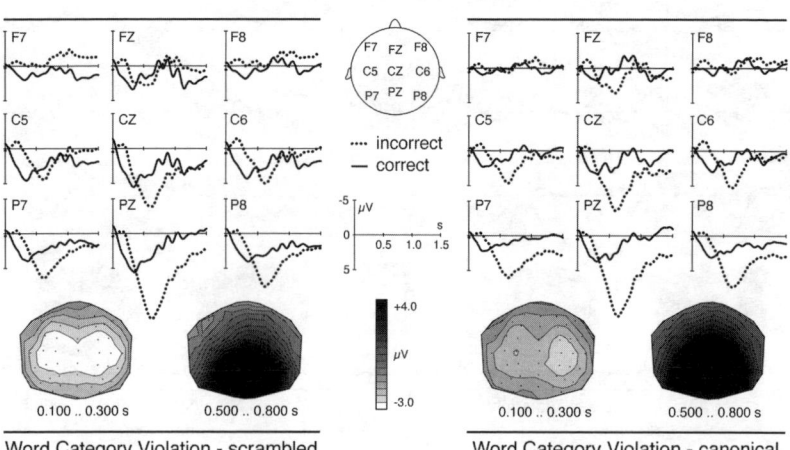

Figure 4. ERPs and isovoltage difference maps of the word category violation conditions in scrambled and canonical sentences for trained nonnative participants.

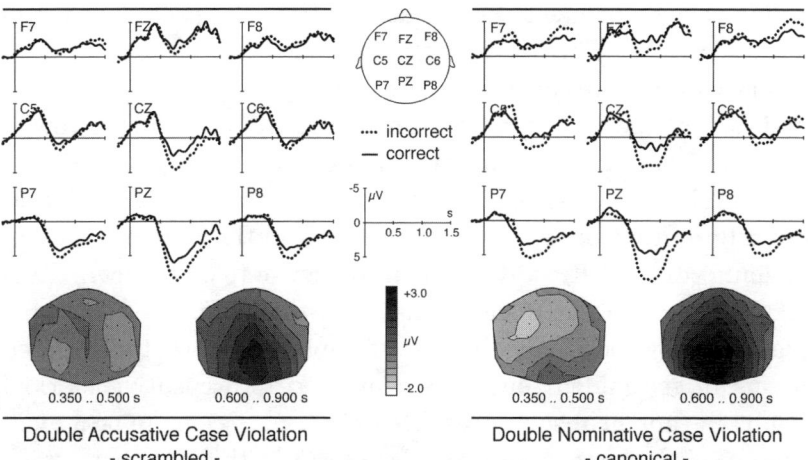

Figure 5. ERPs and isovoltage difference maps of the case violation conditions in scrambled (double accusative violation) and canonical (double nominative violation) sentences for trained nonnative participants.

a temporal overlap of the negativity with the subsequent posteriorly distributed positivity (attenuating the N400 over posterior electrode sites), which was more pronounced in nonnative than in native participants. Thus, it is difficult to decide if the negativities in both groups reflect different functional processes or if they differ only because of a temporal overlap with later processes, which are applied to a different extent in native and nonnative participants. An intriguing result of experiment 2 was the finding of the N400-like negativity for the trained participants in the double nominative case violation only and not for the double accusative case violation. Note that the double nominative case violation was marked by an incorrect "ga" postposition in the NP2 (e.g., *Ichi wa no hato ga ni wa no kamo ga tobikoeru tokoro desu.*), whereas the double accusative case violation was marked by an incorrect "o" postposition (e.g., *Ichi wa no hato o ni wa no kamo o tobikoeru tokoro desu.*). In the case of an accusative case marked noun two consecutive vowels have to be segmented from each other (e.g., *kamo-o*), whereas in the case of a nominative case marked noun, a stop consonant has to be segmented from the preceding vowel (e.g., *kamo-ga*). The spectral changes that mark the transition from a vowel to the stop consonant "g" are potentially more salient and easier to perceive for the trained participants than the vowel-vowel transition (which, in three of four cases, is a duplication of the vowel "o"). Evidence that learners process the two case markers differently from each other and differently from natives can be found in the ERP waveforms in response to correct sentences (cf. Figure 6). Figure 6 shows that only the nonnative group displayed a more positive-going waveform after nominative marked nouns in scrambled sentences compared to accusative marked nouns in canonical sentences. This positivity seems to be a modulation of the general positive waveform at this position, which is interpretable as CPS. The less pronounced CPS in nonnative participants might be related to difficulties in segmenting the "o" postposition, thus leading to a less precise processing of the phrase boundary of an accusative marked nominal phrase. Another possibility would be that the positivity reflects additional syntactic integration processes, whereby the noun phrase is integrated with the preceding context. In both cases, the effect can be

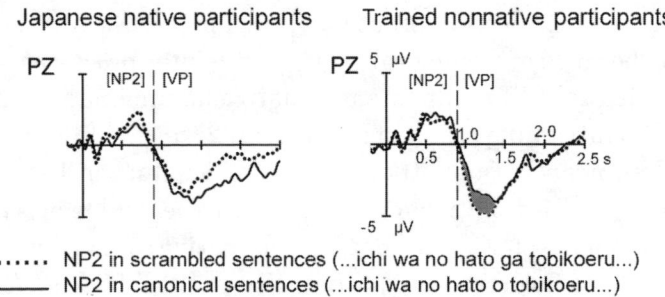

...... NP2 in scrambled sentences (...ichi wa no hato ga tobikoeru...)
⎯⎯ NP2 in canonical sentences (...ichi wa no hato o tobikoeru...)

Figure 6. Global ERPs time-locked to the beginning of NP2 for canonical and scrambled sentences in native Japanese and trained nonnative participants. The mean length of NP2 was exactly 900 ms. The significant difference in the CPS component between accusative and nominative NPs is shadowed gray in the nonnative participants.

seen as indicating more or additional processing after the nominative case marker "ga." Thus, learners potentially rely to a larger degree on the acoustically more salient nominative case marker "ga." The ERP patterns observable in native speakers did not provide indications for similar differences. This finding suggests that the learners have to rely to a large degree on the processing of the more salient one of the two case markers. In our miniature grammar, this would be a sufficient strategy for the comprehension of correct sentences, as the detection of the subject NP implies that the other NP must be the object. Focusing on the "ga" postposition would also be sufficient to detect the errors in both types of case violation condition. In the double nominative violation, the surplus "ga" postposition would be detected immediately, leading to difficulties in the thematic domain, as the hierarchical position, which is indicated by the second "ga" postposition, is already occupied. In the double accusative violation, the error is possibly diagnosed only via absence of the "ga," which is expected to occur after the second noun. Thus, it might be the case that learners conclude the double accusative only after they have checked that there was no nominative case marker at all. In such a sentence, no thematic hierarchy would be specified at all, which would explain the absence of the thematic N400. However, the error leads to structural repair processes, which come into play later, as indicated by the P600. Taken together, the Mini-Nihongo study 2 suggests that changes in the processing requirements of the input together with amount of training make an impact on the applied processing mechanisms in the learners. Under these conditions, an N400-like negativity in response to double nominative case violations emerged in the learners. Relatively automatic processes of phrase structure building, however, seem to be less modifiable and still different from native speakers' processes in the trained nonnative group. Learners' ERP responses reflect a highly strategic use of case marking postpositions in a way that ensures interpretability of the sentences without analyzing elements that are possibly difficult to process and not absolutely necessary for comprehending the sentence.

Conclusion

The miniature language used here, which, in contrast to previously investigated artificial grammars (Friederici et al., 2002; Hoen & Dominey, 2000), consists of natural connected speech, yielded several processing differences between nonnative and native speakers. Particularly, even when nativelike accuracy rates in the grammaticality judgments were achieved, differences in the ERPs were still present. First, these findings suggest, in addition to previously reported findings (e.g., Kotz & Elston-Güttler, 2004), that ERPs can be more sensitive to specific difficulties in L2 speakers than only behavioral measures. Second, due to the differences between native and nonnative speakers, the present findings are interpretable as being in agreement with strong versions of the critical period hypothesis, which assume that syntactic processes are not nativelike in late L2 learners (Bley-Vroman, 1989; Pulvermüller & Schumann, 1994). However, as a very specific learning environment was given and the participants had only relatively limited amount of exposure to the stimuli, it cannot be concluded that the present result is stable over time. Different training methods (e.g., increased focus on formal aspects, such as word category or use of presegmented stimuli), might trigger a more nativelike ERP pattern as well as a higher amount of exposure. Testing the impact of input- and training-related manipulations on the neural processing mechanisms in the model of size-restricted miniature languages seems to be a promising research strategy for the future. Turning away from a deficiency-oriented viewpoint to the resources available to L2 speakers, the two Mini-Nihongo studies shed light also on functions that can be used very efficiently by L2 speakers, namely prosodic phrasing and controlled syntactic processes, as reflected in the early "prosody-related" negativity and the P600 component. These processes were indicated consistently in both experiments and seemed more prominent in the learners than in Japanese native speakers. It might be speculated that those (relatively few) L2 speakers who reach nativelike performance levels use exactly those processes to compensate for their difficulties in other domains, such as syntax and phonology. Hence, even if our brain might use linguistic information in a nonnative manner when processing an L2, we might have access to efficient compensatory mechanisms, leading to the ability to learn and process foreign languages during our whole life.

Notes

[1] Artificial grammars are composed of a limited set of elements and configuration rules, which generate stimulus sequences in any modality.
[2] Intonational phrases are marked by tonal and durational cues, such as phrase final lengthening, pitch variations, rhythmic markers, and pauses.
[3] Finite-state grammars are rule systems that generate sequences that can be exhaustively described by transition probabilities between two succeeding elements. Normally, they are described as nodes that are connected with each other by labeled arrows. Sequences are generated by starting at a particular node and randomly traveling in the direction indicated by the arrows to the final node.
[4] Phrase structure grammars contain rules that define how classes of words can be combined in a hierarchical order, in which the relationship between the elements is characterized by long-distance dependencies.
[5] Japanese mainly uses two script types, namely *kana*, which is a phonographic script (i.e., coding the pronunciation), and *kanji*, which is ideographic (i.e., coding the conceptual meaning). German uses only phonographic script.

References

Ardal, S., Donald, M. W., Meuter, R., Muldrew, S., & Luce, M. (1990). Brain responses to semantic incongruity in bilinguals. *Brain and Language*, *39*, 187–205.

Baldwin, K. B., & Kutas, M. (1997). An ERP analysis of implicit structured sequence learning. *Psychophysiology*, *34*, 74–86.

Barber-Friend, H., Gillon-Dowens, M., & Carreiras, M. (2004). Do late bilinguals learn the L2 over the L1? Electrophysiological correlates of gender and number agreement in highly proficient English-Spanish late bilinguals. *Journal of Cognitive Neuroscience, Supplement*, 210.

Bialystok, E., & Hakuta, K. (1999). Confounded age: Linguistic and cognitive factors in age differences for second language acquisition. In D. Birdsong (Ed.), *Second language acquisition and the critical period hypothesis* (pp. 161–181). Mahwah, NJ: Erlbaum.

Birdsong, D. (1992). Ultimate attainment in second language acquisition. *Language, 68*, 706–755.

Birdsong, D. (1999). Introduction: Whys and why nots of the Critical Period Hypothesis. In D. Birdsong (Ed.), *Second language acquisition and the Critical Period Hypothesis* (pp. 1–22). Mahwah, NJ: Erlbaum.

Birdsong, D., & Molis, M. (2001). On the evidence for maturational constraints in second-language acquisition. *Journal of Memory and Language, 44*(2), 235–249.

Bley-Vroman, R. (1989). What is the logical problem of foreign language learning? In S. Gass & J. Schachter (Eds.), *Linguistic perspectives on second language acquisition* (pp. 41–68). Cambridge: Cambridge University Press.

Brown, C. M., Hagoort, P., & Ter Keurs, M. (1999). Electrophysiological signatures of visual lexical processing: Open- and closed-class words. *Journal of Cognitive Neuroscience, 11*(3), 261–281.

Clahsen, H. (1999). Lexical entries and rules of language: A multidisciplinary study of German inflection. *Behavioral and Brain Sciences, 22*, 991–1060.

Clahsen, H., & Felser, C. (2006). Grammatical processing in language learners. *Applied Psycholinguistics, 27*(1), 3–42.

Conolly, J. F., & Phillips, N. A. (1994). Event-related potential components reflect phonological and semantic processing of the terminal word of spoken sentences. *Journal of Cognitive Neuroscience, 6*, 256–266.

Conolly, J. F., Phillips, N. A., Stewart, S. H., & Brake, W. G. (1992). Event-related potential sensitivity to acoustic and semantic properties of terminal words in sentences. *Brain and Language, 43*, 1–18.

Conolly, J. F., Stewart, S. H., & Phillips, N. A. (1990). The effects of processing requirements on neurophysiological responses to spoken sentences. *Brain and Language, 39*, 302–318.

Coulson, S., King, J. W., & Kutas, M. (1998). Expect the unexpected: Event-related brain response to morphosyntactic violations. *Language and Cognitive Processes, 13*, 21–58.

Donchin, E. (1981). Surprise...surprise? *Psychophysiology, 18*, 493–513.

Eimer, M., Goschke, T., Schlaghecken, F., & Stürmer, B. (1996). Explicit and implicit learning of event sequences: Evidence from event-related brain potentials. *Journal of Experimental Psychology: Learning, Memory and Cognition, 22*(4), 970–987.

Elman, J. L. (1993). Learning and development in neural networks: The importance of starting small. *Cognition, 48*, 71–99.

Eubank, L., & Gregg, K. R. (1999). Critical periods and (second) language acquisition: Divide et impera. In D. Birdsong (Ed.), *Second language* acquisition and the Critical Period Hypothesis (pp. 65–101). Mahwah, NJ: Erlbaum.

Friederici, A. D. (2002). Towards a neural basis of auditory sentence processing. *Trends in Cognitive Sciences*, 6, 78–84.

Friederici, A. D., & Frisch, S. (2000). Verb argument structure processing: The role of verb-specific and argument-specific information. *Journal of Memory and Language*, 43, 476–507.

Friederici, A. D., Pfeifer, E., & Hahne, A. (1993). Event-related brain potentials during natural speech processing: Effects of semantic, morphological and syntactic violations. *Cognitive Brain Research*, 1, 183–192.

Friederici, A. D., Steinhauer, K., & Pfeifer, E. (2002). Brain signatures of artificial language processing: Evidence challenging the Critical Period Hypothesis. *PNAS*, 99(1), 529–534.

Frisch, S., & Schlesewsky, M. (2001). The N400 reflects problems of thematic hierarchizing. *NeuroReport*, 12(15), 3391–3394.

Frisch, S., & Schlesewsky, M. (2005). The resolution of case conflicts from a neurophysiological perspective. *Cognitive Brain Research*, 25, 484–498.

Greenberg, J. (1972). Numeral classifiers and substantival number: Problems in the genesis of a linguistic type. *Working Papers of Language Universals No. 9*, 1–39.

Gunter, T. C., Friederici, A. D., & Schriefers, H. (2000). Syntactic gender and semantic expectancy: ERPs reveal early autonomy and late interaction. *Journal of Cognitive Neuroscience*, 12, 556–568.

Hahne, A. (2001). What's different in second-language processing? Evidence from event-related brain potentials. *Journal of Psycholinguistic Research*, 30, 251–266.

Hahne, A., & Friederici, A. D. (2001). Processing a second language: Late learner's comprehension mechanisms as revealed by event-related potentials. *Bilingualism*, 4, 123–141.

Hahne, A., Mueller, J. L., & Clahsen, H. (2006). Morphological processing in a second language: Behavioral and ERP evidence for storage and decomposition. *Journal of Cognitive Neuroscience*, 18, 121–134.

Hoen, M., & Dominey, P. F. (2000). ERP analysis of cognitive sequencing: A left anterior negativity related to structural transformation processing. *NeuroReport*, 11(4), 3187–3191.

Holcomb, P. J., & Neville, H. J. (1990). Semantic priming in visual and auditory lexical decision: A between modality comparison. *Language and Cognitive Processes*, 5, 281–312.

Johnson, J. S., & Newport, E. L. (1989). Critical period effects in second language learning: The influence of maturational state on the acquisition of English as a second language. *Cognitive Psychology*, 21, 60–99.

Kaan, E., Harris, A., Gibson, E., & Holcomb, P. J. (2000). The P600 as an index of integration difficulty. *Language and Cognitive Processes, 15,* 159–201.

Katsuki-Pestemer, N. (1991). *Grundstudium Japanisch I. Rheinbreitbach: Durr & Kessler.*

Kemler Nelson, D. G., Hirsh-Pasek, K., Jusczyk, P. W., & Cassidy, K. W. (1989). How the prosodic cues in motherese might assist language learning. *Journal of Child Language, 16,* 55–68.

Kotz, S. A., & Elston-Güttler, K. (2004). The role of proficiency on processing categorical and associative information in the L2 as revealed by reaction times and event-related brain potentials. *Journal of Neurolinguistics, 17,* 215–235.

Kutas, M., & Federmeier, K. D. (2000). Electrophysiology reveals semantic memory use in language comprehension. *Trends in Cognitive Neuroscience, 4,* 463–470.

Kutas, M., & Hillyard, S. A. (1980). Reading senseless sentences: Brain potentials reflect semantic incongruity. *Science, 207,* 203–205.

Lenneberg, E. (1967). *Biological foundations of language.* New York: Wiley.

Marslen-Wilson, W. D. (1973). Linguistic structure and speech shadowing at very short latencies. *Nature, 244,* 522–533.

Mitchell, D. C. (1994). Sentence parsing. In M. A. Gernsbacher (Ed.), *Handbook of psycholinguistics* (pp. 375–409). San Diego: Academic Press.

Miyamoto, E. T. (2002). Case markers as clause boundary inducers in Japanese. *Journal of Psycholinguistic Research, 31,* 307–347.

Miyamoto, T., Katayama, J., & Koyama, T. (1998). ERPs, semantic processing and age. *International Journal of Psychophysiology, 29,* 43–51.

Moreno, E. M., & Kutas, M. (2005). Processing semantic anomalies in two languages: An electrophysiological exploration in both languages of Spanish-English bilinguals. *Cognitive Brain Research, 22,* 205–220.

Morgan, J. L., Meier, R. P., & Newport, E. L. (1987). Structural packaging in the input to language learning: Contributions of prosodic and morphological marking of phrases to the acquisition of language. *Cognitive Psychology, 19,* 498–550.

Mueller, J. L. (2005). *Mechanisms of auditory sentence comprehension in first and second language: An electrophysiological miniature grammar study. MPI Series in Human Cognitive and Brain Sciences, Volume 58,* Leipzig: Max Planck Institute for Human Cognitive and Brain Sciences.

Mueller, J. L., Hahne, A., Fujii, Y., & Friederici, A. D. (2005). Native and non-native speakers' processing of a miniature version of Japanese as revealed by ERPs. *Journal of Cognitive Neuroscience, 17*(8), 1229–1244.

Nakagome, K., Takazawa, S., Kanno, O., Hagiwara, H., Nakajima, H., Itoh, K., et al. (2001). A topographical study of ERP correlates of semantic and syntactic violations in the Japanese language using the multichannel EEG system. *Psychophysiology, 38*, 304–315.

Neville, H. J., Mills, D. L., & Lawson, D. S. (1992). Fractionating language: Different neural subsystems with different sensitive periods. *Cerebral Cortex, 2*, 244–258.

Neville, H. J., Nicol, J. L., Barss, A., Forster, K. I., & Garrett, M. F. (1991). Syntactically based sentence processing classes: Evidence from event-related brain potentials. *Journal of Cognitive Neuroscience, 3*, 151–165.

Newport, E. L. (1988). Constraints on learning and their role in language acquisition: Studies of the acquisition of American Sign Language. *Language Sciences, 10*, 147–172.

Newport, E. L. (1990). Maturational constraints on language learning. *Cognitive Science, 14*, 11–28.

Osterhout, L., & Holcomb, P. J. (1992). Event-related brain potentials elicited by syntactic anomaly. *Journal of Memory and Language, 31*, 785–786.

Osterhout, L., Holcomb, P. J., & Swinney, D. A. (1994). Brain potentials elicited by garden-path sentences: Evidence of the application of verb information during parsing. *Journal of Experimental Psychology: Learning, Memory and Cognition, 20*, 786–803.

Pannekamp, A., Toepel, U., Alter, K., Hahne, A., & Friederici, A. D. (2005). Prosody-driven sentence processing: An ERP study. *Journal of Cognitive Neuroscience, 17*(3), 407–421.

Penke, M., Weyerts, H., Gross, M., Zander, E., Münte, T., & Clahsen, H. (1997). How the brain processes complex words: An event-related potential study of German verb inflections. *Cognitive Brain Research, 6*, 37–52.

Pinker, S. (1991). Rules of language. *Science, 253*, 530–535.

Pulvermüller, F., & Schumann, J. H. (1994). Neurobiological mechanisms of language acquisition. *Language Learning, 44*, 681–734.

Rüsseler, J., Henninghausen, E., Münte, T. F., & Rösler, F. (2003). Differences in incidental and intentional learning of sensorimotor sequences as revealed by event-related potentials. *Cognitive Brain Research, 15*, 116–126.

Schlaghecken, F., Stürmer, B., & Eimer, M. (2000). Chunking processes in the learning of event sequences: Electrophysiological indicators. *Memory & Cognition, 28*(5), 821–831.

Schlesewsky, M., & Bornkessel, I. (2004). On incremental thematic interpretation: Degrees of meaning accessed during sentence comprehension. *Lingua, 114*, 1213–1234.

Soderstrom, M., Seidl, A., Kemler Nelson, D. G., & Jusczyk, P. (2003). The prosodic bootstrapping of phrases: Evidence from prelinguistic infants. *Journal of Memory and Language, 49,* 249–267.

Steinhauer, K., Alter, K., & Friederici, A. D. (1999). Brain potentials indicate immediate use of prosodic cues in natural speech processing. *Nature Neuroscience, 2,* 191–196.

Steinhauer, K., & Friederici, A. D. (2001). Prosodic boundaries, comma rules, and brain responses: The closure positive shift in ERPs as a universal marker for prosodic phrasing in listeners and readers. *Journal of Psycholinguistic Research, 30,* 267–295.

Takazawa, S., Takahashi, N., Nakagome, K., Kanno, O., Hagiwara, H., Nakajima, H., et al. (2002). Early components of event-related potentials related to semantic and syntactic processes in the Japanese language. *Brain Topography, 14*(3), 169–177.

Ueno, M., & Kluender, R. (2003). Event-related brain indices of Japanese scrambling. *Brain and Language, 86,* 243–271.

Ullman, M. T. (2001). The neural basis of lexicon and grammar in first and second language: The declarative/procedural model. *Bilingualism: Language and Cognition, 4,* 105–122.

Ullman, M. T. (2004). Contributions of memory circuits to language: The declarative/procedural model. *Cognition, 92,* 231–270.

Weber-Fox, C. M., & Neville, H. J. (1996). Maturational constraints on functional specializations for language processing: ERP and behavioural evidence in bilingual speakers. *Journal of Cognitive Neuroscience, 8*(3), 231–256.

Weber-Fox, C. M., & Neville, H. J. (2001). Sensitive periods differentiate processing of open- and closed-class words: An ERP study of bilinguals. *Journal of Speech, Language and Hearing Research, 44,* 1338–1353.

Weyerts, H., Penke, M., Dohrn, U., Clahsen, H., & Münte, T. (1997). Brain potentials indicate differences between regular and irregular German plurals. *NeuroReport, 8,* 957–962.

White, L., & Genesee, F. (1996). How native is near-native? The issue of ultimate attainment in adult second language acquisition. *Second Language Research, 12,* 233–265.

Yamashita, H. (1997). The effects of word-order and case marking information on the processing of Japanese. *Journal of Psycholinguistic Research, 26,* 163–188.

Cracking the Nutshell Differently. Commentary on Mueller

Monique J. A. Lamers
Radboud University Nijmegen

Since Lenneberg (1967) proposed the critical period hypothesis, nativelike proficiency in late learners of a second language (L2) has been topic of L2 acquisition research. Even with the use of newly developed neuroimaging techniques such as the registration of event-related brain potentials (ERPs) during sentence processing, it remains unresolved whether the differences found between first language (L1) and L2 acquisition are caused by the age of acquisition or by differences in proficiency level. Whereas for semantic processes the difference seems to be mainly quantitative in nature, qualitative differences are reported for syntactic processes almost always involving those processes that are thought to be automatic. These processes are difficult to acquire and are generally learned late in L2 acquisition. In her paper, Mueller takes on the challenge to resolve this issue and focuses on the syntactic processing differences between L1 and L2 acquisition by performing two auditory ERP studies using the miniature language Mini-Nihongo. In this discussion of the study by Mueller in which she performed two auditory experiments investigating the processing of three types of syntactic violation in a miniature

---

Monique J. A. Lamers, Department of Linguistics.

I thank Pascal Brenders and Peter de Swart for providing useful comments on earlier versions of this discussion.

Correspondence concerning this article should be addressed to Monique J. A. Lamers, Radboud University Nijmegen, Postbus 9103, NL- 6500 HD Nijmegen, The Netherlands. Internet: m.lamers@let.ru.nl

language as an L1 or an L2, I will argue that the findings are subject to multiple interpretations.

Mini-Nihongo is a subset of real Japanese and thus the L1 language of native speakers of Japanese. Furthermore, it can also serve as a L2 to be learned by nonnative speakers (i.e., German volunteers) at a high proficiency level in a relatively short amount of time, thus controlling the variability in the age of acquisition. In this discussion, I will argue that using a miniature language, in spite of the advantages, restricts the direct comparison between L1 and L2 users. First, I will point out the difference in language resources between the L2 learners (nonnative participants) and native speakers of Japanese (native participants). Then I will challenge the interpretation of the use of prosodic phrase information as a compensatory mechanism. Additionally, an alternative functional interpretation for the results on the case violation will be presented. Finally, I will point out that establishing the proficiency level of the L2 learners on their behavioural performance on the syntactic violations seems to be insufficient to capture all abilities upon which native speakers can rely.

Differences in Language Resources Between L2 Learners and Native Speakers

Although the limited set of lexical items and grammatical rules in Mini-Nihongo make it, in principle, possible for L2 learners to gain a high level of proficiency, the comparison to native speakers of Japanese is still at odds. After all, in comparison to native speakers of Japanese with the availability of a large, but finite vocabulary and the ability to form and understand an innumerable amount of grammatical sentences, the L2 learners have to cope with only a small number of lexical items, word categories, and variability in sentence constructions (i.e., 2,048 grammatical sentences); for example, in real Japanese, it is possible to leave out the subject (as well as other sentence parts), sentences with a double nominative do occur in certain contexts, and some postpositions are homophones (e.g., *-wa* is ambiguous between a classifier and a topicalization marker). Thus, not withstanding the possibility to gain a high level of proficiency, the discrepancy between the amount of resources available to native speakers of Japanese and the limited resources of the L2 learners to detect and handle a syntactic violation introduces an insuperable difference in language processing. This is exactly what Mueller's data showed.

The Use of Prosodic Phrase Information: Compensatory Mechanism or L2 Learner's Inability?

Mueller reports an early frontal negativity (left anterior negativity; LAN) and a P600 for the Japanese native speakers at the word category violation. Early negativities were also found for nonnatives before training and after training, but with a difference in scalp distribution. She argues that this difference results from the processing of prosodic information in the case of a prosodic phrase boundary, which is present at the second nominal phrase (NP) in the correct but not in the incorrect sentence. She takes this as evidence for the enhanced reliance on prosodic processes as compared to syntactic ones in naive as well as highly proficient L2 learners. Based on the difference in the early time frame as well as a similar P600 indicating the involvement of controlled late processes for the trained L2 learners and the native speakers at the word category violations, Mueller concludes that native speakers and highly proficient L2 learners of Mini-Nihongo use similar as well as different processes. Following up on the suggestion of Mueller that prosodic information provides an important source of information handling in a late acquired L2, it can be argued that the use of prosodic information compensates for the lack of competence of relatively automatic processes, as revealed by the early negativity in the native speakers.

However, in SLA research, it is often reported that L2 learners do not have full capability of using all detailed characteristics of the phonological information of the target language (cf. Gandour et al., 2003). The question arises whether the use of prosodic information should be looked upon as a compensatory mechanism or as a consequence of the inability of the L2 learner in the phonological domain.

Mueller falls back on the difference in use of phonemic information in her explanation of the early negativity found for the native speakers on case violations. In addition to the N400 and the P600 (which will be discussed later), she reports an early negativity for native speakers only. This negativity is interpreted as a phonological mismatch negativity, which seems to be a plausible interpretation, given the construction of the materials. Hence, the absence of an early negativity in nonnative speakers might be due to the *inability* to distinguish and recognize phoneme patterns in detail. Notice that in the second ERP study in which, in addition to subject initial sentences, object initial sentences were used, resulting in a double case violation, Mueller's explanation relates to the inability to make full use of phoneme characteristics. In this study, the nonnative participants actually exploited the more salient pattern of double nominative marking (the double *-ga*) in the comprehension of these sentences, but they were not capable of recognizing the less salient double accusative marking (the double *-o*), resulting in a difference with the native speakers. The absence of the negativity was explained as a result of a difference in the closure positive shift, due to the less salient double accusative violation, which was not detected for the nonnative speakers.

The Functional Interpretation of the Differences in ERP Waveforms at the Case Violation

For the case violation, Mueller reports multiple effects for the native speakers. She adopts the interpretation of the N400 in combination with the P600 from Frisch and Schlesewsky (2001), who used a similar violation in German. Both sentences in (1) and (2) are ungrammatical because they contain two nominative noun phrases when one noun phrase should have been accusative.

Notice that the two types of sentence differ in the animacy of the noun phrases:

(1) *... welcher Bischof...   der Priester   ...
    ... which$_{NOM}$ bishop...   the$_{NOM}$ priest   ...
(2) *... welcher Bischof...   der Zweig   ...
    ... which$_{NOM}$ bishop...   the$_{NOM}$ twig

Frisch and Schlesewsky found an N400 at the second noun phrase in (1), which was lacking in sentence (2). In contrast to Mueller, Lamers and De Hoop (2005) argued that it is not the case-marking violation that elicited the N400, but the lack of the possibility to distinguish the two arguments on semantic/conceptual prominence (i.e, animacy), with typically the subject being the argument highest in prominence (De Hoop & Lamers, 2006). The language comprehender not only expects an accusative case marked argument in sentence (1) but also an argument that is lower in semantic/conceptual prominence than the initial nominative subject argument. Thus, if a nominative case-marked animate argument comes in, not only are the case-marking rules of the language violated, as is reflected by the P600, but also the lexical expectation, resulting in an N400 (Lamers, 2005; Weckerly & Kutas, 1999). A closer look at the examples in Mini-Nihongo gives the impression that this alternative interpretation also applies to the data of Mueller: No disambiguating information becomes available regarding the relative prominence of the two arguments. Especially because real Japanese is known to be a language in which animacy prominence and politeness hierarchies are strongly incorporated (Yamamoto, 1999), it might well be that Japanese native speakers anticipated for a lexical item in the accusative case as well as one being lower in prominence than the initial nominative argument. If the lexicon of Mini-Nihongo could be expanded with nouns that differ in animacy, it could be clarified whether native and nonnative speakers use animacy information to establish a thematic hierarchical structure despite the confounding case information.

The Establishment of the Proficiency Level

Having discussed the possibility of a difference in the use of phonological information, one might question whether the performance on the types of violation used in the above-discussed studies form the right tool to investigate and assess the proficiency level of the learners. Recall that in Mueller's first study, L2 learners in comparison to the native speakers were equally proficient on subcategorization and classifier violations, but not on the case violations (although the difference was small). Brain activity differences were, however, not restricted to the case violations. The results indicated that L2 learners and native speakers were not equally proficient in the use of phonological information, possibly resulting in a difference in the processing of prosodic phrase structure.

Furthermore, the limited set of rules of a miniature language makes it harder to test language irregularities or less well-described linguistic phenomena (e.g., the acquisition of the use of dummy subjects in Dutch; Van Boxtel, 2005). One wonders whether the nonnative participants in Mueller's studies would be highly proficient in learning and applying irregularities, for example, by incorporating regular (morphological) and irregular (lexical) causative formations in Mini-Nihongo. It would be interesting to see whether more training and applying more sensitive methods to measure the proficiency level (i.e., taking into account perception and production time as well as accuracy) makes it possible to select even more nativelike L2 learners, especially on the phonological level, and to compare their performance on different types of syntactic violation as well as their brain activities.

Mueller seems to be well aware of the limitations of her findings. Although at this moment her conclusion that the findings of the ERP studies clearly provide evidence for a strong version of the critical period hypothesis is definitely true, it remains uncertain what exactly triggered the differences between the native and nonnative participants. In this discussion, it was argued that because of the limitations of the miniature language as well as the establishment of the proficiency level, the mechanisms of L2 processing in a nutshell are subject to multiple interpretations. Nevertheless, the approach of Mueller brought a source of information under attention that has not yet been the focus of many studies in SLA, namely prosodic phrase information. Further research will have to show whether this source of information will help us to crack L2 learning not only in a nutshell.

References

Frisch, S., & Schlesewsky, M. (2001). The N400 reflects problems of thematic hierarchizing. *NeuroReport, 12*, 3391–3394.

Gandour, J., Dzemidzic, M., Wong, D., Lowe, M., Tong, Y., Hsieh, L., et al. (2003). Temporal integration of speech prosody is shaped by language experience: An fMRI study. *Human Brian Mapping, 20*, 185–200.

Hoop, De, H., & Lamers, M. (2006). Incremental distinguishability of subject and object. In W. Abraham, N. Noonan (Series Eds.), L. Kulikov, A. Malchukov, & P. de Swart (Vol. Eds.), Studies in Language Companion: Vol. 77. *Series Case, Valency, and Transitivity* (pp. 269–287). Amsterdam: John Benjamins.

Lamers, M. (2005). The on-line resolution of subject-object ambiguities with and without case-marking in Dutch: Evidence from event related brain potentials. In M. Amberber & H. De Hoop (Eds.), *Competition and variation in natural languages: The case for case* (pp. 251–293). Amsterdam: Elsevier.

Lamers, M., & De Hoop, H. (2005). Animacy information in human sentence processing: An incremental optimization of interpretation approach. In H. Christiansen, P. R. Skadhauge, & J. Villadsen (Eds.), *Constraint solving and language processing* (pp. 158–171). Berlin: Springer-Verlag.

Lenneberg, E. (1967). *Biological foundations of language*. New York: Wiley.

Van Boxtel, S. (2005). *Can the late bird catch the worm? Ultimate attainment in L2 syntax*. Unpublished doctoral dissertation, Radboud University Nijmegen, Nijmegen.

Weckerly, J., & Kutas, M. (1999). An electrophysiological analysis of animacy effects in the processing of object relative sentences. *Psychophysiology, 36*, 559–570.

Yamamoto, M. (1999). *Animacy and Reference: A cognitive approach to corpus linguistics*. Amsterdam: Benjamins.

A Meta-analysis of Hemodynamic Studies on First and Second Language Processing: Which Suggested Differences Can We Trust and What Do They Mean?

Peter Indefrey
Max Planck Institute for Psycholinguistics
and
F. C. Donders Centre for Cognitive Neuroimaging

This article presents the results of a meta-analysis of 30 hemodynamic experiments comparing first language (L1) and second language (L2) processing in a range of tasks. The results suggest that reliably stronger activation during L2 processing is found (a) only for task-specific subgroups of L2 speakers and (b) within some, but not all regions that are also typically activated in native language processing. A tentative interpretation based on the functional roles of frontal and temporal regions is suggested.

In recent years, there has been an increasing number of neurocognitive studies investigating language processing in bilingual speakers. Most researchers interested in language are aware of one or another study reporting hemodynamic activation differences between first language (L1) and second language (L2) processing. Given the plethora of experimental details that might lead to signal changes in hemodynamic experiments, however, differences as such might not mean very much as long as they do not overlap across studies with similar paradigms and as long as it is not clear which factors determine the presence or absence

---

Correspondence concerning this article should be addressed to Peter Indefrey, Max Planck Institute for Psycholinguistics, P. O. Box 310, 6500 AH Nijmegen, The Netherlands. Internet: peter.indefrey@mpi.nl

of differences. Finally, many studies reporting *no* differences between L1 and L2 processing tend to mention this negative finding in a small paragraph somewhere between the more interesting significant differences in other comparisons, so that there might be some sort of attentional bias toward overestimating the proportion of experiments reporting differences. In this article, I will focus on the findings obtained in bilingual studies using one or more of five frequently used paradigms that allow an assessment of the agreement between studies. The leading question is: Are there *reliable* neural dissociations between native language processing and processing of an L2? To answer this question, I will use a meta-analysis procedure developed to identify regions of reliable overlap between hemodynamic studies on native language processing (Indefrey, 2004; Indefrey & Cutler, 2004; Indefrey & Levelt, 2000, 2004). Insofar as there are reliable dissociations, we can further ask which processing levels they might reflect and to which L2 speaker characteristics they might be related. The results of the meta-analysis will be presented in sections that are ordered with respect to the language processing levels addressed by the different experimental paradigms. The findings will be discussed by taking into account the relevant findings on L1 processing and bilingual studies that, due to methodological differences, were not included in the meta-analysis as such, but that provide additional information for the questions at hand.

A Meta-Analysis of Bilingual Hemodynamic Activation Experiments

To date, the majority of experimental paradigms that have been used to study bilingual language processing have been taken over from neurocognitive research on monolingual language processing. The paradigms mirror the development in monolingual processing research, which over time proceeded from very general language tasks with low-level control conditions (e.g., storytelling or listening compared to a rest condition), to more specific task and control combinations designed to isolate single processing components. The tasks that to date have been most frequently used are variants of word generation, picture-naming, semantic decision, and sentence or story comprehension. The present analysis is based on 30 bilingual hemodynamic experiments from 24 studies using one or more of these tasks. All tasks have been frequently used in native language experiments, and meta-analyses or reviews identifying reliable activation patterns in these tasks are available. This has the advantage that we do not have to focus on common areas of L1 and L2 processing, which logically have to be subsets of the areas that have been reliably reported in L1 studies, but we will be able to concentrate on *differences* between L1 and L2 activation patterns. As in monolingual studies, there is some variation in the details of the experimental paradigms. Unlike monolingual studies, bilingual studies add considerable variation in the subject populations due to different ages of L2 onset and different levels of L2 proficiency and use. It can be expected that this additional variation will be reflected in more heterogeneous activation patterns and, thus, less overlap between studies. For a meta-analysis of bilingual processing, this means that the possibility of detecting a reliable degree of replication across studies is reduced. Nonetheless, we will attempt to identify areas that have been reliably found to be L1- or L2-specific across studies despite varying designs, languages, and L2 speaker populations.

Procedure

Anatomical Coding

The reported activation foci from hemodynamic activation experiments comparing L1 and L2 language processing were recoded in a descriptive reference system of 114 regions covering the whole brain (Indefrey & Levelt, 2000, 2004). In this system, the cerebral lobes are divided into two or three rostro-caudal or medio-lateral segments of roughly equal size. The segment boundaries are defined in terms of standard brain coordinates (Talairach & Tournoux, 1988). The regions within this gross division are defined in terms of gyri and subcortical structures. Activation foci reported in MNI (Montreal Neurological Institute) coordinates were converted to the Talairach and Tournoux space using the nonlinear algorithm of Brett (1999, available at www.mrc-cbu.cam.ac.uk/Imaging/mnispace.html). Activation foci located near the border of two adjacent regions (<5 mm) were coded in both regions.

Reliability Estimate

The studies included in this meta-analysis were not given any weights reflecting reliability differences due to design or size. This means that a certain degree of overlap of activations between studies was considered reliable, but should not be interpreted as statistically significant. Nonetheless, the notion of "reliability" was not totally arbitrary, but based on the following quasi-statistical estimate: The average number of activated regions per experiment r divided by the number of regions equals the probability for any particular region to be reported in a single experiment if reports were randomly distributed over regions. Assuming this probability, the chance level for a region to be reported n_1 or more times as activated in a number of experiments n was calculated based on a binomial distribution. For every region, the chance level depends on the number of studies reporting the region and on the number of studies that looked at this region whether they found it activated or not. The chance level was calculated separately for every region, such that studies that did not cover the whole brain (e.g., due to technical limitations or a regions of interest [ROIs] data analysis) could be included for the subset of regions that they covered. Regions were considered as reliably replicated if their chance probability to be found at least as often as they were found in independent experiments was less than .05.

Overall Findings

The majority of studies reported no differences in hemodynamic activation between L1 and L2 processing. Because the average number of reported regions was low, the chance level for a coincidental overlap between studies was also low. Regions that were reported in at least two (three for sentence comprehension) independent experiments already passed the reliability criterion.

Table 1 presents the reliably replicated regions for the five tasks. To give some indication of possible factors determining the presence or absence of activation differences, Table 1 also lists L2 onset, proficiency, and exposure of the populations that participated in the experiments.

Word-Level Production

Two types of word production task have been most frequently used in bilingual processing studies: word generation and picture-naming. The two tasks share the core components of word production (lemma retrieval, word form retrieval, syllabification, phonetic encoding) but differ with respect to the processes that are employed to come up with a lexical concept to be produced ("lead-in processes"; Indefrey & Levelt, 2000, 2004). Whether resulting brain activations reflect the complete cascade of word production components depends on the control conditions relative to which increases in cerebral blood flow are measured. Control conditions that involve speech production might obscure some or all core components of word production.

In picture-naming, the lead-in process involves a conceptual representation based on a visual object representation. All five experiments on picture-naming (De Bleser et al., 2003; Hernandez, Dapretto, Mazziotta, & Bookheimer, 2001; Hernandez, Martinez, & Kohnert, 2000; Rodriguez-Fornells et al., 2005; Vingerhoets et al., 2003) used control conditions without word production components. In word generation tasks, the lead-in processes are variable and not very well controlled. In two of the studies analyzed here (Klein, Milner, Zatorre, Zhao, & Nikelski, 1999; Pu et al., 2001), subjects generated verbs from stimulus nouns, in one study (Klein, Milner, Zatorre, Meyer, & Evans, 1995), they generated synonyms for stimulus words, and in two studies (Perani et al.,

Table 1

Overview of regions showing reliable hemodynamic activation differences between L1 and L2 processing in different experimental paradigms

A) Stronger activation in L1 as compared to L2
Semantic decision on written words (6 experiments)

	L2 onset	L2 proficiency	L2 exposure	L anterior/mid middle temporal gyrus
Chee et al. 2001	<5	nondominant	nondominant	
Chee et al. 2001	>12	nondominant	dominant	
Ding et al. 2003	**12**	**high**	**?**	**+**
Illes et al. 1999	12	high	high	
Pillai et al. 2004	**>10**	**moderate/high**	**high**	**+**
Xue et al. 2004	8	low	low	

B) Stronger activation in L2 as compared to L1
Word generation (5 experiments)

	L2 onset	L2 proficiency	L2 exposure	L,R posterior inferior frontal gyrus (BA 47)
Klein et al. 1995	7	?	?	
Klein et al. 1999	12	good	high	
Perani et al. 2003	**3**	**high**	**low/high**	**+**
Pu et al. 2001	8–19	high	low	
Vingerhoets et al. 2003	**10–14**	**mixed**	**low/high**	**+**

Picture naming (5 experiments)

	L2 onset	L2 proficiency	L2 exposure	L posterior inferior frontal gyrus (BA 44, 47)
De Bleser et al. 2003	**10**	**good-very good**	**?**	**+**
Hernandez et al. 2000	<5	high	dominant	
Hernandez et al. 2001	<5	high	dominant	
Rodriguez-Fornells et al. 2005	3	balanced	dominant	
Vingerhoets et al. 2003	**10–14**	**mixed**	**low/high**	**+**

Table 1

Continued

Sentence listening/reading (14 experiments)

	L2 onset	L2 proficiency	L2 exposure	L posterior middle frontal gyrus	L posterior inferior frontal gyrus (BA 44)	L posterior inferior frontal gyrus (BA 47)	L supplementary motor area (SMA)
Chee et al. 1999	<6	high	high				
Frenck-Mestre et al. 2005	>12	high	high				+
Hasegawa et al. 2002	**12**	**high**	**high**	+			
Luke et al. 2002	**>10**	**high**	?	+	+	+	
Nakada et al. 2001	>10	high	?		+		
Nakai et al. 1999	?	?	?				+
Perani et al. 1996	7	moderate	low				
Perani et al. 1998	2	high	high		+		
Perani et al. 1998	10	high	high			+	
Rüschemeyer et al. 2005	?	?	**high**		+		
Vingerhoets et al. 2003	10–14	mixed	low/high			+	
Wartenburger et al. 2003	0	high	high	+	+		
Wartenburger et al. 2003	**19**	**high**	**high**		+		
Wartenburger et al. 2003	**20**	**low**	**high**				

Semantic decision on written words (6 experiments)

	L2 onset	L2 proficiency	L2 exposure	L posterior middle frontal gyrus	L posterior inferior frontal gyrus (BA 45)	L posterior inferior parietal lobule	L anterior cingulate
Chee et al. 2001	<5	*nondominant*	*nondominant*	+	+	+	+
Chee et al. 2001	>12	*nondominant*	*dominant*	+	+		+
Ding et al. 2003	12	high	?				
Illes et al. 1999	12	high	high				
Pillai et al. 2004	>10	moderate/high	high			+	
Xue et al. 2004	8	*low*	*low*				+

Note. For every paradigm, the table lists the reliably activated regions and the experiments that reported them (+ signs). Different subject populations are treated as different experiments. Experiments that found at least one of the regions activated are printed in bold. Not listed are experimental that used the paradigm but did not examine the regions. In addition, the table lists some characteristics reported for the experimental populations. Note that, depending on the study, L2 onset might be given as a mean (e.g., "6"), as a range (e.g., "4–8"), or as an upper or lower end of a range (e.g., "<8," ">4"). L2 proficiency and exposure were reported in various ways that are not comparable across studies. The terms used here should be understood as rough characterizations.

2003; Vingerhoets et al., 2003), they generated words based on a given initial letter. Note that the latter task can be performed by accessing either graphemic or phonological word representations, and in contrast to the other generation tasks, it does not require lemma access from a conceptual representation. Three of the experiments used control conditions such as word repetition or counting that involved lexical or postlexical processing components.

Across word production experiments, no area was replicated as showing stronger activation when the task was performed in the native language as compared to an L2. Whereas for the word generation studies, this finding might be attributed to the heterogeneity of task variants, the same does not hold for the picture-naming studies that had a comparable design.

Stronger activation for L2 word production was found bilaterally in the posterior inferior frontal gyri (BA 47) in the two studies using letter fluency. The left posterior inferior frontal gyrus (BA 44, 47) was also reliably found to be more strongly activated in L2 picture-naming. Due to the complex nature of the letter fluency task, conclusions about the functional role of these regions in letter fluency cannot easily be drawn. Picture-naming is better understood and the neural correlates of L1 picture-naming have been examined and compared to other word production tasks in two comprehensive meta-analyses (Indefrey & Levelt, 2000, 2004). The left posterior inferior frontal gyrus (IFG) is reliably found in L1 picture-naming and seems to support a relatively late processing component in word production, namely postlexical syllabification. This interpretation is in agreement with magnetencephalographic (MEG) data showing activation of this area between 400 and 600 ms after picture onset (Salmelin, Hari, Lounasmaa, & Sams, 1994).

The two bilingual studies reporting stronger activation of the left posterior IFG in L2 as compared to L1 picture-naming (De Bleser et al., 2003; Vingerhoets et al., 2003) both had participants with late L2 onset and variable L2 exposure. By contrast, participants in the three studies reporting no differences between L2 and L1 picture-naming (Hernandez et al., 2000, 2001; Rodriguez-Fornells et al., 2005) had early L2 onset and lived in L2-dominant environments. Both onset and exposure might, therefore, explain the differential findings. Compared to the three other studies, the L2 proficiency was also lower in the De Bleser et al. and Vingerhoets et al. studies. De Bleser et al. tested for a within-subject effect of proficiency by comparing pictures whose names were cognates in the two languages and therefore easier to retrieve ("high proficiency") to pictures with noncognate L2 names ("low proficiency"). The L2 versus L1 difference in the left inferior frontal cortex was only found in the more difficult noncognate condition. Within-subject proficiency differences in L2 word production were also investigated by Briellmann et al. (2004), who compared verb generation in quadrilingual subjects and found increased activation for less fluent languages in a number of areas, including the left IFG (Broca's area), that are also reliably found in L1 word generation (Indefrey & Levelt, 2000, 2004; Poline, Vandenberghe, Holmes, Friston, & Frackowiak, 1996). In sum, L1 and L2 word production seems to engage the same cortical areas. L2 speakers with late L2 onset or lower proficiency might recruit at least the left inferior frontal cortex more strongly.

Hemodynamic activation studies show all areas that are active in a given task compared to a control task. In general, the results reported in these studies are, furthermore, group results that preclude any insight into the individual variability of activation patterns. A recent study by Lucas, McKhann, and Ojemann (2004) used a different technique. In this study, 25 mostly fluent bilingual epilepsy patients underwent language mapping with electrical stimulation of the cortex prior to surgery. While the patients performed an object-naming task in either their L1 or their L2, different cortical sites were electrically stimulated and it was recorded whether the stimulation interfered with object-naming or not. Stimulation-sensitive sites can be interpreted as being necessary for the task at hand, so that for every individual, the procedure resulted in a map of sites that were necessary for L1 picture-naming, L2 picture-naming, or both. Shared sites were found in all left perisylvian regions. L1-specific sites, too, were found in both posterior frontal and temporal regions but more so in the frontal cortex. L2-specific sites were exclusively found in the mid- to posterior temporal cortex and adjacent inferior parietal sites. These data are very important for the interpretation of the two main findings from the hemodynamic studies. First, they show that although there might not be any regions that are exclusively recruited or exclusively necessary for L1 or L2 word production *across* individuals, there seem to be cortical sites in many individuals that are only necessary for word production in one of the languages. Second, the only region that to date has been reliably found to be more strongly activated for L2 word production, the left inferior frontal cortex, contains L1-specific sites but not L2-specific sites. These findings suggest that L2 word production does not, in all individuals, share all processes that are involved in L1 word production, if we assume that L1-specific sites subserve L1-specific processes.

Why would speakers activate a region more strongly for L2 word production that seems to subserve, at least in part, L1-specific processes? A possible interpretation might be that L2 speakers attempt to make use of processes that are in some way tailored to L1 word production (to the extent that L1 but not L2 word production becomes impossible without them). A good candidate for such a process seems to be postlexical syllabification, which is subject to language-specific constraints and seems to engage Broca's area in L1 word production (see previous section).

Word-Level Comprehension

Most studies on bilingual word-level comprehension have used written stimuli and asked their subjects to perform some kind of semantic decision, either in the form of a semantic match-to-sample task (Chee, Hon, Lee, & Soon, 2001; Pillai et al., 2004) or in the form of a semantic judgment task (Ding et al., 2003; Illes et al., 1999; Xue, Dong, Jin, & Chen, 2004). Except for Illes et al., all studies used nonlinguistic stimuli in the control conditions, so that the observed activations might reflect all processing components involved in reading as well as semantic decision. The only areas that were reliably reported to be more strongly activated in semantic decisions on L1 words were the anterior and mid-sections of the left middle temporal gyrus, which were found in two studies on late L2 learners with moderate to high proficiency (Ding et al.; Xue et al.). These subject properties were also represented in some of the studies that did not find activation differences, so that it is unclear which factors might affect middle temporal gyrus activation in L2 semantic decision.

Semantic decisions seem to engage a number of areas more strongly when performed in a language spoken with lower proficiency. The left posterior middle frontal gyrus, the left posterior IFG (BA 45), and the left posterior inferior parietal lobule have been found in two experiments, and the anterior cingulate gyrus has been found in three experiments. Chee et al. (2001) found these areas in two subject groups that differed with respect to the onset and use of the language that was spoken less proficiently, so that these factors do not seem to influence the activation level of the four areas. Xue et al. (2004) found the left posterior inferior parietal lobule and the anterior cingulate in L2 speakers of very low proficiency. In this study, however, the frontal areas were not more strongly activated. None of the areas was found in three studies with late L2 onset but relatively high proficiency and use (Ding et al., 2003; Illes et al., 1999; Pillai et al., 2004).

The left posterior middle frontal gyrus, the left posterior IFG (BA 45), and the left posterior parietal lobe are part of a common (modality unspecific) semantic system that was found to be activated in L1 semantic decisions on both written words and pictures by Vandenberghe, Price, Wise, Josephs, and Frackowiak (1996). Interestingly, stronger L2 activation in this system suggests that "modality-unspecific" might not be the same as "language independent." In addition, Vandenberghe et al. (1996) found word-specific activation in the same region that was more strongly activated in L1 semantic decision (anterior left middle temporal lobe), suggesting that this area might subserve some L1-specific word representation at the graphemic or lemma level.

Sentence- and Discourse-Level Comprehension

Fourteen experiments presented bilingual speakers with spoken (six experiments) or written (eight experiments) sentences. In six of the experiments, the sentences were not isolated but formed a story or a text. Additionally, these were analyzed separately to identify possible language differences at a discourse integration level. Although a number of regions have been reported as being more strongly activated in L1 sentence or story comprehension, there was no region that was reported in more than one study. It must be concluded that, to date, the evidence is not sufficient to conclude that any particular region shows reliable L1-specific activation. Note, however, that, insofar as this information was provided, all of the studies except for Perani et al. (1996) had participants with high L2 proficiency and exposure. The regions found to be more activated in the Perani et al. study when L2 speakers of moderate proficiency and low exposure listened to L1 as compared to L2 stories (bilateral temporal poles and posterior superior temporal gyri, left posterior IFG) might be replicated in future studies with similar populations.

Regions that have been reliably replicated as being more active in L2 than in L1 sentence comprehension concentrate in the left posterior frontal gyrus (middle frontal gyrus, BA 44 and BA 47 of the IFG, and the supplementary motor area [SMA]). None of these regions has been replicated in the subset of studies using story or text stimuli. It can, therefore, be assumed that their activation indeed reflects sentence-level rather than discourse-level processes. Table 1 shows that the agreeing reports on these regions come from only five studies (Hasegawa, Carpenter, & Just, 2002; Luke, Liu, Wai, Wan, & Tan, 2002; Nakai et al., 1999; Rüschemeyer, Fiebach, Kempe, & Friederici, 2005; Wartenburger et al., 2003). Unfortunately, not all of these studies provide details on L2 onset, proficiency, and exposure. Insofar as these data are given, participants in the studies seem to have had late L2 onset (after age 10) and high proficiency and exposure. Among the seven other studies that used sentence-level stimuli and did not find L1/L2 differences in these regions, at least four had similar subject populations. A distinctive feature of the studies reporting differences, however, seems to be the type of sentence material and additional task. Three of the studies (Luke et al.; Rüschemeyer et al.; Wartenburger et al.) presented their subjects with both syntactically correct and incorrect sentences and asked for a grammaticality judgment. Hasegawa et al. used stimulus sentences with conjoined positive and negative clauses followed by a verification task. Taken together, the available evidence suggests that stronger activation for listening or reading L2 sentences is not likely to be found when subjects simply listen or read for comprehension, even when they are of only moderate L2 proficiency, as in Perani et al. (1996). Differences might be found when additional decisions on the sentence material are required, but possibly only for participants with late L2 onset. These results are in line with Wartenburger et al., who directly compared L2 grammatical judgment between highly proficient L2 speakers with early and late L2 onset and found stronger left posterior IFG activation in participants with late L2 onset.

The pattern of results suggests, furthermore, that the left posterior frontal regions are involved in L2 syntactic processing and/or syntactic judgment. In a meta-analysis on native language syntactic processing, Indefrey (2004; see also Kaan & Swaab, 2002) found the left posterior IFG to be reliably replicated for sentence listening and reading with and without additional judgment tasks. The areas were also reliably reported in studies that were well controlled for semantic differences. The left posterior middle frontal gyrus is not typically found in passive L1 sentence comprehension, but it has been found to be activated when syntactic decisions are required (Indefrey, Hagoort, Herzog, Seitz, & Brown, 2001). L2 syntactic processing thus seems to use the same cortical areas as L1 syntactic processing. However, the areas seem to be more strongly recruited than in the native language when there is an extra load on syntactic processing (or when awareness of syntax is required), as in the case of grammaticality decisions.

These considerations raise the question of when syntax-sensitive areas begin to be recruited in the course of L2 acquisition. All of the subjects who participated in the studies analyzed here had been learning their L2 for years, so that the studies cannot provide any evidence with regard to this question. In a recent longitudinal study, Indefrey, Hellwig, Davidson, and Gullberg (2005) conducted a functional magnetic resonance imaging (fMRI) experiment on syntactic processing with native Mandarin Chinese speakers at 3, 6, and 9 months after the onset of learning Dutch in The Netherlands (see Figure 1). The participants were visually presented with scenes of colored geometrical figures performing simple actions. The participants then listened to Dutch descriptions and had to decide whether the descriptions matched the scenes. In one condition, the descriptions were sentences ("The red triangle pushes the blue circle away."). In another condition, the descriptions were syntactically unrelated word sequences ("triangle, red, circle, blue, push away"). In contrast to Dutch native speakers (and the Chinese participants when presented with Mandarin stimuli), who showed significantly enhanced activation of Broca's area when listening to sentences compared to word sequences, the Chinese participants did not show activation differences after 3 months of learning Dutch, although they performed clearly above chance on the task. However, after 6 months, the Chinese learners of Dutch also showed stronger posterior frontal activations for the sentence stimuli and this activation pattern was replicated after 9 months. At this point in time, the participants scored in the low to moderate range of a standardized Dutch proficiency test. These findings suggest that brain regions involved in L1 syntactic processing are relatively soon also recruited for the processing of a new language. A similar time frame has been reported by Osterhout, McLaughlin, Kim, Greenwald, and Inoue (2004; see also Osterhout, McLaughlin, Pitkänen, Frenck-Mestre, & Molinaro, this volume) for the emergence of electrophysiological responses to syntactic violations.

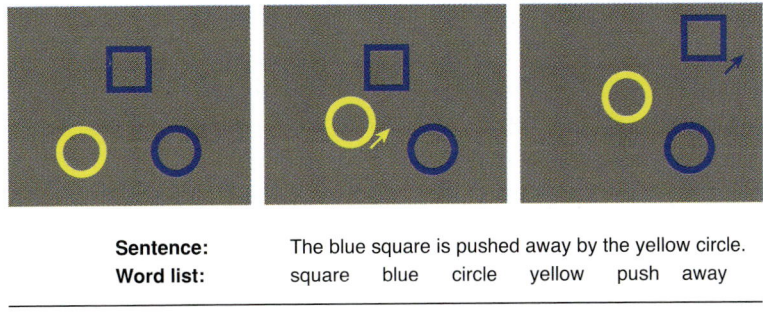

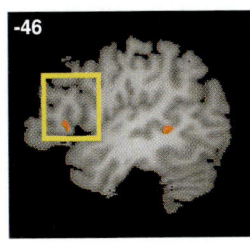
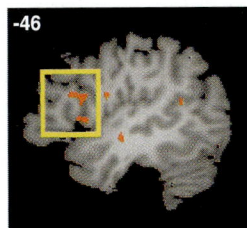

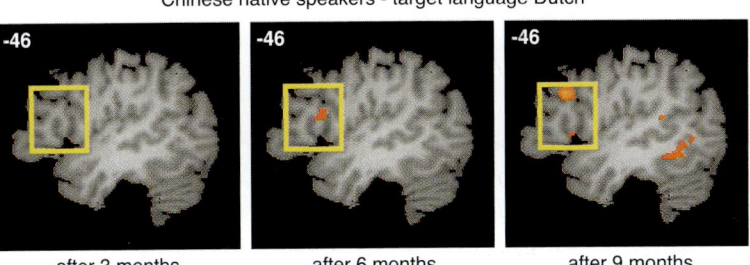

Figure 1. Upper panel: Example of animated visual scenes and corresponding auditory stimulus conditions used in Indefrey et al. (2005). Lower panel: Significant ($p < .05$, small volume correction) left inferior frontal activations for the comparison of sentence and word list conditions. Region of interest indicated in yellow.

Inflectional Morphology

Although stimuli at the sentence and discourse levels involve inflectional morphology (insofar as the languages in question have inflection), the control conditions in the sentence-level studies discussed earlier did not allow to selectively assess activation due to the production or comprehension of inflectional morphemes. There are, however, two recent studies (Sakai, Miura, Narafu, & Muraishi, 2004; Tatsuno & Sakai, 2005) that specifically investigated English past tense inflection in Japanese learners of English. Participants were visually presented with an English verb stem and had to choose between a correct and an incorrect past tense form. Incorrect forms were falsely regularized (catch-catched) or irregularized (smell-smold). In a control condition, participants simply decided which of two verb stems was identical to a previously presented verb stem. Sakai et al. studied the effect of a 2-month training involving explicit instruction on verb inflection in 13-year-old beginners of English. In the past tense matching task, there was stronger activation of the dorsal left posterior IFG (BA 44/45) after training as compared to before training (where no significant activation of this area was found). Activation in this area was higher the better participants performed on the task. The area overlapped with the activation area observed when the same task was conducted with Japanese verbs. In a follow-up study, Tatsuno and Sakai used the same fMRI task on 19-year-old learners of English, who by this age had had 6 years of English instruction. Again, left posterior IFG (together with parieto-occipital and motor areas) was found when comparing past tense form and stem-matching tasks. There were two interesting additional observations. First, a subgroup of high-performing advanced learners showed much weaker activation in this IFG area. Second, activation in the same location was reduced in older as compared to younger participants when performing the task with Japanese verbs. Although the between-group comparisons were conducted in a very small ROI and are therefore susceptible to anatomical alignment differences between groups, the response pattern is suggestive of a change in the linkage between the performance on a certain linguistic task and the accompanying hemodynamic responses. Once an area becomes involved in a linguistic task, its activation level might be initially positively correlated with performance but might show a negative correlation after years of practice. The positive correlation might simply reflect neural and behavioral consequences of the same underlying factor: the degree of effort put into the task. By contrast, a negative correlation might be explained by assuming that the neural structures support the linguistic process involved in the task more effectively, which means that they must have adapted in some way to the process. Note, however, that "adaptation" is not the only option to interpret reduced activation with better performance. Alternatively, a different but more effective brain region might have taken over or such a negative correlation might come about because the task is performed in a different, more effective way that no longer involves the original process.

Do Bilingual Speakers Have a Different Brain Anatomy?

Given that highly practiced cognitive activities have been shown to result in structural brain changes (Draganski et al., 2004; Maguire et al., 2000), one might expect changes in brain structure at least in proficient L2 speakers who must have spoken and heard an L2 frequently and for many years. Anatomical differences between monolingual and bilingual speakers have recently been reported in an inferior parietal region (Mechelli et al., 2004) and in an anterior portion of the mid-body of the corpus callosum (Coggins, Kennedy, & Armstrong, 2004). The latter study compared the cross-sectional areas of different sections of the corpus callosum based on anatomical MRI scans in highly proficient L2 speakers and monolingual speakers. The languages spoken are not reported and the observed structural difference in one of the five callosal regions was small and possibly even nonsignificant, because it is not reported whether the statistical analysis involved appropriate Bonferroni correction for multiple comparisons. Mechelli et al. used voxel-based morphometry, a method that assesses the proportion of gray and white matter in small cubes of brain tissue (voxels) across the whole brain. In L2 speakers of varying L2 onset and proficiency, they found a relative increase in the proportion of gray matter in voxels located in a part of the inferior parietal cortex. The finding indicates some shift of the gray/white matter boundary in the bilingual participants. It does not necessarily mean a thickening of the cortex. Importantly, the structural difference covaried with both L2 onset and proficiency, which makes it unlikely that the finding was due to coincidental anatomical group differences. There is, thus, a piece of evidence that years of bilingual speech processing might indeed lead to structural brain changes. The results of the previous subsections, however, provide little support for the notion that structural changes in the inferior parietal cortex might have been induced by stronger functional recruitment of this cortical region. If at all, stronger L2 activation was mostly observed in the left inferior frontal cortex. The parietal cortex is not reliably activated during word production in the L1 or the L2. To date, there is also no evidence that the left inferior parietal cortex might be more strongly activated in L2 compared to L1 sentence or narrative comprehension. This leaves us with a reliably stronger left inferior parietal activation during L2 semantic decisions on written words. It would certainly be highly speculative to postulate a link between bilingual processing in just this task and structural brain changes.

Summary and Conclusions

The central question of this article was whether there are reliable differences between the hemodynamic activation patterns observed during L1 and L2 processing. Although in most tasks the majority of studies reported no differences, the anatomical overlap of activated regions in those studies that did find differences is unlikely to be due to mere coincidence. The answer is, therefore, yes; there are reliable differences, but only for subgroups of bilingual speakers, and predominantly in the direction of stronger activation during L2 processing. Speaker characteristics that seem to play a role are late L2 onset, low L2 proficiency, and low L2 exposure. In general, this result is in agreement with previous reviews of bilingual brain activation studies (Abutalebi, Cappa, & Perani, 2001; Stowe & Sabourin, 2005). More specifically, the relative influence of the three factors (onset, proficiency, and exposure) seems to differ between the experimental paradigms and, thus, between the language processing components involved. Whereas for word-level production the current evidence is compatible with a role for all three factors, this is different for word-level semantic processing in comprehension, where L2 onset and exposure do not seem to play a major role (see also Stowe & Sabourin, who came to a similar conclusion). By contrast, L2 onset seems to be the most important factor for activation differences related to syntactic processing in sentence comprehension. Note, however, that even in late L2 learners, stronger L2 syntactic processing activations only seem to become visible when subjects are required to make explicit metalinguistic judgments.

In all tasks, stronger L2 activation was confined to regions that are also found when the tasks are performed in the native language, suggesting that there are no L2-specific regions of activation. Note, however, that this statement might only hold at the level of larger regions identified in group analyses of hemodynamic data. At the single-subject level, L2-specific sites within regions might well be found, as suggested not only by the electrical stimulation data discussed earlier but also by single-subject hemodynamic data (Dehaene et al., 1997).

The causes for a stronger recruitment of common L1/L2 regions during L2 processing might differ between tasks, subject groups, and regions. The activation level of the anterior cingulate, for example, is known to depend not only on the attentional demands of tasks but also on the detection of errors. Stronger activation in L2 semantic decision on words might, therefore, be interpreted not only as a higher attentional demand in L2 processing but also as being due to differences in performance. The situation is particularly complex in the case of the left posterior IFG, for which stronger L2 activation was found in all tasks. Considering the heterogeneity of speaker characteristics that seemed to have an influence on left posterior IFG activation and the different language processing components supported by different parts of left posterior IFG in the different tasks, it is very unlikely that the activation level of this region can be traced back to a single factor. A tentative interpretation, which might hold at least for word production and sentence processing, might be based on a distinction between lexical and compositional processes. Whereas lexical processes are mainly supported by temporal lobe areas (Indefrey & Cutler, 2004; Indefrey & Levelt, 2000, 2004), which on the whole do not seem to be more strongly activated in L2 processing, the left posterior IFG is involved in nonlexical compositional processes (postlexical syllabification in word production, syntactic processing in sentence comprehension), which are subject to language-specific rules or constraints. The neuronal organization of the IFG might, therefore, be in some way optimized for the native language and thus less efficient for later learned languages. Stronger activation might then come about by two mechanisms. First, speakers might compensate for lower efficiency by driving this region more strongly. In this case, L2 performance would be expected to covary with activation, as was observed in the Sakai et al. (2004) study on early L2 processing discussed earlier. Second, the activation level of the region might not be modulated by effort but only by the number of neurons needed to perform a task (i.e., by the efficiency of the neuronal organization). In this case, performance can be negatively correlated with the activation level, and the L1 activation level would be intrinsically lower than the L2 activation level. The finding of such a negative correlation in advanced L2 learners (Tatsuno & Sakai, 2005) suggests that the efficiency of the neuronal organization can improve in the course of L2 acquisition.

References

Abutalebi, J., Cappa, S. F., & Perani, D. (2001). The bilingual brain as revealed by functional neuroimaging. *Bilingualism: Language and Cognition, 4*, 179–190.

Briellmann, R. S., Saling, M. M., Connell, A. B., Waites, A. B., Abbott, D. F., & Jackson, G. D. (2004). A high-field functional MRI study of quadri-lingual subjects. *Brain and Language, 89*, 531–542.

Chee, M. W. L., Caplan, D., Soon, C. S., Sriram, N., Tan, E. W. L., Thiel, T., et al. (1999). Processing of visually presented sentences in Mandarin and English studied with fMRI. *Neuron, 23*, 127–137.

Chee, M. W. L., Hon, N., Lee, H. L., & Soon, C. S. (2001). Relative language proficiency modulates BOLD signal change whenbilinguals perform semantic judgments. *NeuroImage, 13*, 1155–1163.

Coggins, P. E., Kennedy, T. J., & Armstrong, T. A. (2004). Bilingual corpus callosum variability. *Brain and Language, 89*, 69–75.

De Bleser, R., Dupont, P., Postler, J., Bormans, G., Speelman, D., Mortelmans, L., et al. (2003). The organisation of the bilingual lexicon: A PET study. *Journal of Neurolinguistics, 16*, 439–456.

Dehaene, S., Dupoux, E., Mehler, J., Cohen, L., Paulesu, E., Perani, D., et al. (1997). Anatomical variability in the cortical representation of first and second language. *NeuroReport, 8*, 3809–3815.

Ding, G. S., Perry, C., Peng, D. L., Ma, L., Li, D. J., Xu, S. Y., et al. (2003). Neural mechanisms underlying semantic and orthographic processing in Chinese-English bilinguals. *NeuroReport, 14*, 1557–1562.

Draganski, B., Gaser, C., Busch, V., Schuierer, G., Bogdahn, U., & May, A. (2004). Neuroplasticity: Changes in grey matter induced by training—Newly honed juggling skills show up as a transient feature on a brain-imaging scan. *Nature, 427*, 311–312.

Frenck-Mestre, C., Anton, J. L., Roth, M., Vaid, J., & Viallet, F. (2005). Articulation in early and late bilinguals' two languages: Evidence from functional magnetic resonance imaging. *NeuroReport, 16*, 761–765.

Hasegawa, M., Carpenter, P. A., & Just, M. A. (2002). An fMRI study of bilingual sentence comprehension and workload. *NeuroImage, 15*, 647–660.

Hernandez, A. E., Dapretto, M., Mazziotta, J., & Bookheimer, S. (2001). Language switching and language representation in Spanish-English bilinguals: An fMRI study. *NeuroImage, 14*, 510–520.

Hernandez, A. E., Martinez, A., & Kohnert, K. (2000). In search of the language switch: An fMRI study of picture naming in Spanish-English bilinguals. *Brain and Language, 73*, 421–431.

Illes, J., Francis, W. S., Desmond, J. E., Gabrieli, J. D. E., Glover, G. H., Poldrack, R., et al. (1999). Convergent cortical representation of semantic processing in bilinguals. *Brain and Language, 70*, 347–363.

Indefrey, P. (2004). Hirnaktivierungen bei syntaktischer Sprachverarbeitung: eine Meta-Analyse. In H. M. Müller & G. Rickheit (Eds.), *Neurokognition der Sprache* (pp. 31–50). Tübingen: Stauffenburg Verlag.

Indefrey, P., & Cutler, A. (2004). Prelexical and lexical processing in listening. In M. S. Gazzaniga (Ed.), *The cognitive neurosciences III* (pp. 759–774). Cambridge, MA: MIT Press.

Indefrey, P., Hagoort, P., Herzog, H., Seitz, R. J., & Brown, C. M. (2001). Syntactic processing in left prefrontal cortex is independent of lexical meaning. *NeuroImage, 14,* 546–555.

Indefrey, P., Hellwig, F., Davidson, D., & Gullberg, M. (2005). *Native-like hemodynamic responses during sentence comprehension after six months of learning a new language.* Poster presented at the 11th Annual Meeting of the Organization for Human Brain Mapping, Toronto.

Indefrey, P., & Levelt, W. J. M. (2000). The neural correlates of language production. In M. S. Gazzaniga (Ed.), *The new cognitive neurosciences* (2nd ed., pp. 845–865). Cambridge, MA: MIT Press.

Indefrey, P., & Levelt, W. J. M. (2004). The spatial and temporal signatures of word production components. *Cognition, 92,* 101–144.

Kaan, E., & Swaab, T. Y. (2002). The brain circuitry of syntactic comprehension. *Trends in Cognitive Sciences, 6,* 350–356.

Klein, D., Milner, B., Zatorre, R. J., Meyer, E., & Evans, A. C. (1995). The neural substrates underlying word generation: A bilingual functional-imaging study. *Proceedings of the National Academy of Sciences of the United States of America, 92,* 2899–2903.

Klein, D., Milner, B., Zatorre, R. J., Zhao, V., & Nikelski, J. (1999). Cerebral organization in bilinguals: A PET study of Chinese-English verb generation. *NeuroReport, 10,* 2841–2846.

Lucas, T. H., McKhann, G. M., & Ojemann, G. A. (2004). Functional separation of languages in the bilingual brain: A comparison of electrical stimulation language mapping in 25 bilingual patients and 117 monolingual control patients. *Journal of Neurosurgery, 101,* 449–457.

Luke, K. K., Liu, H. L., Wai, Y. Y., Wan, Y. L., & Tan, L. H. (2002). Functional anatomy of syntactic and semantic processing in language comprehension. *Human Brain Mapping, 16,* 133–145.

Maguire, E. A., Gadian, D. G., Johnsrude, I. S., Good, C. D., Ashburner, J., Frackowiak, R. S. J., et al. (2000). Navigation-related structural change in the hippocampi of taxi drivers. *Proceedings of the National Academy of Sciences of the United States of America, 97,* 4398–4403.

Mechelli, A., Crinion, J. T., Noppeney, U., O'Doherty, J., Ashburner, J., Frackowiak, R. S., et al. (2004). Structural plasticity in the bilingual brain: Proficiency in a second language and age at acquisition affect grey-matter density. *Nature, 431,* 757.

Nakada, T., Fujii, Y., & Kwee, I. L. (2001). Brain strategies for reading in the second language are determined by the first language. *Neuroscience Research, 40,* 351–358.

Nakai, T., Matsuo, K., Kato, C., Matsuzawa, M., Okada, T., Glover, G. H., et al. (1999). A functional magnetic resonance imaging study of listening comprehension of languages in human at 3 tesla-comprehension level

and activation of the language areas. *Neuroscience Letters*, 263, 33–36.
Osterhout, L., McLaughlin, J., Kim, A., Greenwald, R., & Inoue, K. (2004). Sentences in the brain: Event-related potentials as real-time reflections of sentence comprehension and language learning. In M. Carreiras & J. Clifton (Eds.), *The on-line study of sentence comprehension: Eyetracking, ERP, and beyond* (pp. 271–308). London: Psychology Press.
Perani, D., Abutalebi, J., Paulesu, E., Brambati, S., Scifo, P., Cappa, S. F., et al. (2003). The role of age of acquisition and language usage in early, high-proficient bilinguals: An fMRI study during verbal fluency. *Human Brain Mapping*, 19, 170–182.
Perani, D., Dehaene, S., Grassi, F., Cohen, L., Cappa, S. F., Dupoux, E., et al. (1996). Brain processing of native and foreign languages. *NeuroReport*, 7, 2439–2444.
Perani, D., Paulesu, E., Galles, N. S., Dupoux, E., Dehaene, S., Bettinardi, V., et al. (1998). The bilingual brain: Proficiency and age of acquisition of the second language. *Brain*, 121, 1841–1852.
Pillai, J. J., Allison, J. D., Sethuraman, S., Araque, J. M., Thiruvaiyaru, D., Ison, C. B., et al. (2004). Functional MR imaging study of language related differences in bilingual cerebellar activation. *American Journal of Neuroradiology*, 25, 523–532.
Poline, J. B., Vandenberghe, R., Holmes, A. P., Friston, K. J., & Frackowiak, R. S. J. (1996). Reproducibility of PET activation studies: Lessons from a multi-center European experiment—EU concerted action on functional imaging. *NeuroImage*, 4, 34–54.
Pu, Y. L., Liu, H. L., Spinks, J. A., Mahankali, S., Xiong, J. H., Feng, C. M., et al. (2001). Cerebral hemodynamic response in Chinese (first) and English (second) language processing revealed by event-related functional MRI. *Magnetic Resonance Imaging*, 19, 643–647.
Rodriguez-Fornells, A., Van der Lugt, A., Rotte, M., Britti, B., Heinze, H. J., & Münte, T. F. (2005). Second language interferes with word production in fluent bilinguals: Brain potential and functional imaging evidence. *Journal of Cognitive Neuroscience*, 17, 422–433.
Rüschemeyer, S. A., Fiebach, C. J., Kempe, V., & Friederici, A. D. (2005). Processing lexical semantic and syntactic information in first and second language: fMRI evidence from German and Russian. *Human Brain Mapping*, 25, 266–286.
Sakai, K. L., Miura, K., Narafu, N., & Muraishi, Y. (2004). Correlated functional changes of the prefrontal cortex in twins induced by classroom education of second language. *Cerebral Cortex*, 14, 1233–1239.
Salmelin, R., Hari, R., Lounasmaa, O. V., & Sams, M. (1994). Dynamics of brain activation during picture naming. *Nature*, 368, 463–465.

Stowe, L., & Sabourin, L. (2005). Imaging the processing of a second language: Effects of maturation and proficiency on the neural processes involved. *International Review of Applied Linguistics in Language Teaching, 43*, 329–354.

Talairach, J., & Tournoux, P. (1988). *Co-planar stereotaxic atlas of the human brain.3-dimensional proportional system: An approach to cerebral imaging*. Stuttgart: Thieme.

Tatsuno, Y., & Sakai, K. L. (2005). Language-related activations in the left prefrontal regions are differentially modulated by age, proficiency, and task demands. *Journal of Neuroscience, 25*, 1637–1644.

Vandenberghe, R., Price, C., Wise, R., Josephs, O., & Frackowiak, R. S. J. (1996). Functional anatomy of a common semantic system for words and pictures. *Nature, 383*, 254–256.

Vingerhoets, G., Van Borsel, J., Tesink, C., Van den Noort, M., Deblaere, K., Seurinck, R., et al. (2003). Multilingualism: An fMRI study. *NeuroImage, 20*, 2181–2196.

Wartenburger, I., Heekeren, H. R., Abutalebi, J., Cappa, S. F., Villringer, A., & Perani, D. (2003). Early setting of grammatical processing in the bilingual brain. *Neuron, 37*, 159–170.

Xue, G., Dong, Q., Jin, Z., & Chen, C. S. (2004). Mapping of verbal working memory in nonfluent Chinese-English bilinguals with functional MRI. *NeuroImage, 22*, 1–10.

When Does the Neurological Basis of First and Second Language Processing Differ? Commentary on Indefrey

Laurie A. Stowe
University of Groningen

A large number of studies have been carried out over the last few years investigating whether the neurological representation of language differs between first (L1) and second (L2) language. In a recent review, Stowe and Sabourin (2005) concluded that both L1 and L2 typically activate the same areas, particularly the typical language areas (i.e., that there is no consistent qualitative difference between the neural architecture supporting processing of the two languages). The evidence suggested that this was so even when the L2 was learned relatively late.

On the other hand, the extent and intensity of the activation quite frequently differed between the two languages, typically with increased activations for the L2. This was true for word paradigms, phonological processing, and sentence comprehension. This suggests, instead, a quantitative difference between the processing resources used by the two languages. Particularly, it suggests that L2 frequently cannot be processed as efficiently as L1, requiring more "work" from the same processing resources.

Laurie A. Stowe, Department of Linguistics and Neuroimaging Center.
Correspondence concerning this article should be addressed to Laurie A. Stowe, Department of Linguistics, Faculteit der Letteren, Rijksuniversiteit Groningen, Postbus 716, 9700 AS Groningen, The Netherlands. Internet: l.a.stowe@rug.nl

Factors That Contribute to Finding a Difference Between L1 and L2

Determining *when* L2 is processed differently remains difficult. Trying to disentangle the effects of age and proficiency on language processing requires finding groups of speakers with well-defined onsets of L2 or clearly distinguishable levels of proficiency. The degree of difference between the L1 and L2 might also be an important factor. Indefrey (this volume) has carried out a systematic and impressive review of studies that showed differences between L1 and L2, employing similar tasks, in order to investigate which areas are reliably replicated. The results of this meta-analysis led to a conclusion similar to that drawn by Stowe and Sabourin (2005): Only areas that are typically involved in language processing in general—and are typically activated in monolingual subjects carrying out the same task—showed increased activation for L2. Even for these areas, however, such differences do not always replicate. Indefrey further analyzed the variations in subject groups and tasks, which might explain why some studies found differences and others did not. He concludes that differences are probably easier to find with late acquisition or less proficient speakers or when the task increases in its linguistic demands. There seems little reason to doubt that these factors do play a role in how readily differences between L1 and L2 can be demonstrated.

Age of acquisition and proficiency are probably the most interesting factors for theories of how the brain supports L2 acquisition. The word-naming studies summarized by Indefrey showed statistical differences in the left inferior frontal gyrus (LIFG, a portion of which is called Broca's area) for two studies with late learners, whereas no significant difference was found in two studies with early learners. This suggests that early learners do not require additional resources in order to carry out these tasks.

Indefrey notes that another factor that appears to affect whether differences are found between L1 and L2 is the complexity of the linguistic materials being employed and the task that the subjects are asked to carry out. In L1, sentence and story comprehension typically activate the LIFG and the posterior temporal lobe, with the extent of the activations somewhat dependent on the baseline condition employed. Increased frontal activation for these areas in L2 were found in four out of nine studies, suggesting that this is reliable. These studies differed from the studies in which no difference was found, in that they employed particularly complex sentences (Hasegawa, Carpenter, & Just, 2002) or an additional task such as grammaticality judgment, which required conscious introspection or deeper processing of the materials (Rüschemeyer, Fiebach, Kempe, & Friederici, 2005). As long as these activations are due to language processing rather than task processing, this is not problematic, but it complicates the interpretation of such results. Such task effects are not unexpected; Stowe, Haverkort, Maguire, Wijers, and Paans (1999) demonstrated that although complex sentences typically activate the LIFG, simple sentences show relatively little activation relative to simply fixating on a central point on a screen unless a task is required of subjects. As the same area seems to be involved, the area nevertheless probably contributes to the same aspect of processing in both complex sentences and when carrying out an additional task.

Missing Differences Between L1 and L2

A less interesting reason why differences do not show up between L1 and L2 might simply be lack of statistical power. Most neuroimaging L2 studies have relatively small numbers of subjects. Further, because the areas that show differences are typically activated in *both* languages, we are in fact attempting to show an interaction between the factors L1 versus L2 and language task versus control. Under these circumstances, it is not very surprising that it is sometimes difficult to show a difference between L1 and L2. This suggests some caution in interpreting nonreplications.

Hernandez, Martinez, and Kohnert (2000), who tested early learners of their L2 on word-naming, did not replicate the LIFG activation found for late learners. Only six subjects were included in the study. Although there was a clear tendency toward increased activation in Broca's area for English over Spanish in both volume and intensity of activation (see Figure 2 in their article), this failed to reach significance. Given the activations seen in the other studies, one cannot readily dismiss the trend as truly nonsignificant. Similarly Vingerhoets et al. (2003) do not report significant LIFG differences between L1 and L2 for reading sentences without any extra task, unlike studies with more complex materials or additional tasks. Nevertheless, this area is significantly activated in L2 (French) but not L1 (Dutch), again suggesting a trend toward more activity, even though no extra task was superimposed.

A second factor might be important in detecting differences between L1 and L2: individual variability. Most group statistics assume that the voxels (points in three-dimensional space) within the brain are not only at the same *anatomical* location of the brain for all (or at least most) subjects but also that they have the same cytoarchitectural structure (i.e., the proportion of types of cell and receptor, as well as the organization of the cells into layers). This is necessary for the assumption that the area has the same neurocomputational characteristics in all the subjects. However, there is considerable variability in the location of specific cytoarchitectures across individuals (see Uylings, this volume). When individuals are superimposed onto a template brain using the same method as in group statistics, there is virtually no voxel at which 10 random individuals all have Brodmann Area 44 or 45, which are two types of cytoarchitecture found in the IFG bilaterally. This can obviously result in variability in the exact place in the brain in which various individuals process some particular aspect of language and in which they thus show a difference between L1 and L2. Hasegawa et al. (2002) showed activations for L1 and L2 per individual (cf. Figure 1 in their article). Although these appeared consistently larger for L2 than L1, the locations within the frontal cortex on which these differences were found were quite diverse. Hasegawa et al. might have been able to show differences primarily because they compared the number of activated voxels per subject rather than using group statistics.

Optimizing the Chances of Finding L1 Versus L2 Differences

At least one study (Perani et al., 2003) has suggested that even very early L2 acquisition does not lead to totally equivalent levels of activation in L1 and L2. Similarly, a number of event-related potential (ERP) studies demonstrate that processing does not become equivalent to L1 processing even when a relatively good level of proficiency has been achieved (cf. reviews in Sabourin & Stowe, submitted; Stowe & Sabourin, 2005). One goal of investigating L1 and L2 differences is clearly to determine if late acquisition is qualitatively or quantitatively different than early acquisition; another, to determine the role of proficiency in neural organization. Before we draw conclusions based on lack of differences in more proficient or early learners, it is important to be sure that these results are not simply due to a lack of statistical power. Clearly, a subthreshold difference is not the same as no difference.

The above discussion suggests two factors that increase the likelihood of finding a difference between L1 and L2. First, a relatively difficult aspect of L2 processing should be tested (e.g., comprehension of complex sentences). Second, analysis techniques that focus on individuals rather than groups might also optimize the design. If even this sort of design, using a fairly sizable group of subjects, failed to show any sign of a difference between L1 and L2, the negative results should be convincing.

Interpreting L2 Activations

One last issue that I would like to bring up is whether age of acquisition effects should necessarily be regarded as due to differences in processing between L1 and L2. This is probably the normal assumption for this sort of research. Indefrey (this volume) gives a version of this hypothesis when he suggests that the LIF lobe is more activated in L2 because it is an area that has learned early to deal with particular patterns of syllabification and cannot be used efficiently to deal with others. However, we might want to question the validity of this assumption. Fiebach, Friederici, Müller, Von Cramon, and Hernandez (2003) investigated age of acquisition effects on activation elicited by L1 words (i.e., words that do not presumably differ in syllable structure or spelling patterns, merely in how early they were learned). They found that late age of acquisition led to increases in activation in an area of the LIFG similar to that found for L2. This suggests that later learning even within L1 cannot always benefit from a general processing efficiency for L1.

References

Fiebach, C. J., Friederici, A. D., Müller, K., Von Cramon, D. Y., & Hernandez, A. W. (2003). Distinct brain representations for early and late learned words. *NeuroImage, 19*, 1627–1637.

Hasegawa, M., Carpenter, P. A., & Just, M. A. (2002). An fMRI study of bilingual sentence comprehension and workload. *NeuroImage, 15*, 647–660.

Hernandez, A. E., Martinez, A., & Kohnert, K. (2000). In search of the language switch: An fMRI study of picture naming in Spanish-English bilinguals. *Brain and Language, 73*, 421–431.

Perani, D., Abutalebi, J., Paulesu, E., Brambati, S., Scifo, P., Cappa, S., & Fazio, F. (2003). The role of age of acquisition and language usage in early, high-proficient bilinguals: An fMRI study during verbal fluency. *Human Brain Mapping, 19*, 170–182.

Rüschemeyer, S. A., Fiebach, C. J., Kempe, V., & Friederici, A. D. (2005). Processing lexical semantic and syntactic information in first and second language: FMRI evidence from German and Russian. *Human Brain Mapping*, *25*, 266–286.

Sabourin, L. L., & Stowe, L. A. (submitted). Second language processing: Circumstances under which L1 and L2 are similar. *Brain and Language*.

Stowe, L. A., Haverkort, M., Maguire, R. P., Wijers, A. A., & Paans, A. M. J. (1999). Effects of secondary tasks on language activations. *NeuroImage*, *9*, S988.

Stowe, L. A., & Sabourin, L. L. (2005). Imaging the processing of a second language: Effects of maturation and proficiency on the neural processes involved. *International Review of Applied Linguistics*, *43*, 329–353.

Vingerhoets, G., Van Borsel, J., Tesink, C., Van den Noort, M., Deblaere, K., Seurinck, R., et al. (2003). Multilingualism: An fMRI study. *NeuroImage*, *20*, 2181–2196.

Summing up: Some Themes in the Cognitive Neuroscience of Second Language Acquisition

John H. Schumann
University of California, Los Angeles

From this excellent set of presentations, it seems to me that three themes emerge: (a) Can the brain address all relevant linguistic issues? (b) What are the sources of individual differences within and across brains? (c) What areas of the brain subserve language?

To address the first question, which was raised in Wolfgang Klein's opening remarks at the conference from which this volume emerged, it might be useful to distinguish between phenomena that can be understood by indexical reference versus those that are understood by symbolic reference (Deacon, 1997). In order to make this distinction, let us imagine an evolutionary scenario in which a group of hominids developed the ability to produce particulate sounds and to concatenate those sounds into words (Lee & Schumann, 2003).[1] They then attempt to communicate various meanings by combining these words. They do this in conversational interaction where certain structures emerge as part of a complex adaptive system generated by the speakers' efforts to make meanings in consistent ways over time. In such a system, words are not simply indexical (i.e., they do not simply refer to things in the world); instead, they refer to other words. This

---

John H. Schumann, Department of Applied Linguistics.

Correspondence concerning this article should be addressed to John H. Schumann, Dept. of TESL/Applied Linguistics, University of California, Los Angeles, 3300 Rolfe Hall, P. O. Box 951531, Los Angeles, CA 90095-1531. Internet: schumann@humnet.ucla.edu

emergent system now exists in the environment as a cultural artifact or technology, and to a large extent, it can take on a life of its own. Some of the structure that is characteristic of this system might not be controlled directly via the brains of its speakers but might be the product of how the elements of the language must interact among themselves once some initial steps have been taken in structuring the system. Although the forms might not be reducible to the brain, they are nevertheless mediated by the brain. However, now that the system exists *between* and *among* brains, the language itself can no longer be considered essentially *in* or *of* the brain. In other words, there might be facts about the structure of the language that are products of how simpler structures are designed and how that design constrains the form of subsequent structures. Thus, the emergent linguistic system might be subject to constraints that are independent of, but not entirely outside of, neural constraints.

Another way to look at this is to ask the question "Where is the waltz in the brain?" Physical movements that constitute the waltz might not be reducible to motor systems in the brain, but they nevertheless are constrained by them. Thus, motor systems that allow us to waltz are brain based, but the specific movements that comprise the waltz are products of the dance's choreography in relation to the music. Another example might be revealed by asking the question, "Where is the lifting in a crane?" (Asif Agha, personal communication). Does the lifting occur in the shovel, in the motor, in the cables, in the operator, or in the orders given to the operator by the foreman?

Thus, Klein's skepticism about the degree of reduction possible in brain-language relationships might be well placed. There might be constraints on the structure of the emergent system that cannot be reduced to neural structures and still remain meaningful. The second issue—concerning individual differences (IDs)—came up in the talks by Birdsong (IDs in second language [L2] proficiency), Uylings (IDs in neural structure), and Green et al. (IDs resulting from degeneracy within the brains). Edelman (1987, 1989, 1992; see also Schumann, 1997) has identified several sources of variation in brains. The first is the shuffling of parental genes during fertilization. Each child receives approximately 50% of his genes from his father and 50% from his mother, but the mix of genes in each child is different. The second is called developmental selection and involves stochastic processes in cell migration during embryology. Genes do not specify the targets of all neurons; therefore, neural migration is subject to some variability caused by differences in the chemical milieu in which cell migration takes place. The third source of variation is called experiential selection and occurs as a result of an organism's experience in the world. The result of this selection is that some neuronal groups, ensembles, and circuits are strengthened and others are weakened and die out. Degeneracy refers to the situation in which two structurally different neural systems can produce the same output. The degenerate structures might be different in different brains. These four processes combine to produce brains that are as different as faces at the level of microstructure.

Differences in brains are very important when considering critical period issues in SLA. The fact that some older learners might achieve native or near-native proficiency in the L2 says nothing about critical periods. It might be that late starters who are high achievers simply have different brains than late starters who are low achievers. The high achievers might have certain neural hypertrophies that provide advantages in postcritical period SLA.

An example of how such hypertrophies can advantage a brain comes from research on Einstein's brain, which was preserved after his death. Diamond, Scheibel, Murphy, & Harvey (1985) have shown that Einstein had many more glial cells (support structures for neurons) than age-matched controls. Witelson, Kigar, & Harvey (1999, as summarized in Schumann et al., 2004) showed that

> the posterior limbs of the Sylvian fissure do not exist and the fissure joins postcentral sulcus. This architecture eliminates the parietal operculum which, in normal brains, lies between the postcentral sulcus and the posterior segment of the lateral sulcus. In addition, Einstein's parietal lobes

were symmetrical whereas those of most humans lack this symmetry. Each hemisphere of his brain was 15% larger than in controls, and his parietal lobes were wider and more spherical than normal. The elimination of the parietal operculum expanded the area of the inferior parietal lobule in which visual, somatosensory, and auditory stimuli are integrated and where visuospatial and mathematical cognition as well as movement imagery are processed. (Schumann et al., 2004, p. 12)

Witelson et al. (1999) suggested that this hypertrophy had functional consequences that allowed Einstein to cognize creatively in domains related to his scientific contributions. Einstein evidently had an extensive hypertrophy in neural structure. Less profound hypertrophies in the areas of the brain subserving L2 acquisition might allow certain older individuals to achieve very high proficiency in an L2.

The third issue emerging from the presentations concerns which areas of the brain subserve language. Rodriguez-Fornells et al.'s research showed that the dorso-lateral prefrontal cortex and the anterior cingulate cortex are involved in language to facilitate the processes of selection, switching, and inhibition/suppression. Indefrey demonstrates that the areas of the brain activated in L2 processing are very similar to those involved in first language (L1) processing, and those areas include large parts of the temporal lobe and the frontal lobe. Given this extensive substrate for language acquisition, it might be enlightening to ask what parts of the brain can be shown to play NO role in language. Another way to express this view might be to hypothesize that there are no language areas in the brain; there are simply areas that process language. This position becomes even stronger when one considers issues such as the neural substrate for social affiliation. If one believes that the caregiver-child bond in L1 learning or integrative motivation in L2 learning are important in language acquisition, then one would have to argue that language acquisition is also subserved by the extended amygdala, basolateral amygdala, the nucleus accumbens shell, the medial orbital area, and the hypothalamus for the production of oxytocin and vasopressin (Depue & Morrone-Strupinsky, 2005; Luciana, 2001). Because language involves both declarative knowledge and procedural knowledge, both the hippocampus and the basal ganglia must also be involved (Schumann et al., 2004).

Summing up the summary, we note that not all structures found in target languages or interlanguages can be traced to particular structures or functions in the brain; they might be the result of how language, as a cultural artifact, develops as it is used between and among brains. Brains vary substantially across individuals. This variation might put some L2 learners at an advantage in acquiring an L2 after the close of any critical/sensitive period. Finally, it would appear that the brain areas involved in L1 acquisition are very similar to those involved in L2 acquisition; these areas are extensive and might not constitute a dedicated module for language.

Note

[1] The position developed by Lee & Schumann (2003) follows on earlier work by Deacon (1997), Larson-Freeman (1997), Batali (1998), Kirby (1998), Steels (1998), De Boer (2000), MacNeilage and Davis (2000), Weber and Deacon (2000), and Studdert-Kennedy (2000).

References

Batali, J. (1998). Computational simulations of the emergence of grammar. In J. M. Hurford, M. Studdert-Kennedy, & C. Knight (Eds.), *Approaches to the evolution of language* (pp. 405–426). Cambridge: Cambridge University Press.

De Boer, B. (2000). Emergence of sound systems through self-organization. In C. Knight, M. Studdert-Kennedy, & J. M. Hurford (Eds.), *The evolutionary emergence of language: Social function and the origins of linguistic form* (pp. 177–198). Cambridge: Cambridge University Press.

Deacon, T. W. (1997). *The symbolic species: The co-evolution of language and the brain*. New York: Norton.

Depue, R. A., & Morrone-Strupinsky, J. V. (2005). A neurobehavioral model of affiliative bonding: Implications for conceptualizing a human trait of affiliation. *Behavioral and Brain Sciences, 28,* 313–350.

Diamond, M. C., Scheibel, A. B., Murphy, G. M., & Harvey, T. (1985). On the brain of a scientist: Einstein, Albert. *Experimental Neurology, 88,* 198–204.

Edelman, G. M. (1987). *Neural darwinism: The theory of neuronal group selection.* New York: Basic Books.

Edelman, G. M. (1989). *The remembered present: A biological theory of consciousness.* New York: Basic Books.

Edelman, G. M. (1992). *Bright air, brilliant fire: On the matter of mind.* New York: Basic Books.

Kirby, S. (1998). Fitness and the selective adaptation of language. In J. M. Hurford, M. Studdert-Kennedy, & C. Knight (Eds.), *Approaches to the evolution of language* (pp. 359–383). Cambridge: Cambridge University Press.

Larsen-Freeman, D. (1997). Chaos/complexity science and second language acquisition. *Applied Linguistics, 18*(2), 141–165.

Lee, N., & Schumann, J. H. (2003, March). *The evolution of language and of the symbolosphere as complex adaptive systems.* Paper presented at the conference of the American Association for Applied Linguistics, Arlington, VA.

Luciana, M. (2001). Dopamine-opiate modulations of reward-seeking behavior: Implications for the functional assessment of prefrontal development. In C. A. Nelson & M. Luciana (Eds.), *Handbook of developmental cognitive neuroscience.* Cambridge, MA: MIT Press.

MacNeilage, P. F., & Davis, B. L. (2000). Evolution of speech: The relation between ontogeny and phylogeny. In C. Knight, M. Studdert-Kennedy, & J. M. Hurford (Eds.), *The evolutionary emergence of language: Social function and the origins of linguistic form* (pp. 146–160). Cambridge: Cambridge University Press.

Schumann, J. H. (1997). *The neurobiology of effect in language.* Malden, MA: Blackwell.

Schumann, J. H., Crowell, S. E., Jones, N., Lee, N., Schuchert, S. A., & Wood, L. A. (2004). *The neurobiology of learning: Perspectives from second language acquisition.* Mahwah, NJ: Erlbaum.

Steels, L. (1998). Synthesizing the origins of language and meaning using coevolution, self-organization and level formation. In J. M. Hurford, M. Studdert-Kennedy, & C. Knight (Eds.), *Approaches to the evolution of language* (pp. 384–404). Cambridge: Cambridge University Press.

Studdert-Kennedy, M. (2000). Evolutionary implications of the particulate principal: Imitation and the dissociation of phonetic form from semantic function. In C. Knight, M. Studdert-Kennedy, & J. M. Hurford (Eds.), *The evolutionary emergence of language: Social function and the origins of linguistic form* (pp. 161–176). Cambridge: Cambridge University Press.

Weber, B., & Deacon, T. W. (2000). Thermodynamic cycles, developmental systems, and emergence. *Cybernetics & Human Knowing, 7*(1), 21–43.

Witelson, S. F., Kigar, D. L., & Harvey, T. (1999). The exceptional brain of Albert Einstein. *Lancet, 353*, 2149–2153.

Glossary of Neuroanatomical Terms and Phrases

(Note. The asterisk (*) refers to other index words.)

Acetylcholine
A neurotransmitter (see synapse*)

Amygdala
See Figure 1B

Associational cortices
Cortical regions that do not directly receive sensory input, typically performing higher order cognitive functions

Axonal connections / axonal fibers
The axon is the fiber of a neuron that transports the neuron's outgoing signals to other neurons or muscle cells.

Basal ganglia
Summary term for caudate nucleus, putamen, globus pallidus, and nucleus accumbens (see Figures 1B and 1D)

Brainstem
The definition of the summary term "brainstem" varies. Most commonly, the brainstem has been defined as midbrain, pons, and medulla oblongata. A few authors include also the thalamus and hypothalamus (see Figure 1B).

Broca's area
Anterior language area, Brodmann areas* 44 and 45 of the left inferior frontal gyrus (see Figure 1A)

Brodmann area, BA
Based on a microscopic analysis of the whole cortex, the neurologist K. Brodmann created a map of the cortex, in which he distinguished and numbered areas of similar cytoarchitecture*. Referring to this map, Brodmann areas are frequently used as convenient macroscopic anatomical labels. When used in this way, Brodmann areas have lost their original cytoarchitectonic meaning. See Figure 1A for BAs mentioned in this volume.

Caudate nucleus
See Figure 1B

Cingulate cortex
See Figure 1B

Cytoarchitecture
Spatial arrangement of different nerve cell groups as seen in stainings, which make the cell bodies of all neurons and glia* cells visible (e.g., Nissl staining)

Dendrite
Dendrites are fibers of neurons receiving signals of other neurons.

Dentate gyrus, fascia dentata
Narrow gyrus* in the medial temporal lobe near the hippocampus (see Figure 1D)

Dopamine
A neurotransmitter (see *Synapse**)

Dorsal
Toward upper end of the brain (see Figure 1C)

Dorsolateral prefrontal cortex, DLPC, DLPFC
Part of the prefrontal* cortex that is known to have important working memory, executive, and attentional functions. It encompasses the middle and superior frontal gyri*.

Ectopic cells
Cells that have migrated to the wrong place

Entorhinal cortex
Part of the anterior parahippocampal gyrus (see Figure 1B)

Fascia dentata
See *Dentate gyrus**

Frontal cortex, frontal lobe
See Figure 1C

Fusiform face area
A cortical area in the fusiform gyrus* that is sensitive to faces

Fusiform gyrus
See Figure 1B

Ganglionic eminence
An embryonic cluster of nerve cells near the ventricle, where neurons are generated and migrate toward different parts of the brain (e.g., the neocortex*, the thalamus, and the amygdala*)

Glia fibers
Extensions of glia cells. Glia cells are part of the neural tissue. They separate nerve cells from blood vessels and also produce the myelin* sheath of axons*.

Gyrus, (pl.) gyri
Convex fold of the surface of the brain

Inferior frontal gyrus
See Figure 1A

Insular cortex, insula
See Figure 1D

Lateral
Toward the right or left side of the brain (see Figure 1D)

Medial
Toward the midline of the brain (see Figure 1D)

Middle frontal gyrus
See Figure 1A

Myelin
Substance that forms a sheath around axons*, which allows for faster signal propagation. Myelin is produced by glia* cells,

Neocortex, neocortical
Phylogenetically younger, largest part of the cortex with six nerve cell layers. Phylogenetically older parts of the cortex (olfactory cortex, hippocampus*, dentate gyrus*) are referred to as allocortex.

Neurogenesis
The formation of nerve cells (neurons)

Nigrostriatal
Part of a pathway connecting the substantia nigra (see Figures 1B and 1D) and the striatum*

Nucleus
A group of nerve cells that forms an anatomical entity

Occipital cortex, occipital lobe
See Figure 1C

Olfactory bulb
End of an anterior extension of the brain (see Figure 1B), containing smell-sensitive cells that reach into the nasal cavity

Orbitomedial frontal cortex
See Figure 1B

Parahippocampal place area
In rats, possibly also in humans, a cortical area containing nerve cells encoding the animal's position; see parahippocampal gyrus in Figure 1B

Parietal lobe, parietal cortex
See Figure 1C

Perisylvian areas
Cortex adjacent to the Sylvian fissure, a deep sulcus* between the frontal and the temporal lobe (see Figure 1A). The term is frequently used as a summary term for the language-related areas in the left hemisphere.

Prefrontal cortex, PFC
Summary term for the frontal cortex without the motor and premotor cortex

Primary sensory and motor areas
The first cortical areas that receive input from sensory organs (visual: posterior occipital cortex, auditory: mid superior temporal gyri; touch: postcentral gyri) and the last cortical areas that generate output to the musculature (precentral gyri)

Putamen
See basal ganglia* and Figure 1D

Pyramidal dendrites
Dendrites* of pyramidal cells, a type of nerve cell in the cortex

Radial glia fibers
Glia* fibers across cortical layers

Radial migration
Nerve cell migration from the proliferative zone (i.e., the subventricular and ventricular zone) around the ventricle into the cortex

Rostral
Toward the front end of the brain (see Figure 1C)

Supplementary motor area, SMA
A premotor area (see Figure 1A)

Spina bifida
A developmental malformation of the spine

Striatum, striatal
Summary term for caudate nucleus*, putamen*, and globus pallidus (see basal ganglia*)

Sulcus, (pl.) sulci
Concave fold of the surface of the brain

Superior temporal gyrus
see Figure 1A

Supramarginal gyrus
See Figure 1A

Synaptogenesis
The formation of synapses*

Synapse
Contact point between nerve cells, where the electrical signal of one nerve cell's axon* induces an electrical signal in another nerve cell's dendrite* through the release of chemical substances (transmitters)

Tangential migration
Nerve cell migration from a far distance (e.g., the ganglionic eminence*) in layers or zones parallel to the brain surface and/or the ventricle

Temporal lobe, temporal cortex
See Figure 1C

Temporal poles
The most anterior part of the temporal lobes

Thalamic nuclei
Groups of nerve cells in the thalamus (see Figure 1B). The thalamus is an important relay station for all input and output of the brain.

Ventral
Toward the lower end of the brain (see Figure 1C)

Ventricle
The ventricles form a system of liquid-filled cavities in the brain.

White matter
Brain tissue beneath the cortex (gray matter) that appears lighter than the cortex because it mainly contains myelinated axons* (i.e., axons with a myelin* sheath)

328   *Glossary of Neuroanatomical Terms and Phrases*

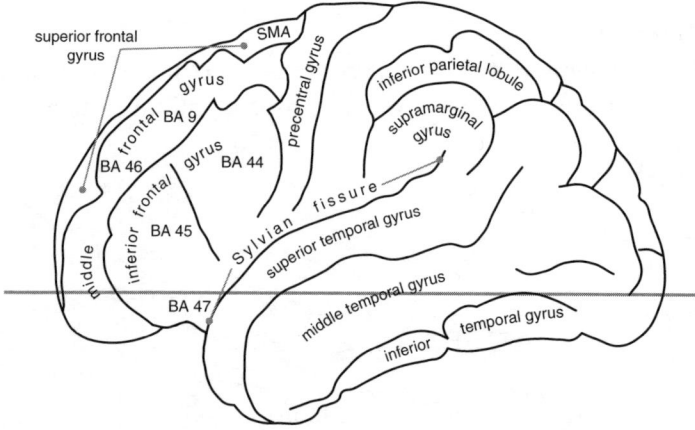

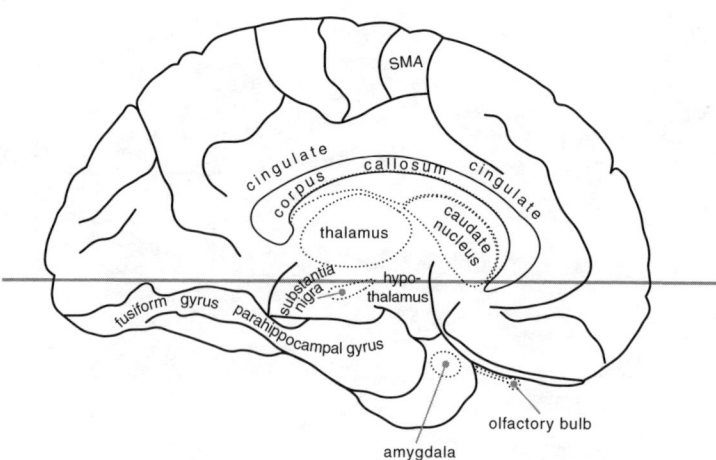

Figure 1. Schematic drawings of the brain showing brain structures mentioned in this volume. (A, B) lateral and medial views of the left hemisphere indicating gyri and Brodmann areas; (C) lateral view of the left hemisphere indicating the cortical lobes and the cerebellum; (D) horizontal slice of the brain indicating subcortical structures. The horizontal gray line in parts A and B shows the level of the slice (D).

Glossary of Neuroanatomical Terms and Phrases 329

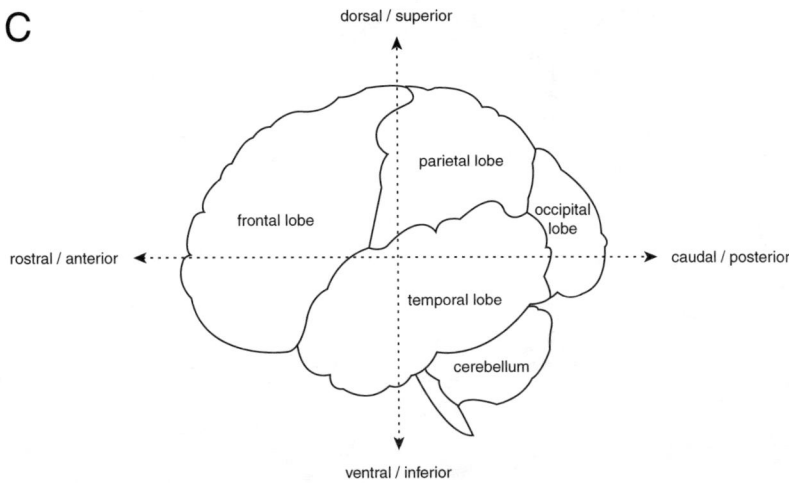

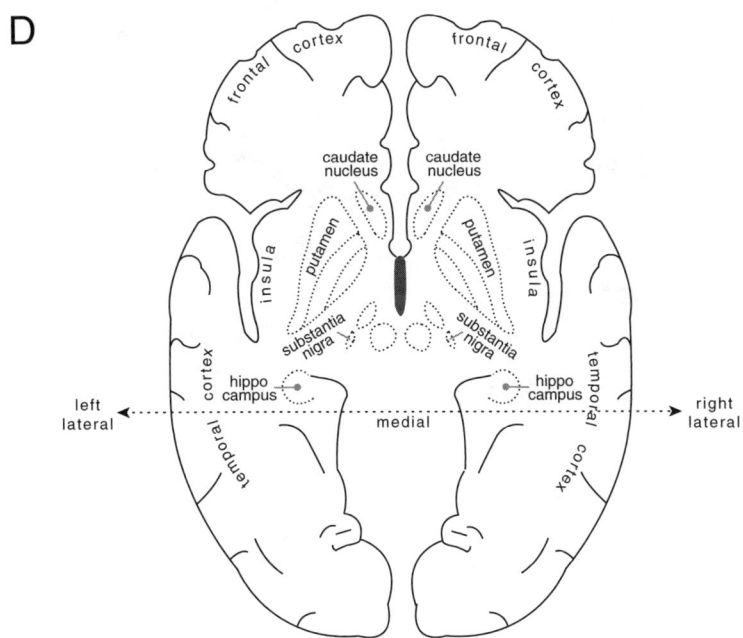

Figure 1. (continued)

Author Index

Abrahamsson, N., 19, 20, 21
Abutalebi, J., 24, 105, 110, 116, 136, 299
Adelstein, A., 78
Ader, R., 78
Adleman, N. E., 145
Aglioti, S., 151
Ainsworth-Darnell, K., 226
Akahane-Yamada, R. A., 20
Akiyama, M., 70
Albert, M. L., 139
Albin, R. L., 158
Ali, S. O., 153
Allen, M. D., 205, 226
Allison, J., 124
Allison, J. T., 303
Allison, T., 172, 173
Allport, A., 140, 141
Alston, R. L., 30
Alter, K., 241,
Amaral, D. G., 93
Ambrose, S. H., 120
Amedi, A., 94
Amunts, K., 115, 129
Anderson, M. C., 38
Anderson, S. W., 77
Anthony, J. C., 71
Arbib, M. A. 118
Ardal, S., 242
Armstrong, T. A., 297
Aron, A. R., 145
Arteaga, D., 215
Aslin, R. N. 211
Avery, D., 152

Bäckman, L., 28, 29, 32, 33
Baddeley, A., 153
Baddeley, A. D., 113
Badecker, W., 205

Badre, D., 158
Baker, M. C., 100
Balcom, P., 22
Baldwin, K. B., 245
Bandstra, E. S., 71
Banfield, J., 147
Barber-Friend, H., 239
Barcelo, F., 171, 172
Barch, D. N., 170
Barker, R. A., 116
Barnea-Goraly, N., 67
Barr, H. M., 70
Barr, R. G., 86
Barss, A., 205, 241
Bartels, A., 93
Batali, J., 317
Bavelier, D., 79
Bechara, A., 77
Behrens, T. E. J., 110,
Benet-Martinez, V. 119
Bentin, S., 203, 209
Bergan, J. F., 73
Berger, M. S., 137
Berman, K. F., 145
Bernard-Bonnin, A. C., 82
Bersick, M., 226
Bialystok, E., 14, 16, 17, 109, 130, 138, 141, 192, 238
Birdsong, D., 11, 12, 13, 14, 16, 17, 18, 19, 20, 21, 36, 37, 51, 52, 60, 79, 131, 166, 236, 237, 238
Bley-Vroman, R., 19, 264
Bloom, F. E., 69
Bobb, S., 108
Boland, J., 226
Bongaerts, T., 20, 138, 139
Bookheimer, S., 136, 283
Bornkessel, I., 251

Bornstein, M. H., 15, 17
Boronat, C. B., 22
Bosch, L., 170
Bott, S., 22
Botvinick, M. 153, 162, 170
Botvinick, M. M., 153, 170
Boustagui, E., 21
Bowden, H. W., 36, 38
Bradlow, A. R., 20
Brake, W. G. 254
Braver, T. S., 170
Brenders, P., 193
Briellmann, R. S., 289
Brinke, S., 166
Brodmann, K., 93
Brown, A., 89
Brown, C. M., 143, 241, 293
Brown, R. M., 68
Brown, C., 226
Bryck, R. L., 158
Bryden, M. P., 78
Buchsbaum, B. R., 145
Bullock, B. E., 4
Bunge, S. A., 153
Butler, J., 76

Cabeza, R., 35
Cadoret, G., 94
Callan, D. E., 116
Cameron, J. L., 69
Cappa, S. F., 24, 105, 299
Caramazza, A., 138, 139
Carpenter, P. A., 136, 292, 307
Carreiras, M., 239
Carter, C. S., 153, 162, 170
Caspi, A., 78
Cassidy, K. W., 248
Cazzaniga, L., 77
Chang, W. L., 145
Chanquoy, L., 214
Chao, L. L., 172
Chapron, C., 80
Chechik, G., 129
Chee, M. W. L., 3, 25, 26, 103, 113, 115, 136, 201, 290, 291
Chen, C. S., 290

Chomsky, N., 37, 120,
Chugani, H. T., 69, 78
Chwilla, D. J., 193
Clahsen, H., 21, 167, 168, 226, 238, 244
Clyne, M. G., 4
Coggins, P. E., 297
Cohen, D. 78
Cohen, J. D., 153, 162, 170
Cohen, N., 78
Cok, B., 27
Coles, M. G. H., 158, 170, 176
Colomé, A., 139, 149, 156, 170, 193
Conolly, J. F., 240, 254
Conway, M., 20
Cook, V., 22
Cooney, J. W., 172
Cooper, J. R., 69
Coopmans, P., 37
Corbetta, M., 171
Cornelissen, K., 174
Corsellis, J. A., 30
Costa, A., 115, 138, 139, 141, 164, 170
Coulson, S., 168, 176, 241
Courchesne, E., 29
Craik, F. I. M., 28, 109, 141
Cranshaw, A., 20
Crinion, J., 104, 113
Crosson, 32
Cruess, D. G., 78
Curtis, C. E., 153
Cutler, A., 280, 300,

Dabholkar, A. S., 67, 75
Daems, F., 215
Dale, A. M., 232, 234
Damasio, H., 77
D'Amato, C. J., 70
Dapretto, M., 136, 283
Darcy, I., 20
Davidson, D., 27, 294
Davis, B. L., 317
De Bleser, R., 25, 105, 283, 288, 289

De Boer, B., 317
De Bot, K., 116, 138, 139
De Bruijn, E., 193
De Bruijn, E. R. A., 193
De Bruin, J. P. C., 68
De Courten, C., 67
De Diego Balaguer, R., 133, 135, 161, 162, 168, 169, 170
De Groot, A. M. B., 176
De Hoop, H., 275
De Smedt, K., 191
Deacon, T. W., 313, 317
Dehaene, D., 25, 137, 299
Dehaene, S., 103
DeKeyser, R., 12, 16, 60, 78, 79
DeKeyser, R. M., 13, 17, 37
Dekydtspotter, L., 53, 54, 55
Delalle, I., 68
Dell, G. S., 151
Denburg, N. L., 77
Depue, R. A., 317
Desimone, R., 172
Desmond, J. E., 153
D'Esposito, M., 153, 172
Devlin, J. T., 106, 108
Dhond, R. P., 234
Di Biase, B. 10
Diamond, M. C., 66, 315
Diamond, A. 138
Diaz, B., 170
Dijkstra, T., 7, 22, 116, 140, 165, 166, 191, 193
Ding, G., 26, 284, 287, 190, 291
Dobmeyer, S., 171
Dohrn, U., 168, 244
Dominey, P. F., 246, 247, 264
Donald, M. W., 242
Donchin, E., 158, 170, 240
Dong, Q., 26, 90
Dong, Y., 176
Donkers, F. C., 145
Dörnyei, Z., 12, 36,
Doughty, C. J., 12
Douglas, R. J., 94
Dove, A., 153

Draganksi, B., 122
Dreher, J. C., 153
Dumaret, A. C., 80
Dumay, N., 174
Duncan, J., 153, 172
Dupoux, E., 20, 21
Dussias, P. E., 21
Duyme, M., 80

Echevarria, S., 170
Edelman, G. M., 112, 314
Eimer, M., 242, 245
El Tigi, M., 21
Elman, J. L., 238
Elston-Güttler, K., 264
Emslie, H., 153
Eubank, L., 238
Evans, A. C., 25, 283
Evans, D. L., 78
Evers, P., 68

Fabbro, F., 114, 137, 151, 152
Fabiani, M., 176
Fallon, J. H., 88
Farde, L., 29, 33
Farkas, I., 142
Farmer, J., 80, 152
Fayol, M., 215
Federmeier, K. D., 158, 240
Feenstra, M. G. P., 68
Feldman, L. B., 209
Felser, C., 21, 238
Felten, D. L., 78
Felton, C. V., 76
Fiebach, C. J., 292, 310
Fiez, J. A., 20
Finkbeiner, M., 103
Fissell, K., 153
Fitch, W. T., 120
Fize, D., 145
Flechsig, P., 67
Flege, J. E., 20, 22, 38, 60, 79, 166
Fodor, J. D., 92
Ford, J. M., 143

Forgie, M., 77
Forster, K., 103
Forster, K. I., 103, 205, 241
Frackowiak, R. S., 162
Frackowiak, R. S. J., 101, 289, 291
Franceschina, F., 213
Franceschini, R., 136, 175
Frank, D. A., 71
Freeman, G. B., 34
Frenck-Mestre, C., 21, 103, 286, 294
Friederici, A. D., 27, 32, 116, 174, 175, 205, 224, 225, 226, 232, 233, 237, 239, 240, 241, 242, 243, 246, 247, 251, 255, 256, 264, 292, 307, 310
Frisch, S., 237, 240, 254, 275
Frisson, S., 215
Friston, K. J., 112, 162, 289
Friston, K. I., 117
Frith, C. D. 162
Fujii, Y. , 27, 251
Fuster, J. M., 145

Gabrieli J. D. F., 153
Gally, J. A., 112
Gandour, J., 273
Garavan, H., 145, 153
Garey, L. G., 68
Garrett, M., 205, 241
Gaskell, M. G., 174
Gathercole, S. E., 113
Gazzaley, A., 172
Gehring, W. J., 158, 170
Gemba, H., 143, 145
Genesee, F., 236
Gentilucci, M., 118
Gerard, L. D., 166
Gerken, L., 175
Gess, R., 215
Gibb, R., 77
Gibson, G. E., 34
Gibson, E., 167, 241
Giedd, J. N., 64, 67
Gillon-Dowens, M., 239

Givon, T., 225
Glaser, D., 78
Glover, G. H., 145, 153
Golato, P., 38
Goldman-Rakic, P. S., 63, 68
Golestani, M., 115, 129
Gollan, T. H., 38
Golomb, J., 31
Gomez, R. L., 175
Gorny, G., 77
Gorski, R. A., 75
Gortmaker, S., 71
Goschke, T., 245
Gosling, S. D., 119
Gould, E., 62
Grafman, J., 153
Grainger, J., 103, 193
Gratton, G., 176
Graziano, M. S., 62
Green, D. W., 6, 24, 27, 100, 101, 108, 110, 111, 116, 117, 119, 129, 136, 139, 160, 314
Greenberg, J., 250
Greenough, W. T., 66, 79, 80
Greenwald, R., 166, 202, 294
Greer, S., 145
Gregg, K. R., 238
Grimshaw, G. M., 78
Grönholm, P., 174
Grosjean, F., 22, 38, 117, 138, 139, 141, 142, 175, 192, 195, 201
Gross, C. G., 62
Gross, B., 170
Gui, S. S., 176
Gullberg, M., 27, 294
Gunter, T. C., 241
Gutchess, A. H., 26, 38

Hadland, K. A., 153
Hagoort, P., 111, 120, 135, 143, 194, 226, 241, 293
Hahm, J., 130
Hahne, A., 27, 167, 168, 207, 224, 226, 237, 241, 242, 243, 244, 247, 251

Hakansson, G., 10
Hakuta, K., 14, 16, 238
Halgren, E., 232, 234
Hari, R., 288
Harley, T. A., 192
Harlow, H. F., 77
Harris, A., 167
Harvey, T., 315
Hasegawa, M., 136, 286, 292, 293, 307, 309
Hauser, M. D., 120
Haverkort, M., 117, 307
Hawkins, R., 213
Hay, D. F., 78
Hein, G., 116
Heinze, H. J., 153, 191
Hellwig, F., 27, 111, 294
Heminger, G. R., 76
Henninghausen, E., 245
Hensch, T. K., 73, 75
Herlenius, E., 68
Hermans, D., 138, 156
Hernandez, A. E., 136, 149, 283, 285, 289, 308
Hernandez, A. W., 310
Herschensohn, J., 36, 37, 215
Herzog, H., 111, 293
Hicks, S. P., 70
Hillenbrand, J., 22
Hillyard, S. A., 172, 203, 205, 209, 239
Hink, R. F., 172
Hirsch, J., 24, 137, 201
Hirsh-Pasek, K., 248
Hochstetter, F., 61
Hoen, M., 246, 247, 264
Hofman, M. A., 75
Holcomb, P., 167, 203, 205, 240, 241
Holmes, A. P., 289
Holroyd, C. B., 170
Holtmaat, A. J., 68
Holtzheimer, P., 152
Hon, N., 25, 136, 290

Hooper, J., 216
Hulstijn, J., 29
Huttenlocher, P. R., 67, 75
Huttunen, M. O., 71
Hyltenstam, K., 19, 20, 21
Hyvärinen, J., 75

Iarossi, E., 22
Ikonomidou, C., 72
Illes, J., 102, 136, 284, 287, 290, 291
Indefrey, P., 24, 27, 111, 239, 280, 281, 283, 288, 289, 293, 294, 295, 300, 306, 310
Innocenti, G. M., 67
Inoue, K., 27, 166, 202, 294
Ioup, G., 21
Ito, C., 70

Jahanshahi, M., 153
Janer, K. W., 162
Jansma, B. M., 143, 156
Janssen, N., 165, 193
Jescheniak, J. D., 155
Jessell, T. M., 75
Jevtovic-Todorovic, V., 72
Jiang, N., 103
Jin, Z., 26, 290
Jobert, A., 103
Johnson, E. K., 211
Johnson, J. S., 13, 14, 15, 16, 17, 18, 52, 60, 79, 166, 236
Johnson-Frey, S. H., 117
Jokela, M., 221
Jonides, J., 158
Josephs, O., 101, 291
Judáš, M., 88
Juraska, J. M., 75, 80
Jusczyk, P. W., 92, 248
Just, M. A., 136, 292, 307

Kaan, E., 167, 241, 293
Kaes, T., 67
Kakigi, R., 27
Kalsbeek, A., 68
Kamada, N., 70

Kandel, E. R., 75
Kastner, S., 172
Katayama, J., 248
Katsuki-Pestemer, N., 248
Kaushanskaya, M., 119
Kawaguchi, S., 10
Keenan, R. A., 116
Kemler Nelson, D. G., 248
Kempe, V., 292, 307
Kempermann, G., 62
Kennedy, T. J., 297
Kerkhofs, R., 193, 194
Kiehl, K. A., 145
Kigar, D. L., 315
Kim, K. H. S., 24, 25, 137, 201
Kim, A., 27, 166, 179, 202, 205, 209, 210, 226, 294
Kim-Cohen, J., 78
Kimura, D., 34
King, J. W., 168, 176, 241
Kirby, S., 317
Klein, D., 25, 36, 136, 201, 283, 285
Klein, R., 109, 141
Klein, W., 10, 175
Kline, J., 71
Kluender, R., 248
Knight, R. T., 171, 172
Knudsen, E. I., 2, 18, 73
Koda, K., 114
Koechlin, E., 153
Koenderink, M. J., 64, 66, 68
Kohnert, K., 136, 283, 308
Kok, A., 143
Kolb, B., 76, 77
Kolodny, J., 153
Konishi, S., 145
Kooistra, L., 75
Kopell, B. S., 143
Kopra, M., 222
Korr, H., 62
Koskinas, G. N., 93
Kosofsky, B. E., 70, 71
Kostović, I., 62, 63, 64
Kotz, S. A., 32, 116, 264
Kouider, S., 103

Koyama, T., 248
Kozorovitskiy, Y., 62
Kroll, J., 208
Kroll, J. F., 38, 108, 165, 174, 176, 193, 209
Kuhl, P., 225
Kuhl, P. K., 23
Kuperberg, G., 226
Kushnir, S. L., 137
Kutas, M., 135, 143, 158, 163, 168, 176, 203, 205, 209, 239, 240, 241, 242, 245, 275
Kuypers, K., 66

La Heij, W., 115
Lagercrantz, H., 68
Laine, M., 167, 168, 174
Lamers, M., 275
Landing, B. H., 88
Laplante, D. P., 72
Largy, P., 215
Larsen, c. c., 62
Larsen-Freeman, D., 11
Larson-Hall, J., 12, 16, 60, 78, 79
Laufer, B., 22
Lawson, D. S., 241
Le Bihan, D., 103
Lecours, A. R., 67
Lee, H. L., 3, 25, 136
Lee, H. W., 103, 136, 313
Lee, K. M., 24, 137, 201, 290
Lee, N., 32
Leino, S., 221
Lenneberg, E. H., 18, 52, 238
Letinic, K., 63
Levelt, W. J. M., 139, 143, 146, 151, 155, 175, 280, 281, 283, 288, 289, 300
Lewis, D. A., 68
Li, P., 141, 142, 192
Li, S. C., 33
Li, S. J., 153
Liddle, P. F., 145, 162
Lindenberger, U., 33
Liu, H. L., 292

Liu, S., 20, 166
Long, M. H., 10, 18, 20
Longworth, C. E., 116
Lounasmaa, O. V., 288
Lucas, T. H., 111, 136, 289
Luce, M., 242
Luciana, M., 317
Luke, K. K., 286, 292, 293
Lupien, S., 34
Luria, A. R., 145
Lutz, T., 155, 161

MacDonald, A. W., 153, 162
Mack, M., 22
MacKay, I. R., 20, 60
Mackey, S., 94
MacKinnon, G. E., 78
Macnamara, J., 137
MacNeilage, P. F., 317
MacWhinney, B., 23, 36, 176
Maguire, E., 127, 128
Maguire, E. A., 109, 297
Maguire, R. P., 307
Mai, H., 167
Maki, P. M., 34
Malach, R., 94
Malanga, C. J., 70, 71
Marian, V., 119
Marin, O., 63
Marinova-Todd, S., 21
Marin-Padilla, M., 62
Marlot, C., 145
Marshall, D. B., 10
Marslen-Wilson, W. D., 114, 116, 237
Martin, K. A., 94
Martínez, A., 136, 283, 308
Matsuzaki, R., 143
Matthews, P. M., 106
Matzke, M., 167
Mazziotta, J., 136, 283
McCandliss, B. D., 20, 225
McCarthy, G., 172, 173, 203
McClelland, J. L., 20
McEvoy, K., 172
McEwen, B. S., 75

McGregor, A. M., 76
McKhann, G. M. M., 111, 136, 289
McKinnon, R., 205, 226
McLaughlin, B., 209
McLaughlin, J., 27, 166, 174, 202, 205, 209, 210, 226, 194
Meador, D., 60
Mechelli, A., 3, 5, 109, 127, 128, 129, 297
Mehler, J., 92
Meier, R. P., 248
Meilijson, I., 129
Menon, V., 145
Mestres, A., 174
Meuter, R. F. I., 140, 141,
Meuter, R., 242
Meyer, A. S., 139
Meyer, D. E., 170
Meyer, E., 25, 283
Michael, E. B., 38
Miezin, F. M., 171
Miller, A. K., 30
Miller, B., 17
Miller, E. K., 153, 162
Miller, J. L., 141
Miller, M. W., 70
Mills, D. L., 241
Milner, B., 25, 76, 136, 201, 283
Miozzo, A. 105
Miozzo, M., 138
Mitchell, D. C., 237
Mitchell, R., 216
Miura, K., 296
Miyamoto, E. T., 251
Miyamoto, T., 248
Miyashita, Y., 184
Mobley, L. A., 226
Moffitt, T. E., 78
Molis, M., 13, 14, 17, 18, 166, 238
Möller, J., 143
Moran, J., 172
Moreno, E. M., 158, 163, 242
Morgan, J. L., 248, 254
Moro, A., 32

Morrone-Strupinsky, J. V., 317
Morrow, C. E., 71
Morton, J., 92
Moselle, M., 21
Moyer, A., 19, 36
Mrzljak, L., 63, 64, 65
Mueller, J. L., 27, 167, 174, 244, 251, 252, 255, 257, 275, 277
Muldrew, S., 242
Müller, K., 310
Mummery, C. J., 105
Münte, T., 133, 135, 143, 147, 153, 155, 161, 167, 168, 174, 191, 192, 244, 245
Muraishi, Y., 296
Murphy, G. M., 315
Myles, F., 216

Nager, W., 167
Nagy, M. E., 234
Nakada, T., 286
Nakagome, K., 248
Nakai, T., 286, 292
Nakajima, H., 269, 270
Nakajima, K. 184,
Nakamura, K., 103, 104
Nakata, H., 27
Nambu, A., 143
Narafu, N., 296
Negro, I., 214
Neville, H. J., 32, 79, 166, 175, 205, 225, 240, 241, 242, 243, 247, 250
Newman, A. J., 32
Newport, E. L., 13, 14, 15, 16, 17, 18, 52, 60, 79, 166, 175, 211, 236, 238, 248
Newton, M., 78
Nicol, J., 103,
Nicol, J. L., 205, 241
Nicolaides, K. H., 76
Nieuwenhuis, S., 145
Nikelski, J., 136, 201, 283
Niskanen, P., 71
Nitsch, C., 136
Norman, D. A., 139

Nösselt, t., 153, 191
Nulman, i., 71
Nuñez, J. L., 75
Nyberg, L., 35
Nystrom, L. E., 153

Obenauer, H. G., 54
Obler, L. K., 139
Ochsner, K. N., 153
Ojemann, G. A., 111, 136, 289
Ojima, S., 27
O'Leary, D. D. M., 60, 67, 73
Olney, J. W., 70, 72
Opitz, B., 232, 233
Osinsky, R., 147
Osterhout, L., 27, 166, 167, 174, 202, 203, 204, 205, 209, 210, 225, 226, 241, 294
Ozawa, K., 32

Paans, A. M. J., 307
Paavilainen, P., 222
Pallas, S., 130
Pallier, CC., 3, 23, 113, 170
Palolahti, M., 221
Palomero-Gallagher, N., 93
Pancheva, R., 32
Pandya, D. N., 171
Pannekamp, A., 241, 253
Papadopoulou, D., 21
Papagno, C., 113
Paradis, M., 24, 110, 117, 130, 137, 139, 142
Pardo, J. V., 162
Pardo, P. J., 162
Park, 26, 35, 38
Park, D. C., 28
Parnavelas, J. G., 63
Parodi, T., 226
Passingham, R. E., 153
Patkowski, M. S., 17, 37
Paus, T., 64, 115, 129
Pavlenko, A., 119
Pawlby, S., 78
Penfield, W., 136, 137

Author Index

Penke, M., 168, 226, 244
Pennebaker, J. W., 119
Peperkamp, S., 20, 21
Perani, D., 24, 25, 105, 113, 136, 201, 283, 285, 286, 292, 293, 299, 309,
Perfetti, C. A., 114
Perner, J., 13
Perry, B. D., 78
Pessoa, L., 172
Petanjek, Z., 68
Petersen, S. E., 172
Peterson, R., 151
Petitto, J. M., 78
Petrides, M., 94
Pfefferbaum, A., 29, 143
Pfeifer, E., 174, 225, 237
Phelps, M. E., 82
Phillips, N. A., 240, 254
Pianka, P., 94
Picton, T. W., 172
Pienemann, M., 10
Pillai, J., 102, 284, 287, 290, 291
Piller, I., 36
Pinker, S. 15, 18, 38, 244
Piske, T., 20
Pisoni, D. B., 20
Poldrack, R. A., 135, 145, 192
Poline, J. B., 269, 289
Pollmann, S., 153
Portin, M., 167, 168
Posner, M. I., 225
Post, B., 114
Potter, J. P., 119
Potter, M. C., 209
Poulisse, N., 138, 139
Pouratian, N., 137, 175
Price, C.,
Price, C. J., 105, 110, 112, 116, 117, 136, 149, 291
Price, D. J., 67
Protopapas, A. 20
Proverbio, A. M., 27
Pu, Y. L., 283, 285

Puce, A., 172, 173
Pulvermüller, F., 236, 238, 264

Quinones-Hinojosa, A., 137

Raboyeau, G., 209
Radford, A., 212
Rafii, 88
Raichle, M. E., 162
Rakic, P. S., 60, 62, 63, 68
Ramirez-Esparza, 125
Randall, B., 114
Rasmussen, T., 76
Raz, N., 29, 30, 31, 94
Reeves, A. J., 62
Reiss, A. L., 145
Relkin, N. R., 24, 137, 201
Repetto, M. J., 78
Resnick, S. M., 34
Reuter-Lorenz, P. A., 31
Ridderinkhof, K. R., 145
Rinne, J. O., 174
Ritter, W., 143
Riva, D., 77
Ro, D., 73
Ro, P., 73
Robbins, T. W., 145
Roberts, L., 136
Rodriguez-Fornells, A., 133, 135, 143, 144, 145, 146, 147, 153, 154, 155, 156, 160, 161, 162, 165, 170, 174, 191, 193, 283, 285, 289, 316
Roe, A., 130
Roelofs, A., 115, 139, 193
Rogers R.D., 153
Rogers, W. A., 29
Rösler, F., 245
Ross, L. L., 153
Ross T.J., 145
Roth, R. H., 69
Rothwell, J. C., 153
Rotte, M., 133, 153, 160, 191
Roux, F. E., 137

Rowntree, S., 77
Rubenstein, J. L. R., 60, 63
Rubia, K., 145
Rugg, M. D., 234
Ruiz-Marcos, 90
Ruppin, E., 129
Rüschemeyer, 286, 292, 293
Rushworth, M. F. S., 106, 153
Rüsseler, 167, 245
Ryan, L., 36, 37, 52

Sabourin, L., 24, 26, 27, 60, 79, 299, 305, 306, 309
Saffran, J. R., 211
Sakai, K. L., 78, 174, 296, 300
Salmelin, R., 288
Salthouse, T. A., 28
Sampson, P. D., 70
Sams, M., 288
Sanders, L. D., 175
Sandra, D. 215
Santesteban, M., 141
Sanz, C., 36
Sasaki, H., 70,
Sasaki, K., 143, 145
Savoy, P., 151
Scabini, D., 172
Scarbourough, D. L., 166
Schaie, K. W., 13
Scheibel, A. B., 315
Schiltz, 188
Schlaghecken, F., 245
Schleicher, A., 115, 129
Schlesewsky, M., 237, 251, 254, 274, 275
Schmitt, B. M., 133, 143, 147, 156
Schmitz, C., 62
Schreuder, R., 116, 138
Schriefers, H. J., 155, 165, 193, 241
Schubert, T., 153
Schumann, J. H. 32, 236, 238, 264, 313, 315, 316, 317
Schwartz, B. D., 53
Schwent, V. L., 172

Scovel, T., 18
Sebastian-Galles, N., 139, 170
Segalowitz, N., 29
Seidl, A., 248
Seitz, R. J., 111, 293
Sekihara, 184
Seliger, H. W., 18
Selinker, L., 19
Sesack, S. R., 68
Shallice, T., 105, 139
Shankle, W. R., 62
Shigematsu, I., 70
Shors, T. J., 62
Shulman, G. L., 172,
Shulman R., 226
Sidman, R. L., 62
Sikström, 33
Simson, R., 143
Singleton, D., 18, 36, 37, 52
Skehan, P., 12, 36
Skrap, M., 151
Slotkin, T.A., 71
Small, B. J., 28, 69
Smith, A. M., 145
Snow, C. E., 10
So, K. F., 209
Sobol, A., 71
Soderstrom, M., 248
Sodhi, J., 75
Soon, C. S., 3, 25, 103, 113, 136, 290
Spencer, D. D., 172
Sperry, R., 115
Spitz, R. A., 77
Sporns, O., 112
Sprouse, R. A., 53
Stafford, C. A., 36
Stamatakis, E. A., 114
Steels, L., 317
Stein, Z., 71, 145, 153
Steinhauer, K., 174, 225, 239, 241, 253
Stellakis, N., 22
Stemberger, J. P., 151
Stenger, V. A., 153
Stewart, E., 174,

Stewart S. H., 240, 254
Stowe, L. A., 24, 26, 27, 60, 79, 117, 299, 305, 306, 307, 309
Strafella, A. P., 152
Streissguth, A. P., 70
Stryker, M., 130
Studdert-Kennedy, M., 317
Stürmer, 245
Stuss, D. T., 172
Sunderman, G., 208, 209
Sur, M., 60, 130
Susser, M., 71
Susser, E., 71
Suwazono, S., 171
Svoboda, K., 68
Swaab, D. F., 75, 293
Swanson, K. A., 54
Swinney, D. A., 241

Takagi, N., 20
Takazawa, S., 248
Talairach, J., 281
Tan, E. W. L., 25, 114, 136, 201, 292
Tatsuno, Y., 296
Taupin, P., 62
Taylor, A., 78
Teichman, 48
Ter Keurs, M., 241
Teruel, E., 155
Teuber, H. L., 77
Thiel, A., 113
Thiel T., 25, 136, 201
Thorpe, S., 76, 145
Thorpe-Beeston, J. G., 76
Thyre, R., 53, 54
Toepel, U., 241
Tohkura, Y., 20
Tokowicz, N., 38, 176
Tokumaru, Y., 22
Tomasello, M., 216
Tomie, J. A., 77
Tomkiewicz, S., 80
Tononi, G., 112
Toribio, A. J., 4
Tournoux, P., 218, 282

Townsend, D. A., 62
Trachtenberg, J. T., 68
Tranel, D., 77
Trémoulet, 137
Trevarthen, C., 78
Treves, A., 120
Tsujimoto, T., 145
Tyler, L. K., 92

Uchida, 184
Ueno, M., 248
Ullman, M. T., 32, 38, 111, 114, 175, 238, 250
Ullsperger, M., 170
Ungerleider, L. G., 172
Unsworth, S., 55, 56
Uylings, H. B. M., 2, 59, 60, 61, 62, 63, 64, 65, 66, 68, 79, 308, 314

Vainikka, A., 216, 226
Vallar, G., 113
Van Boxtel, G. J. M., 145, 276
Van den Wildenberg, W., 145,
Van der Loos, H., 68
Van der Lugt, A., 133, 147
Van Eden, C. G., 63, 64, 65
Van Hell, J. G., 22, 176, 193
Van Heuven, W. J. B., 7, 116, 140, 165, 191, 193,
Van Jaarsveld, H., 166
Van Pelt, 90
Van Turennout, M., 143, 156
Vandenberghe, R., 101, 106, 107, 289, 291
Vaughan, H. G., 143
Veltman, W. A. M., 66,
Ventureyra, V. A., 23
Vingerhoets, G., 283, 285, 288, 289, 308
Viswanathan, M., 109, 141
Vogel, A. L., 71
Vogt, C., 93
Vogt, O., 93
Volkow, N. D., 33

Von Cramon, D. Y., 116, 153, 170, 310
Von Eckardt, 229
Von Economo, C., 93
Von Studnitz, R., 110, 136
Vorobyev, V., 174

Wager, T. D., 162
Wagner, A. D., 135, 192
Wahlin, A., 28
Wai, Y. Y., 292
Walker, J. A., 137
Wallace, C. S., 66
Wan, Y. L., 292
Wang, Y, 26
Wang, M., 114
Wartenburger, 25, 26, 286, 292, 293
Weber, B., 317
Weber-Fox, C. M., 166, 242, 243, 247, 250
Weckerly, J., 275
Weinert, F. E., 13
Weitzman, M., 71
Weller, B. J., 143
Werheid, K., 116
Weyerts, H., 168, 176, 244,
Wheeldon, L. R., 146
Whishaw, I. Q., 76
Whitaker, H. A., 111, 136
White, L., 11, 53, 213, 236
White C.D., 145
Wiggins, C. J., 153
Wijers, A. A., 307
Wilson, C., 152
Wise, R., 101, 291
Witelson, S. F., 315, 316
Withers, G. S., 66

Witzel, T., 234
Wodniecka, Z., 108
Wolpert, L., 117
Wong-Fillmore, L., 216
Wood, C., 203
Woods, B. T., 77
Woods, D. L., 172
Woolley, C. S., 75
Wozniak, D. F., 72
Wray, A., 216

Xue, G., 26, 284, 287, 290, 291
Xue, L., 71

Yakovlev, P. I., 67
Yamaguchi, S., 172
Yamamoto, M., 275
Yamashita, H., 251
Yeni-Komshian, G. H., 20, 166,
Yeterian, E. H., 171
Yeung, N., 145, 170
Young, C., 72
Young-Scholten, M., 216, 226
Yudkoff, M., 76

Zaake, 188
Zani, A., 27
Zappatore, D., 136
Zatorre, R. J., 25, 115, 129, 136, 201, 283
Zeki, S., 93
Zhang, L., 26
Zhao, M., 62
Zhao V., 136, 201, 283
Zilles, 64, 93, 115, 129
Zito, K., 68
Zohary, E., 94
Zoncu, R., 63
Zwarts, F., 117

Subject Index

acetylcholine, 34
age of acquisition, AoA, 4, 11, 12, 13, 14, 15, 16, 17, 18, 20, 21, 23, 24, 25, 26, 36, 37, 52, 236, 242, 243
age of exposure, AoE, 11, 12, 26, 27, 225, 242
agreement, 27, 64, 138, 155, 167, 176, 213, 214, 216, 217, 219, 220, 224, 241, 255, 264, 280, 288, 298
American Sign Language, 78, 243
amygdala, 33, 317, 323
anterior cingulate cortex, ACC, 149, 151, 152, 153, 154, 160, 161, 162, 165, 170, 195, 316
articulation, 15, 108
artificial language, 225, 250, 255
attainment, 1, 2, 10, 11, 12, 15, 17, 18, 19, 20, 22, 28, 29, 36, 37, 51, 52, 113, 131, 238
attention, 10, 28, 32, 33, 35, 53, 71, 109, 113, 158, 172, 192, 215, 277
automaticity, 7, 29
axonal connections, 80

BA 44, 94, 95, 144, 285, 286, 288, 292, 296
BA 45, 95, 153, 287, 291
BA 47, 285, 286, 288
basal ganglia, 32, 116, 317, 325, 326
BIA model, 140, 193–195
brain volume, 29
Broca's area, 26, 62, 95, 115, 289, 290, 294, 306, 308, 321

Brodmann Area 44, BA 44, 93, 308
Brodmann Area 45, BA 45, 95, 153, 287, 291
Brodmann Area 47, BA 47, 285, 286, 288

case, 2, 7, 13, 19, 23, 26, 29, 31, 34, 35, 38, 56, 60, 92, 102, 105, 110, 114, 130, 137, 139, 140, 145, 149, 151, 152, 158, 165, 166, 168, 170, 176, 206, 213, 215, 217, 221, 222, 240, 243, 245, 250, 251, 252, 254, 256, 257, 259, 260, 261, 262, 263, 272, 273, 274, 275, 276, 294, 299, 300
Catalan, 133, 141, 149, 151, 153, 155, 160, 161, 162, 163, 164, 168, 169, 170, 176
caudate nucleus, 30, 31, 33, 105, 321, 326
cingulate, 33, 113, 149, 194, 195, 287, 291, 299, 316
closed-class words, 243
Closure Positive Shift, CPS, 241, 253, 262, 274
code-switch, 158, 159
code-switching, 137, 138
cognate, 25, 164, 165, 169
cognitive aging, 28, 52
cognitive control, 6, 134, 138, 142, 153, 158, 162, 163, 165, 166, 174, 175, 191, 193, 196
compositional, 167, 300
computational model, 193
conceptual representation, 139, 288

343

control, 1, 4, 6, 7, 27, 99, 100, 102, 105, 107, 110, 115, 116, 117, 118, 119, 120, 129, 134, 138, 139, 140, 141, 142, 152, 153, 156, 158, 160, 162, 163, 164, 168, 171, 173, 174, 191, 192, 193, 194, 195, 196, 202, 204, 209, 280, 283, 288, 289, 290, 296, 307
convergence, 1, 4, 5, 27, 100, 111, 114, 116
corpus callosum, 67, 70, 297
cortical stimulation mapping, 111
cortisol, 34
covert naming, 155, 161
critical period, 1, 15, 18, 36
Critical Period Hypothesis, CPH, 36, 53
cytoarchitecture, 115, 308, 322

dendrite maturation, 59
discourse, 21, 292, 295
dominance, 20, 23, 37, 38, 75
dominant language, 23, 159
dopamine, 30, 32, 34, 35, 68
dorsolateral prefrontal cortex, DLPFC, 65, 195
Dutch, 27, 55, 56, 105, 131, 165, 194, 215, 248, 276, 294, 308

early learners, 306, 309
early left anterior negativity, ELAN, 241, 243, 244, 246, 273
EEG, 232, 236, 239
electrical stimulation, 289
end state, 10, 11, 12, 19, 53, 202
English, 13, 21, 22, 23, 26, 27, 32, 54, 55, 56, 79, 100, 102, 103, 104, 105, 106, 107, 108, 109, 113, 114, 119, 128, 137, 140, 141, 149, 152, 158, 167, 194, 205, 209, 211, 213, 214, 215, 216, 221, 222, 223, 225, 242, 243, 248, 249, 296, 308

entorhinal cortex, 30, 31
environmental enrichment, 66, 68, 73, 79, 80
error-related negativity, ERN, 170
errors, 12, 37, 155, 162, 214, 236, 241, 263, 299
event related brain potential, ERP, 143, 199, 202
executive control, 133
executive functions, 134, 135, 175
experience learning, 79

feral children, 78
Finnish, 167, 221, 222, 223
first language acquisition, L1 acquisition, 9, 53, 78, 248, 253, 317
fMRI, 23, 25, 27, 92, 103, 104, 107, 136, 143, 144, 145, 146, 149, 152, 158, 160, 165, 172, 175, 193, 194, 196, 200, 201, 202, 206, 232, 236, 294, 296
French, 9, 23, 54, 55, 105, 113, 140, 155, 166, 167, 209, 210, 211, 213, 214, 215, 216, 217, 218, 221, 223, 225, 308
Friulian, 151, 152
functional degeneracy, 111
fusiform gyrus, 26, 172, 177, 323

gender, 148, 155, 156, 157, 158, 159, 161, 165, 169, 176, 212, 213, 214, 241
genetic, 59, 66, 78, 80
genotype, 80, 81
German, 25, 27, 55, 104, 105, 133, 146, 147, 148, 149, 153, 154, 155, 156, 159, 161, 165, 167, 168, 207, 243, 244, 248, 251, 254, 255, 256, 257, 265, 272, 274

gonadal hormones, 75
grammaticality, 6, 12, 13, 25, 27, 246, 252, 257, 264, 293, 294, 307
grammaticalization, 212, 213
graphemic, 288, 292
gray matter, 4, 29, 64, 109, 110, 127, 128, 129, 130, 201, 298, 327

higher order function, 117, 118
hippocampus, 28, 30, 31, 61, 62, 75, 177, 317, 322, 324

IC model, 139, 140, 141, 160
implicit memory, 28
incongruence, 147, 158
individual differences, 4, 36, 100, 101, 112, 114, 119, 129, 163, 175, 217, 233, 313, 314
individual variability, 289, 308
inferior frontal gyrus, 106, 144, 285, 288, 306, 321
inflectional
  morphemes, 217
  morphology, 215, 295
inhibition, 7, 69, 138, 140, 141, 142, 143, 145, 147, 153, 155, 160, 171, 173, 175, 176, 195, 196, 316
inhibitory mechanisms, 140, 171, 173, 175
innate language faculty, 51, 52
interference, 6, 23, 135, 137, 138, 142, 143, 146, 147, 148, 149, 150, 151, 152, 153, 155, 156, 157, 158, 159, 162, 163, 164, 165, 166, 169, 173, 175, 176, 195
interlingual homographs, 165, 194
intersubject variation, 13
Italian, 9, 22, 25, 109, 128, 152, 155, 164

Japanese, 3, 99, 103, 104, 105, 205, 235, 243, 247, 248, 250, 251, 253, 254, 255, 256, 257, 259, 262, 264, 272, 273, 275, 296

language nonselective lexical access, 7, 192, 195
language schema, 140, 153
language-specific access, 193
language-specific neuronal sites, 111
late acquisition, 25, 306, 309
late learners, 14, 15, 21, 24, 25, 37, 236, 271, 306, 308
left anterior negativity, LAN, 163, 240, 241, 255, 273
lemma, 139, 140, 146, 151, 155, 156, 193, 283, 288, 292
lesion, 76, 77, 92, 113, 142, 151, 201
lexicon, 32, 135, 139, 140, 160, 166, 171, 175, 208, 249, 275
localization, 24, 135, 199, 200

Mandarin Chinese, 294
maturation, 15, 16, 18, 19, 35, 51, 52, 61, 67, 68, 69, 79, 138
MEG, 92, 232, 236, 288
memory, 7, 12, 28, 29, 31, 32, 33, 34, 35, 38, 91, 94, 113, 153, 158, 163, 174, 175, 176, 192, 195, 255, 322
meta-analysis, 5, 14, 279, 280, 281, 282, 293, 306
metabolic rate, 69
metalinguistic judgments, 299
miniature language, 235, 236, 237, 246, 248, 251, 264, 272, 276
monitor, 176, 192
monitoring, 7, 145, 152, 155, 162, 196

morphology, 94, 127, 142, 161, 168, 169, 202, 208, 212, 213, 250
morphosyntactic acquisition, 166
morphosyntactic processes, 244, 255
morphosyntax, 12, 20, 205, 212, 213, 221, 223, 226
motivation, 11, 20, 32, 36, 201, 316
myelin development, 67

N200, 144, 145, 146, 147, 155, 158, 172, 173, 240
N400, 27, 163, 166, 169, 194, 202, 203, 205, 206, 209, 210, 211, 212, 216, 217, 218, 220, 221, 222, 232, 239, 240, 242, 244, 248, 254, 256, 257, 260, 261, 263, 274, 275
narrative, 21, 298
nativelike, 11, 19, 20, 21, 27, 36, 37, 52, 53, 56, 60, 79, 236, 239, 250, 251, 254, 256, 264, 271, 276
natural selection, 112
Nature-Nurture, 59
neocortex, 32, 60, 61, 62, 64, 94, 323
neurogenesis, 28, 61, 69, 70, 72, 80
neuronal adaptation, 101, 103, 104, 105
neuronal migration, 59, 70
number, 4, 19, 23, 37, 52, 54, 61, 62, 67, 68, 72, 75, 79, 80, 81, 94, 99, 102, 109, 110, 117, 127, 129, 138, 143, 164, 167, 168, 170, 175, 176, 212, 213, 214, 217, 219, 221, 226, 240, 241, 247, 248, 249, 272, 279, 282, 283, 289, 291, 292, 300, 305, 309

optical imaging, 136, 137
orthography, 91, 142, 214

P300, 158, 240
P3b, 245
P600, 5, 27, 167, 168, 202, 205, 206, 207, 212, 216, 217, 218, 219, 220, 221, 222, 226, 232, 241, 242, 243, 244, 246, 248, 253, 254, 255, 256, 257, 260, 263, 264, 273, 274, 275
parahippocampal cortex, 172, 177
parahippocampal cortex, 172, 177
parietal cortex, 64, 75, 109, 113, 116, 128, 298, 325
past tense inflection, 296
PET, 25, 27, 33, 92, 104, 105, 136, 202, 236
phenotype, 81
phenylalaline, 76
phenylketonuria, 76
phonetic encoding, 283
phonological mismatch negativity, 240, 274
phonological retrieval, 106, 108
phonology, 18, 32, 142, 208, 213, 214, 215, 221, 223, 236, 237, 248, 256, 265
picture-naming, 105, 108, 146, 281, 283, 288, 289
plasticity, 59, 60, 62, 69, 73, 74, 76, 77, 94, 128, 130, 131, 236, 238
poverty-of-the-stimulus, 52, 54, 56
practice, 20, 26, 36, 73, 166, 297
prefrontal cortex, PFC, 325
proficiency, 1, 3, 4, 11, 21, 25, 26, 27, 37, 38, 53, 100, 104, 106, 109, 113, 116, 117, 118, 119, 128, 130, 138, 141, 142, 159, 163, 175, 200, 201, 202, 224, 225, 236, 237, 238, 239,

Subject Index

242, 243, 244, 246, 252, 255, 257, 271, 272, 273, 276, 277, 283, 289, 291, 292, 293, 294, 297, 298, 299, 306, 309, 314, 315, 316
pronunciation, 12, 14, 20, 21, 22, 23, 79, 265
prosodic phrase, 273
prosody, 6, 100, 115, 237
putamen, 30, 33, 110, 321, 326

recovery patterns, 119, 142
revised hierarchical model, 174
rote-memorization, 216
rules, 32, 38, 60, 166, 167, 171, 205, 212, 213, 215, 216, 219, 221, 224, 237, 246, 248, 265, 272, 275, 276, 300
Russian, 119, 167, 168, 207, 243

scrambling, 56
　direct object scrambling, 56
　object scrambling, 55, 56
　scrambled indefinite objects, 56
second language learning, L2 learning, 4, 22, 27, 35, 36, 37, 200, 202, 208, 215, 216, 225, 226, 238, 239, 245, 250, 253, 277, 316
second language onset, L2 onset, 3, 281, 284, 285, 286, 287, 288, 289, 291, 292, 293, 297, 298, 299
second language proficiency, L2 proficiency, 12, 24, 27, 38, 201, 231, 236, 281, 284, 285, 286, 287, 289, 292, 293, 298
second language use, L2 use, 7, 10, 13, 20, 27, 29, 34
selection, 6, 7, 100, 112, 115, 116, 117, 120, 134, 139, 140, 141, 146, 151, 152, 153, 160, 168, 171, 173, 175, 193, 196, 315, 316

semantic decision, 106, 143, 290, 291, 299
semantic retrieval, 106, 107, 108
sensitive period, 2, 60, 61, 69, 70, 242
sentence processing, 3, 21, 219, 231, 232, 235, 237, 248, 251, 300
sequence learning, 245
simultaneous interpreter, 110
simultaneous interpretation, 110
serial models, 151
social neglect, 59, 77
Spanish, 13, 17, 102, 133, 137, 141, 146, 147, 148, 149, 151, 153, 154, 155, 158, 159, 160, 161, 162, 163, 164, 165, 168, 169, 170, 308
spreading activation model
statistical power, 5, 307
stereoscopic vision, 59, 75
story comprehension, 281, 292, 307
stress, 21, 34, 59, 69, 71, 72, 100, 238
striatum, 28, 30, 32, 324
Stroop task, 162
supplementary motor area, SMA, 149, 152, 154, 195, 286, 292, 325
Swedish, 167
switching, 33, 105, 119, 134, 138, 140, 141, 142, 151, 152, 153, 156, 158, 175, 192, 316
syllabification, 283, 288, 290, 300, 310
syntactic, 3, 22, 27, 54, 55, 91, 95, 108, 134, 143, 155, 156, 159, 166, 167, 175, 205, 206, 207, 212, 216, 217, 219, 221, 224, 226, 236, 237, 238, 239, 240, 241, 242, 243, 244, 245,

246, 247, 248, 250, 251, 253, 254, 255, 256, 260, 262, 264, 271, 272, 273, 276, 293, 294, 299, 300
syntax, 32, 56, 95, 212, 223, 225, 236, 237, 238, 242, 265, 294

temporal cortex, 30, 172, 290, 326
thyroid hormone, 75
transcranial magnetic stimulation, TMS, 113, 152
transmitter, 68

ultimate attainment, 11, 12, 17, 18, 28, 29, 36, 131

Universal Grammar, UG, 10, 37, 51, 52, 53, 54, 55, 56

voxel-based morphometry, 108, 297

white matter, 30, 67, 108, 115, 129, 201, 297, 298
word-form retrieval, 283
word generation, 25, 102, 281, 283, 289
word production, 196, 283, 288, 289, 290, 298, 300
word recognition, 174, 193
working memory, 7, 12, 28, 29, 32–35, 38, 113, 153, 163, 176, 192